OUR UNCERTAIN HERITAGE

Genetics and Human Diversity   Second Edition

DANIEL L. HARTL
James S. McDonnell Professor of Genetics
James S. McDonnell Department of Genetics
Washington University School of Medicine
St. Louis, Missouri

1817

HARPER & ROW, PUBLISHERS, New York
Cambridge, Philadelphia, San Francisco,
London, Mexico City, São Paulo, Singapore, Sydney

For Dana Margaret and Theodore James

Sponsoring Editor: Claudia M. Wilson
Project Editor: Bob Greiner
Text Design Adaptation: Robert Sugar
Cover Design: 20/20 Services Inc.
Cover Photos: © Copyright 1984 Mark W. Berghash. All rights
            Reserved
Photo Research: Mira Schachne
Production: Debra Forrest Bochner
Compositor: Ruttle, Shaw & Wetherill, Inc.
Printer and Binder: R. R. Donnelley & Sons Company

**Our Uncertain Heritage: Genetics and Human Diversity,
Second Edition**

Library of Congress Cataloging in Publication Data

Hartl, Daniel L.
    Our uncertain heritage.

    Includes bibliographical references and index.
    1. Human genetics.   I. Title.
QH431.H296   1985   573.2′1   84-10759
ISBN 0-06-042684-5

41,731

84 85 86 87   9 8 7 6 5 4 3 2 1

Contents

iii

3 Mendel's Laws and Dominant Inheritance 63

4 Mendel's Laws and Recessive Inheritance 94

5 The Genetic Basis of Sex 118

6 Abnormalities in Chromosome Number

153

7 Abnormalities in Chromosome Structure

180

11 Viruses and Cancer   299

12 Immunity and Blood Groups   324

13 Population Genetics   357

14 Quantitative and Behavior Genetics   394

Preface

Our Uncertain Heritage is a textbook of human genetics for nonscience and nonbiology majors. It is written for students who wish to gain some familiarity with human genetics either to satisfy their undergraduate distribution requirements or to satisfy their own curiosity. The book's aim is to provide an understanding of the principles and implications of human genetics without getting bogged down in nonessential details of cellular, molecular, or mathematical biology. The book has no prerequisites, and where the discussion involves concepts that are likely to be unfamiliar to the nonscientist, the necessary background has been provided. However, this background material has been presented in a fresh framework so as to make it interesting even to those who may have encountered it in other courses or in their own reading. This approach is necessary because human genetics has proved to be a popular course with students in sciences other than biology, such as physics, chemistry, geology, mathematics, and the large number of agricultural and engineering fields.

Human genetics, as might be expected, has proved to be a suitable unifying theme for teaching virtually all aspects of genetics. The scope and fascination of human genetics serve to focus and bind together students whose interests are as diverse as art, languages, literature, history, philosophy, anthropology, sociology, and psychology. This popularity of human genetics is due in part to the natural curiosity that nearly everyone shares about the human body and how it works. Most of us are also concerned with one or more hereditary traits, particularly genetic disorders, that may occur in ourselves or among our relatives. Apart from these personal considerations is the fact that new findings in genetics appear almost weekly in popular news magazines, newspapers, or on television—human insulin produced in bacteria, giant mice produced by injection of growth-hormone genes, genetically engineered plants with superior characteristics. Genetics, which in its widest sense includes parts of cell biology, cytology, molecular biology, microbiology, immunology, virology, population biology, and other fields, has become part of everyday life in our high-tech society, and many students perceive that a knowledge of basic human genetics is an important part of their general education.

Human genetics is an appropriate theme for teaching genetics for another reason: It is a lively, challenging field, marked by exciting current research such as recombinant DNA and enormously complex legal, moral, ethical, and social problems. Many of these issues inevitably involve a conflict of rights, interests, or values, and reasonable people may disagree as to how these should be weighed and balanced. In many of these cases, an understanding of the scientific issues is prerequisite for discussion of their social implications. Problems related to the social or ethical implications of human genetics are discussed throughout the text in their appropriate context. I have attempted to give a fair and balanced presentation of each of these controversial topics, having adopted the view that, given the facts, students should deliberate the issues and make their own judgments.

Readers familiar with the first edition will note many changes. This is a different book in many respects. The original 23 chapters have been pared to 14, and some material in the first edition, such as human population growth and an outline of human evolution, has had to be eliminated, whereas other subjects, such as recombinant DNA and cancer-causing viruses, have had to be expanded. These changes in part reflect changes in the field, particularly in the development of the powerful new procedures that have come to be called recombinant DNA and in the growth of understanding of gene structure and function that has been achieved through their application. There are also changes in emphasis in this edition. For example, the discussion of population and quantitative genetics has been rewritten to make it easier, and the discussion of various aspects of behavior genetics has been reorganized into one section in Chapter 14.

This second edition also has several other features that students and instructors had requested. Each chapter is provided with a comprehensive summary for purposes of emphasis and review. Key words are highlighted in the text where they are first defined, a list of key words appears at the end of each chapter, and there is a glossary at the end of the book. Each chapter is provided with a set of problems, graded in difficulty, that challenge the students' understanding and test their skills, and answers are worked out in full at the back of the book. End-of-chapter lists of annotated references for further reading, chosen for accessibility as well as for subject and level of presentation, are also provided. I have emphasized such sources as *Scientific American, Annual Review of Genetics, Science, Nature,* and, for some subjects, *Cell.*

Some parts of this edition of *Our Uncertain Heritage* are an abbreviated version of the corresponding material in *Human Genetics* by D. L. Hartl (Harper & Row, 1983), and students may consult this more technical book for further detail. Other parts of *Our Uncertain Heritage* are wholly new, such as the discussion of population genetics in Chapter 13 or the new approach to heritability in Chapter 14. One of the advantages of writing a book for nonmajors is that the author is less constrained to cover material in a standard manner or from a particular point of view. Thus, I have used analogy rather freely and have incorporated illuminating historical detail, such as the story, in Chapter 5, of hemophilia in Tsarevich Alexis.

The organization of the book is straightforward and proceeds from the less abstract to the more abstract. Chapters 1 through 7 cover classical Mendelian genetics with a focus on cells (Chapter 1), chromosomes and gamete formation (Chapter 2), autosomal dominance (Chapter 3), autosomal-recessive inheritance and linkage (Chapter 4), sex linkage (Chapter 5), and abnormalities in chromosome number (Chapter 6) and structure (Chapter 7). Somatic cell genetics is included in Chapter 4 and general aspects of sexual differentiation in Chapter 5. Some instructors may wish to begin their course with Chapters 8 through 12, which cover DNA structure and replication (Chapter 8), transcription and translation (Chapter 9), mutation (Chapter 10), viruses and cancer (Chapter 11), and immunity and blood groups (Chapter 12). Recombinant DNA is included in Chapter 8, aspects of gene regulation in Chapter 9, transposable elements in Chapter 10, and retroviruses in Chapter 11. The final chapters deal with population genetics (Chapter 13) and quantitative genetics and behavior (Chapter 14), and there is a section on genetic counseling in Chapter 14.

Special thanks should go to the reviewers—Dr. Carl Huether of the University of Cincinnati, Dr. Andrew Clark of The Pennsylvania State University, Dr. Sydney S. Y. Young of Ohio State University, Dr. Wendell McKenzie of North Carolina State University, Dr. Frank J. Rice of Fairfield University, and Dr. James Bowman of Utah State University—whose careful reading and constructive criticisms did much to improve the book. Scarcely a page remains that has not had some significant improvement as a result of their efforts.

My laboratory associates and colleagues—Daniel Dykhuizen, Nancy Scavarda, Ray Miller, Ann Honeycutt, Louis Green, Jean de Framond, James Jacobson, and Antony Dean—also deserve credit for their patience and understanding. Special thanks go to Nicki Glaser for her help in preparing the final draft of the manuscript. Washington University in St. Louis has continued to provide a nearly perfect atmosphere for serious research and scholarship.

No project of this magnitude is ever completed without support from one's family, and I wish to thank my wife, Christine, for her patience through much frustration and hard work. Dana and Ted continue as an inspiration in trying to convey the excitement and elegance of genetics in simple terms.

The entire Harper & Row staff has once again achieved an outcome we can all be proud of. In particular, I'd like to thank Claudia Wilson, Bob Greiner, Robert Sugar, Mira Schachne, and Jeanette Perez. The credit for an attractive book should go to them. I would also like to thank the copyeditor, Marymae Klein, who did an excellent job in straightening my often snarled syntax.

A few comments about the wordplay in the title are perhaps in order, lest the title be mistaken as an author's failed attempt to be cute. The title does not imply any questioning of the validity of Mendel's laws or associated genetic principles. The play is on the word "uncertain," referring, on the one hand, to the unpredictability of the outcome of segregation and recombination in the transmission of genes from one generation to the next. Prospective parents who decide to have a baby are undertaking a genetic experiment the outcome of which is by no means entirely predictable, and hence is uncertain. A second connotation of uncertain is "dubious," and in this sense the term alludes to the perhaps surprising number of harmful genes hidden in normal individuals—all of us— which, when they come together in unfavorable combinations, cause severe genetic disorders and much unhappiness. The title also plays on "heritage," meaning primarily genetic inheritance in the present context, but also pointing to the social heritage of today's newborns, an uncertain heritage of fearsome problems in a divided and dangerous world.

Daniel L. Hartl

chapter 1

Cells and Cell Division

Genetics is the study of the fundamental units of inheritance called *genes,* particularly in relation to their function in the organism and how they are passed from generation to generation. This book is an introduction to *human genetics,* the branch of genetics that deals primarily with human beings, so its focus is on the human organism—human genes, human cells, human development, human behavior, human populations. Human genetics is an exciting scientific discipline because it is a lively, challenging field, marked by discoveries of great practical importance, such as the prenatal diagnosis of genetic disease or the recent applications of genetic technology to the large-scale production of such drugs as insulin and growth hormone. These new discoveries in human genetics are often accompanied by enormously complex legal, moral, ethical, and social problems.

Human genetics is a subject of personal interest to many people, especially those who are affected with some inherited disorder or whose relatives are affected. Disorders that can be traced in part to genetic factors are relatively common. Indeed, few families are entirely free of such conditions as high blood pressure, diabetes, schizophrenia, mental retardation, epilepsy, stuttering, deafness, and blindness—conditions that are caused in part by genetic factors. Certain inherited disorders that have occurred among royalty have become notorious, such as hemophilia (bleeder's disease) among several of Queen Victoria's male descendants, or porphyria, another inherited blood disorder, which afflicted King George III with severe episodic pain and mental derangement and may have contributed to his mishandling of the unrest in the American Colonies that ultimately led to the American Revolutionary War.

Whether among royalty or commoners, the occurrence of an inherited disorder in a family often evokes feelings of guilt or personal inadequacy on the part of the parents, and it often creates feelings of anxiety in the unaffected relatives because they may be carriers of the harmful genes. Such guilt and anxiety can be counteracted only by an adequate understanding of the principles of human inheritance together with the facts pertaining to the genetic basis of the particular condition in question.

Not only is human genetics of interest to each of us as individuals, but the subject also has wider social implications. In the early years of this century, genetic knowledge was misused. In those years, there developed a politically powerful **eugenics** movement—a pseudoscientific movement that aimed to "improve" the "genetic quality" of human beings. The eugenics movement demonstrated its political clout in the United States with the passage of the Immigration Restriction Act of 1924, which was designed to limit the immigration of ethnic minorities from southern and eastern Europe on the grounds that the "Nordic" types from northern and western Europe were genetically superior. In state legislatures, eugenicists pushed for the passage of laws for the forcible sterilization of "hereditary defectives," "sexual perverts," "drug fiends," "drunkards," "prostitutes," and others. In 1907, Indiana became the first state to adopt such a compulsory sterilization law, and by 1930 the majority of states had adopted such measures. Luckily, these laws were largely ignored by public officials in most of the United States. In California, however, where the eugenics movement was particularly influential, at least 10,000 men and women were involuntarily sterilized.

The eugenics movement was worldwide, with centers of activity in Japan and western Europe as well as in the United States. The lunacy got entirely out of hand in Nazi Germany, where the Eugenic Sterilization Law of 1933 ultimately led to the compulsory sterilization of at least 250,000 men and women deemed "hereditarily defective." Toward the end of the Nazi regime, compulsory sterilization was dispensed with, and millions of "undesirables" were simply murdered.

In the United States and elsewhere, reputable geneticists were either cowed into silence by the power of the eugenics movement or were simply ignored. Ordinary citizens were largely ignorant of the principles of human genetics, so the eugenicists controlled events. Even today, a surprising amount of eugenic nonsense is found in the public press, but it is hoped that an educated populace will prevent a recurrence of the tragedies that grew out of the eugenics movement.

In short, a knowledge of human genetics serves a social function in allowing people to recognize and thwart those who would misuse genetics for their own ends and prejudices.

As mentioned earlier, knowledge of human genetics also satisfies a private interest. Who has never wondered (or, indeed, worried) about his or her own personal hereditary endowment? If we are cigarette smokers, might we not wonder whether we may have a hereditary predisposition toward lung cancer and so may be dooming ourselves? If we sometimes overindulge in alcohol, might we not wonder about a possible hereditary predisposition toward alcohol abuse? As it happens, much of what is known about human genetics has come from the study of genetic abnormalities and diseases. As will become evident in later chapters, no family is free of actual or potential genetic misfortune, and some of the examples discussed may remind you of situations in your own family, among your relatives, or among your friends.

CELLS

To a considerable extent, the study of modern genetics involves the study of **cells**—the smallest units of living matter. Cells are living units because they take in nutrients, conduct chemical activities, move, reproduce, and do all the other things that are considered to be part of being alive. Cells are fundamental units in biology because animals, plants, and other forms of life are composed of cells. Cells are particularly important in genetics because inheritance is basically a cellular phenomenon. Sexually mature organisms form specialized reproductive cells called **gametes**. (In animals, the male gamete is the sperm and the female gamete is the egg.) Gametes from a male and female individual then unite in the process of fertilization to form a cell called the **zygote**, which undergoes successive cycles of cellular reproduction to form the embryo and ultimately the newborn organism. This process of gamete formation and fertilization is the basic one in genetics because the gametes carry the hereditary material—the genes—from generation to generation. Figure 1.1 is a photograph of a human egg cell surrounded by sperm.

Cells are of interest in modern genetics for practical reasons, too. In many cases, inherited diseases can be traced to defects in certain types of cells in the body. For example, the inherited disorder **sickle cell anemia** has many complex symptoms, including lack of stamina, susceptibility to infections, recurrent pain in skeletal joints, and often heart and kidney abnormalities. Yet all these symptoms have a single underlying cause—a defect in the oxygen-carrying protein **hemoglobin**, which is a normal constituent of the red blood cells. The seemingly unconnected symptoms of sickle cell anemia all relate to the reduced supply of oxygen to the tissues brought on by the hemoglobin defect. When the cellular basis of a disease is known, it is often possible to compensate for the defect directly rather than to settle for mere treatment of the symptoms.

Most cells are too small to be visible to the naked eye. A convenient measure of size for such small objects is the **micrometer** (or **micron**), abbreviated by the symbol μm (μ is the Greek letter *mu*). There are 1000 μm to the millimeter, and about 25,000 μm to the inch. The unaided human eye can resolve objects no

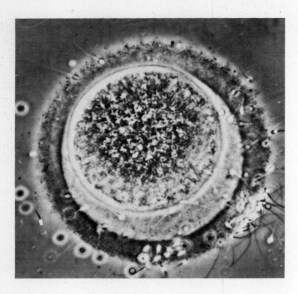

Figure 1.1 A human egg surrounded by human sperm.

smaller than about 80 μm in diameter. With an ordinary light microscope, which consists of series of magnifying and focusing lenses, objects as small as about 0.5 μm can be resolved. Finer resolution requires an **electron microscope**, which uses an arrangement of magnets to focus a beam of electrons. Because the waves in an electron beam are much shorter than those of visible light, an electron microscope can resolve objects as small as about 0.001 μm.

Figure 1.2 illustrates several cellular-sized objects drawn approximately to scale. The innermost large circle represents the limit of resolution of the unaided human eye. Within this circle are objects that require a light microscope or an electron microscope to be seen. Among the smallest cells are bacteria, which range in diameter from 0.2 to 5 μm. The smallest bacteria would require an electron microscope to be seen, and, in the scale of Figure 1.2, are represented as a mere speck. The much-studied intestinal bacterium *Escherichia coli* is in the middle of the bacterial size range.

Another tiny object represented in Figure 1.2 is a smallpox virus (a virus is an internal parasite of cells), which is among the largest of viruses and is approximately the same size as a small bacterium. At the other end of the cellular size range is the ostrich egg, which is 75,000 μm (about 3 inches) in diameter and is the largest known cell. (The ostrich egg owes much of its size to its yolk and other stored foods for nourishment of the ostrich embryo.) The size of a small bacterium is in about the same ratio to the size of an ostrich egg as a thimble is to a domed football stadium.

Among plants and animals, an "average" cell is roughly spherical and about 10 μm in diameter, although there is very great variation in size and shape. In the human body, the smallest cell is a type of white blood cell that is about the same size as a bacterium. The largest cells are certain nerve cells, which have slim projections that may extend more than 3 feet. The human egg, one of the

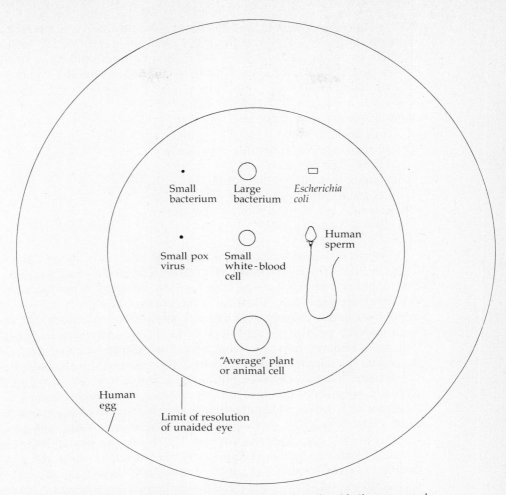

Figure 1.2 Various cellular-sized objects drawn to approximately the same scale.

largest cells in the body, is spherical and about 130μm in diameter. The egg is therefore visible to the naked eye and is roughly the size of the dot over this i. As indicated in Figure 1.2, the human sperm is shaped something like a polliwog. Dimensions as small as these are hard to make vivid because they are so much smaller than those encountered in everyday life. It may help to note that all the sperm that gave rise to all the people that ever lived could be carried in a teaspoon; the eggs, being larger, would require a small bucket.

Human reproduction is accomplished through the production of sperm and eggs, which, recall, are collectively known as **gametes**. The genes in an individual are distributed to the gametes in a highly orderly fashion that will be discussed in Chapter 2. For now it is sufficient to note that a normal human female releases one egg about once a month from one or the other of her ovaries. Rarely will two eggs be ovulated in the same month, very rarely three, and almost never four or more—unless the ovaries have been artificially stimulated by fertility drugs or hormones. A normal male, by contrast, will expel 350 million sperm in a single ejaculate. Thus, any particular sperm is very unlikely to fertilize an egg. This consideration led Aldous Huxley to comment as follows on his own chances of ever being born:

A million million spermatozoa,
    All of them alive:
Out of their cataclysm but one poor Noah
    Dare hope to survive.
And among that billion minus one
    Might have chanced to be
Shakespeare, another Newton, a new Donne—
    But the one was Me.*

When a sperm penetrates an egg in the process of fertilization (see Figure 1.1), the resulting cell is known as a **zygote**. The single-celled zygote then undergoes a process of cell division called **mitosis**, whereby it produces two daughter cells that are genetically identical. (The process of mitosis will be discussed in detail later in this chapter.) The two daughter cells produced by the first division each undergo mitosis, giving a total of four genetically identical cells. The four cells each undergo mitosis, giving a total of eight genetically identical cells. So on and on the number of cells increases. By means of signals that are still largely mysterious, the cells organize into a recognizable human embryo. Although the cells are genetically identical, they nevertheless undergo changes in form and function—a process called **differentiation**—whereby they become specialized into heart cells, liver cells, brain cells, lung cells, muscle cells, and so on. As the embryo grows, cell divisions continue. At birth, the human infant consists of about 1 trillion (a million million) cells. This would correspond to roughly 40 doublings of the fertilized egg and its products if the cells in the embryo had all divided in unison. They do not divide synchronously, however. Some of the cells lag behind others. Some cease division entirely. Therefore, it is only approximately true to say that about 40 cell doublings occur between fertilization and birth. The growth of a child into an adult is partly due to an increase in the number of cells. A fully grown adult has about 10 trillion cells. Some of the cells in an adult are specialized to form gametes, and so the cycle of life continues.

CELL STRUCTURE

In addition to wide-ranging sizes and numbers, cells are also found in a variety of shapes (see Figure 1.3 for several examples). Often we picture cells as spheres, but this does an injustice to their variety. The nerve cell has long, thin projections; the muscle cell resembles an earthworm; and the pigment cells (found just beneath the outer layer of skin and containing the pigments responsible for skin color)

* Aldous Huxley, from the *Fifth Philosopher's Song*.

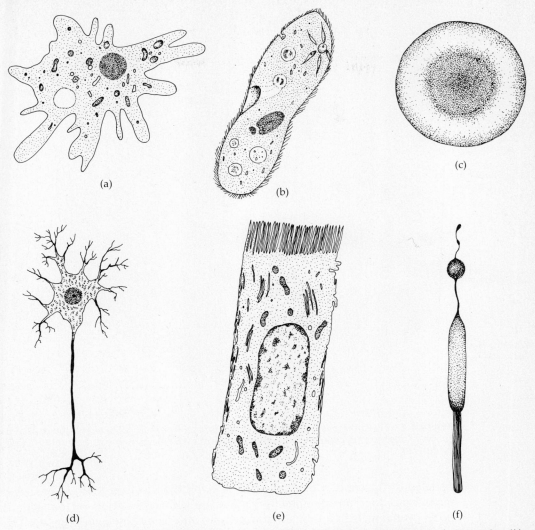

Figure 1.3 Specialized cells have a great variety of shapes: (*a*) a single-celled amoeba, (*b*) a paramecium, (*c*) a red blood cell, (*d*) a nerve cell, (*e*) an intestinal cell, and (*f*) a rod cell from the retina of the eye.

look very much like splotches of ink. Some cells can change their shape. Muscle cells are long and thin when relaxed, short and fat when contracted. The variety of cell shapes in the human body only hints at the many and often bizarre shapes found among other forms of life.

For many purposes, the cell may be thought of as a little package of specialized structures and molecules that carries out the functions of life in a coordinated manner. Figure 1.4 illustrates the three-dimensional organization of a "typical" animal cell. The outer boundary of the cell is a thin film called the **cell membrane**. This membrane is a flexible structure about 0.01 μm thick, so its

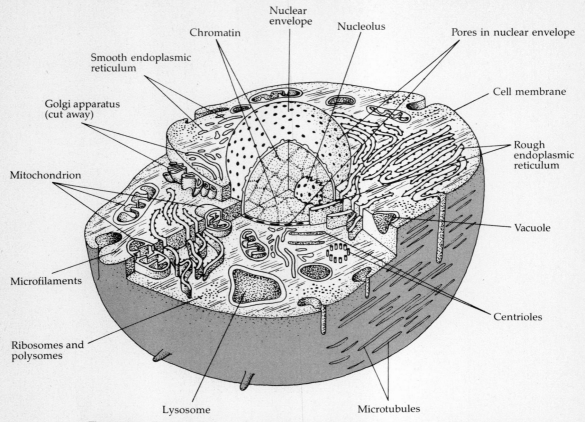

Figure 1.4 Three-dimensional drawing of a "typical" animal cell showing the spatial organization of the various cellular components. Note the extensive membrane structures within the cytoplasm.

detailed structure can be seen only through the electron microscope. The cell membrane does not completely isolate the cell from the world outside. It has holes or channels in it that allow water and other small molecules to pass through easily. Large molecules and important charged atoms or molecules known as **ions**, such as sodium ions, cannot get through the membrane easily, however. The movement of these large molecules and ions through the cell membrane is accomplished by energy-requiring "pumps," the nature of which is still not understood.

Most cells manufacture substances that surround or coat the cell membrane. These are usually too porous to hinder the movement of even very large molecules, but they do serve to protect and support the cells. Plant cells have a rigid outer coating called the **cell wall** (see Figure 1.5*b*), which is made up of a fibrous material called **cellulose**. Wood is mostly cellulose, and cotton is more than 90 percent pure cellulose.

Inside, the cell contains not only molecules but also internal structures of

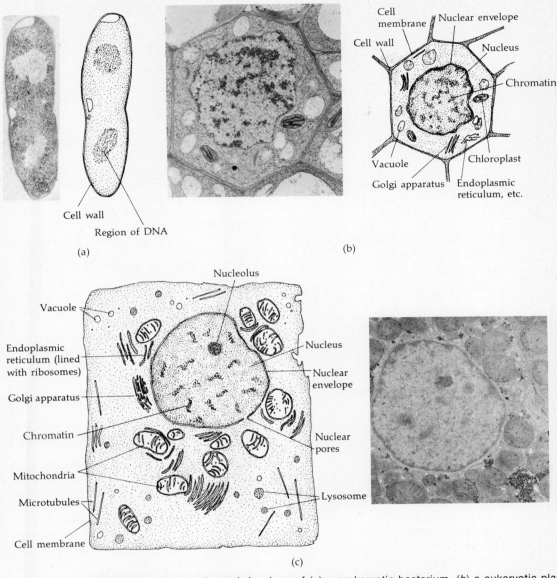

Cell wall

Region of DNA

(a)

Cell
membrane | Nuclear envelope

Cell wall

Nucleus

Chromatin

Vacuole

Chloroplast

Golgi apparatus | Endoplasmic
reticulum, etc.

(b)

Nucleolus

Vacuole

Endoplasmic
reticulum (lined
with ribosomes)

Golgi apparatus

Chromatin

Mitochondria

Microtubules

Cell membrane

Nucleus

Nuclear
envelope

Nuclear
pores

Lysosome

(c)

Figure 1.5 Micrographs and drawings of (*a*) a prokaryotic bacterium, (*b*) a eukaryotic plant cell, and (*c*) a eukaryotic animal cell. Note the absence of a nuclear envelope in the prokaryotic cell. Eukaryotic cells are conveniently considered to consist of two parts—the nucleus, which contains the chromosomes (although only chromatin is visible at this stage of the cell's life cycle), and the cytoplasm, which includes everything outside the nucleus. The photos are: (*a*) the human intestinal bacterium *Escherichia coli* (20,000×); (*b*) a cell from the stem of a pea seedling, *Pisum sativum* (7000×); and (*c*) a bat liver cell (6000×).

**Table 1.1   FUNCTIONS OF ORGANELLES AND SOME OTHER STRUCTURES FOUND IN
EUKARYOTIC CELLS**

Nucleus

Nuclear envelope	Membranous boundary of nucleus
Nuclear pores	Holelike gaps in nuclear envelope; function uncertain
Nucleolus	Dense spherical body within nucleus; site of synthesis of certain constituents of ribosomes (see below)
Chromatin	Contains genetic material DNA (deoxyribonucleic acid) and other constituents; becomes organized into visible **chromosomes** at certain stages in dividing cells

Cytoplasm

Ribosomes	Tiny spherical particles containing special proteins and RNA (ribonucleic acid); synthesis of all proteins occurs on ribosomes; frequently found in chains of 5 to 10 ribosomes called **polysomes,** actively engaged in protein synthesis
Endoplasmic reticulum (ER)	Network of extensively folded internal membranes; two principal categories are **rough ER** (surface peppered with ribosomes) and **smooth ER** (free of ribosomes)
Golgi apparatus	Collection of smooth membranous structures of characteristic appearance near the nucleus; important in storage and secretion of certain substances
Mitochondria	The "power plants" of eukaryotic cells; site of chemical transformations resulting in the high-energy molecule ATP (adenosine triphosphate), which is a source of energy for other chemical transformations in the cell
Chloroplasts (plants only)	Site of photosynthesis, in which the energy of sunlight is used to combine carbon dioxide and water into the sugar glucose
Lysosomes	Membranous sacs for storage of powerful digestive enzymes.
Vacuoles	Membranous sacs serving as storage reservoirs of diverse substances; some contain only water, others dissolved sugars, still others pigments or engulfed food particles
Microtubules and microfilaments	Cablelike aggregates of special proteins; important in cell rigidity, movement, and (particularly microtubules) cell division
Centrioles (animals only)	Bodies with a characteristic arrangement of 27 microtubules; important in chromosome movement during cell division
Cilia and flagella (some cells only)	Short hairlike (cilia) or long whiplike (flagella) projections from the cell surface; important in cell movement; flagellated cells usually have one or two such projections, whereas cilia are numerous

Cell membrane	Membranous outer boundary of cell
Cell wall (plants only)	Rigid boxlike structures enclosing cell membrane; composed primarily of network of tough cellulose molecules

various kinds that carry out specialized functions or activities. These intracellular structures are called **organelles**, or little organs. About a dozen classes of organelles are found in typical animal and plant cells (see Figures 1.4, 1.5a and 1.5b for illustrations and Table 1.1 for a list).

One organelle, the **nucleus**, is of particular importance in genetics because it contains molecules that carry the hereditary information in the cell—the genetic information that makes the cells of a particular species unique. This information gives the cell its identity as a human cell, a toad cell, an elm cell, or an apple cell. Genetic information in the nucleus is present in molecules of a substance known as **DNA** (for **deoxyribonucleic acid**). The nucleus of nondividing cells contains strands of a material called **chromatin**, which is dispersed throughout the nucleus and, because it is more dense in some regions than others, gives the nucleus an uneven or granular appearance. The chief chemical constituent of chromatin is DNA. In dividing cells, the chromatin strands undergo a little-understood process of coiling and so become visible through the light microscope as discrete entities called **chromosomes**.

The nucleus is the control center of the cell because hereditary instructions in the nucleus are "read" and they trigger the rest of the cell to manufacture certain things or to carry out certain tasks. For this reason, it is customary to consider the cell as consisting of two major regions—the nucleus and the cytoplasm. The **cytoplasm** includes everything in the cell other than the nucleus. The distinction between nucleus and cytoplasm turns out to be useful because most cells have just one nucleus, and this is set off from the cytoplasm by its own membranous **nuclear envelope**. Large molecules can move between the nucleus and the cytoplasm, very likely by means of **nuclear pores** in the nuclear envelope.

Bacteria and blue-green algae have no well-defined nucleus. This creates the major distinction that divides all cells, and indeed all forms of life, into two groups. Organisms represented by cells with a nucleus are called **eukaryotes** (see Figure 1.5b and c); relatively primitive cells such as bacteria and blue-green algae with no nuclear envelope surrounding the hereditary material are called **prokaryotes** (see Figure 1.5a). In accord with their comparatively simple structure, prokaryotic cells lack most of the complex organelles associated with eukaryotic cells. Moreover, cell division in prokaryotes is quite different from that in eukaryotes. Although the genetic material in prokaryotes is DNA, as it is in eukaryotes, the DNA in prokaryotic cells is not organized into chromatin, so prokaryotic cells do not possess true chromosomes. Nevertheless, the DNA in prokaryotic cells is often called a "chromosome," although this terminology is, strictly speaking, incorrect.

Next in prominence to the nucleus in most eukaryotic cells are the bacteria-sized **mitochondria**. These organelles range in size from 0.2 to 3 μm and in shape from spheres to sausages. An active cell like a liver cell may contain up to 1000 mitochondria. The covering of the mitochondrion is a double membrane, each layer like the cell membrane, and the inner layer is wrinkled or corrugated, with its numerous folds extending into the interior of the mitochondrion (Figure 1.6a). This folding provides a large surface area inside the mitochondrion to support the chemical processes that occur there—vital processes that generate most of

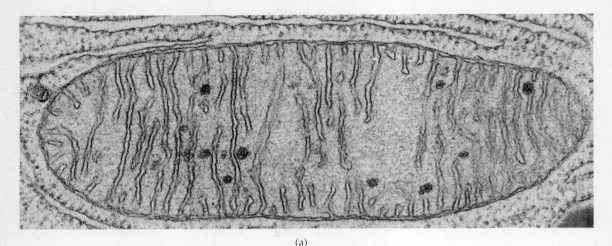

(a)

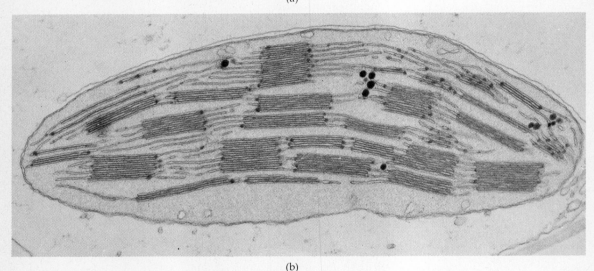

(b)

Figure 1.6 Micrographs of (*a*) a mitochondrion, found in all eukaryotic cells—this one is from a bat pancreatic cell (61,600×); and (*b*) a chloroplast, found only in plant cells—this one is from a mature lettuce leaf (25,550×).

the energy in the cell. (Many of the chemical processes in a cell take place on the surface of membranes.) Thus, the mitochondrion may be thought of as the cell's power plant, coverting the energy in the chemical bonds of various substances into a supercharged chemical bond that is stored in a molecule known by the initials **ATP** (for **adenosine triphosphate**). In loose terms, the mitochondrion ''cascades'' an electron down a special sequence of molecules, and at every step the electron gives up a pulse of energy to form a molecule of ATP. Finally, the electron has very little energy left, and it is incorporated into a molecule of water. The process of releasing energy to ATP in the mitochondrion is known as **cellular respiration**.

The storage battery of the cell is the high-energy bond in ATP. In the same way that some of the energy required to compress a spring is regained when the spring jumps back, some of the energy required to make a chemical bond between atoms is regained when the bond is cleaved. All chemical bonds release some energy when broken, but the high-energy bond of ATP is unusual in that it releases more energy than most chemical bonds. The high-energy bond of ATP provides the energy for most cellular activities.

The original source of energy in the chemical bonds of biological molecules is the sun. Green plants are almost alone in being able to convert the energy in sunlight into chemical bonds usable by living things; green plants are the original source of energy for almost all other forms of life, including humans. The cellular locations of the energy-binding process in eukaryotic plant cells are in organelles called **chloroplasts**, and the process is known as **photosynthesis**. Chloroplasts are present in only certain cells of green plants; they make green plants green. Found in a variety of shapes, often disk-shaped particles 5 to 8 μm in diameter, chloroplasts contain, among other things, the green pigments collectively called chlorophyll (see Figure 1.6b). Chlorophyll has an electron that becomes very excited (it gains energy) when struck by sunlight. The chloroplast can then convert the trapped energy in this excited electron into the chemical bonds of molecules. The chloroplast, therefore, does roughly the opposite of the mitochondrion. Whereas the chloroplast acquires energy by building chemical bonds during photosynthesis, the mitochondrion releases the energy by breaking them down during respiration.

Mitochondria and chloroplasts contain small ring-shaped molecules of DNA, very similar to the DNA in prokaryotic cells. These organelles also contain **ribosomes** (small granular particles on which proteins are manufactured) and certain other constituents that are very similar to their prokaryotic counterparts. The contribution of organelle DNA to the cell's heredity is minor, however; the DNA in chromosomes in the cell nucleus is vastly more important in heredity than is the DNA in mitochondria and chloroplasts. Moreover, the bulk of protein synthesis occurs on ribosomes located in the cytoplasm, not on ribosomes in these organelles. Nevertheless, many biologists believe that the ancestors of mitochondria and chloroplasts were once prokaryotic organisms in their own right, not parts of cells, and that the DNA and ribosomes in these organelles today are vestiges of this ancient past. As evolution proceeded, these primitive organisms came to be associated with cells and, along the way, lost their ability to survive independently. The eukaryotic cell, in turn, has become dependent on them for its own survival; the mitochondrion is necessary for respiration and the chloroplast for photosynthesis.

Convoluted throughout the cytoplasm of a eukaryotic cell is the **endoplasmic reticulum**, a maze of membranes that folds back and overlaps, providing a large surface area that supports the numerous chemical reactions that occur on them (see Figures 1.4 and 1.5). The interior surface of parts of the endoplasmic reticulum is often dotted with strings of five or six **ribosomes**. These tiny, almost spherical structures are the workers on the cellular assembly lines. Only 0.015 μm in diameter and numbering in the hundreds of thousands in a typical eukaryotic cell, ribosomes assemble the countless protein molecules needed for specific chemical reactions to occur in the cell. (In general, one particular kind of protein is able to aid one particular reaction, although some proteins, rather than accelerating chemical reactions, become parts of intracellular structures.)

The cell nucleus usually contains one or more dense areas, each called a **nucleolus** (see Figure 1.4). The nucleolus is involved in the production of the **RNA** (for **ribonucleic acid**) constituent of ribosomes. Even prokaryotic cells contain ribosomes; a typical prokaryotic cell may contain 20,000 to 30,000 ribosomes.

Among the thousands of proteins produced in cells are some that form the **microtubules** and others that form the **microfilaments**. These long, cablelike aggregates of protein provide structural support for the cell. The cell is evidently able to assemble these supporting structures whenever they are needed and to disassemble them later. A most important function of microtubules is to form the football-shaped network of fibers that arches through the cell during the middle phases of cell division and guides the movement of the chromosomes; this network is called the **spindle**.

Eukaryotic cells also contain other types of specialized structures, whose functions are outlined in Table 1.1. Included here are the *Golgi apparatus,* a series of folds of membranes serving as a reservoir of materials to be secreted from the cell; *lysosomes,* which are small spherical sacs containing powerful digestive enzymes; and *vacuoles,* which are globules or blebs of various kinds and sizes that may contain water, pigments, stored food, or substances temporarily stored for transport into or out of the cell.

Many cells also carry small appendages or hairlike structures that enable movement. Among these are small hairs called *cilia,* such as those lining the throat and the bronchial tubes leading to the lungs, which normally have a back-and-forth brushlike movement that occurs in waves, sweeping out trapped dust, dirt, and small organisms that are inhaled. Another example of a movable cellular appendage is the tail of a sperm, which whips and propels the sperm through its liquid surroundings. Such long whiplike appendages are known as *flagella*.

All cells, and particularly eukaryotic cells, are highly complex living things, and something has to coordinate a cell's many activities. The coordinator of the eukaryotic cell is the nucleus. More precisely, it is the genetic or hereditary material in the nucleus—the celebrated molecule known as DNA. One key to understanding the cell—and the whole organism—is to understand the nature and behavior of the genetic material in the nucleus.

CELL DIVISION

As noted earlier, the genetic material resides in the chromosomes found only within the nucleus of eukaryotic cells. In this section, we focus on the behavior of chromosomes during cell division. Cell division is a central aspect of genetics, because it is the process by which chromosomes are transmitted from one generation to the next.

Actually, two types of cell division occur in humans and most other eukaryotes. Figure 1.7 is an outline of the human life cycle showing the role played by each type of division. One type of division, which occurs only in the ovaries

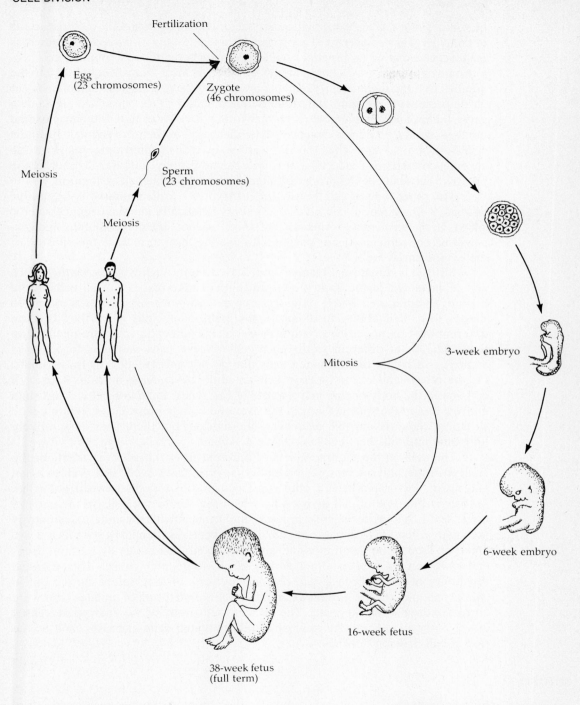

Fertilization

Egg
(23 chromosomes)

Zygote
(46 chromosomes)

Meiosis

Sperm
(23 chromosomes)

Meiosis

Mitosis

3-week embryo

6-week embryo

16-week fetus

38-week fetus
(full term)

Figure 1.7 The human life cycle.

of females and the testes of males, is concerned exclusively with the production of gametes (eggs or sperm). This gamete-producing type of division is known as **meiosis**, and one of its principal features is that it reduces the number of chromosomes from 46 (the characteristic number in most human cells) to 23 (the number in human gametes). Gametes are said to carry a **haploid** complement of chromosomes; the zygote and all subsequent stages of the life cycle are said to carry the **diploid** complement of chromosomes. Details of how the chromosome number is reduced during meiosis will be discussed in the next chapter. For now it is sufficient to note that the fertilization of a 23-chromosome egg by a 23-chromosome sperm produces a 46-chromosome zygote; thus, fertilization is the process that restores the proper chromosome number in each generation.

The other type of cell division in eukaryotes is called **mitosis**. By means of mitosis, a single parent cell gives rise to two genetically identical daughter cells. Although the term *mitosis* is technically used to denote the division of the nucleus, it will be convenient for our purposes to use the term to include the division of the cytoplasm as well.

Mitosis is the normal manner of cell division through which growth, repair, and replacement of the body's cells and tissues take place. Not all cells in the body of an adult are capable of undergoing mitosis, however. The cells of certain tissues lose their ability to divide in the embryo or child when they become differentiated by acquiring or expressing specializations in structure or function that make them different from other cells. Nerve cells and muscle cells are examples of highly differentiated cells that do not undergo mitosis in the adult. On the other hand, some types of cells in adults do undergo mitosis. Rapid cell division in the body occurs in a number of cell types, including cells in the bone marrow and in the spleen. Cells in the liver and kidney represent an intermediate situation; they are capable of undergoing division, but there are comparatively long time intervals between successive divisions.

The cells in the embryo are very different from those in the adult. In the embryo, especially in very young embryos, practically every cell divides again and again very rapidly with only short pauses for growth between divisions. Mitosis is the type of cell division by which the number of cells in the embryo increases. It is a matter of great practical importance that cells undergoing mitosis are highly sensitive to damage by radiation and certain chemicals; for this reason, embryos tend to be much more severely damaged than adults by harmful drugs and radiation as well as other forms of environmental stress. This extreme sensitivity of the embryo is one reason why tests of drug safety carried out in animals should include careful examination of the offspring of females who were given the drug when pregnant. The extreme sensitivity of the embryo decreases as the embryo ages, as the number of differentiated cells increases, and as the rate of cell division declines.

MITOSIS

Mitosis lends itself to an easy summary: the chromosomes become double (replicated) and the two halves pull apart as the cell pinches into two. The process is that simple, but it does not occur without considerable ceremony. In this section we take a closer look at mitosis, paying particular attention to the behavior of the chromosomes during the process. Mitosis is a cellular *process* in which the events that occur at any one moment blend imperceptibly into those that occur at the next moment. Nevertheless, for purposes of discussion, mitosis is usually described in terms of landmark stages known as *interphase*, *prophase*, *metaphase*, *anaphase*, and *telophase*. These landmark stages are illustrated by the sketches in Figure 1.8 and in the photographs of nuclei of cells of the desert locust *Schistocerca gregaria* in Figure 1.9. Each stage of mitosis warrants a brief discussion.

As is apparent in Figure 1.8, mitosis is a cyclical process, with the products of one division being able to recycle into yet another division. It will be convenient for discussion to break into the cycle at **interphase**, as this is the stage of mitosis mainly concerned with cellular growth. During interphase, the cell usually doubles, or nearly doubles, in size; it increases the number or amount of virtually all the cytoplasmic components—mitochondria, ribosomes, endoplasmic reticulum, and so on—but not the nucleus. The nucleus remains single. During this period, however, within the nucleus, the cell makes an exact copy or replica of each chromosome.

Chromosome replication, as the chromosome-copying process is called, is not apparent in the nuclei of cells in interphase because at this stage the chromosomes are not visible through the light microscope. All that can be seen in the nucleus is one or more nucleoli and the diffuse granular substance called *chromatin* mentioned earlier (see Figures 1.8 and 1.9). In the electron microscope, however, the chromosomes are seen as slender, very thin, coiled threads that meander throughout the nucleus, their tips attached to the inner surface of the nuclear envelope.

During interphase, the cell produces DNA, RNA, protein, lipids, and several other kinds of molecules, and it grows rapidly as the various constituents of the cytoplasm, such as mitochondria and ribosomes, are increasing in number. The production of molecules of certain kinds may be restricted to specific places in the cell or to particular times. Almost all DNA synthesis, for example, is confined to one period approximately in the middle of interphase. The synthesis of DNA is part of chromosome replication, during which the cell makes an exact copy of each of its chromosomes.

The chromosomes that have been replicated develop more coils. Like a string that becomes shorter and thicker by being twisted into coils and twisted further into more coils, each chromosome becomes shorter and thicker by coiling. As the chromosomes thicken, they provide more surface area for absorbing the dye added to stain the cells; eventually their thickness concentrates enough dye that the chromosomes become visible through the light microscope. The chromosomes are first barely visible as long, slender threads convoluted throughout the nucleus. The first visible appearance of the chromosomes marks the beginning of **prophase**.

During prophase, the chromosomes continue to shorten and thicken, presumably by further coiling. Gradually the chromosomes become short enough so

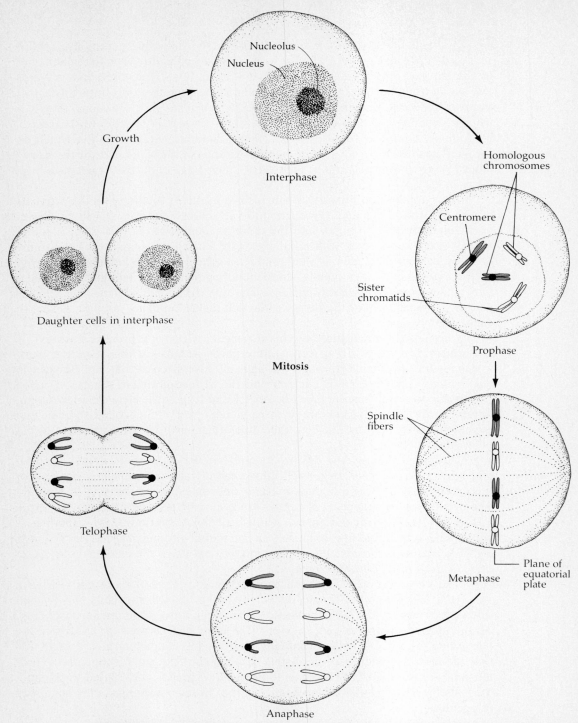

Figure 1.8 Mitosis in a somatic cell of a hypothetical organism that has two pairs of chromosomes.

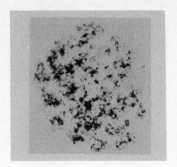

Interphase

Prophase

Mitosis

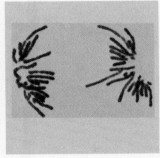

Telophase

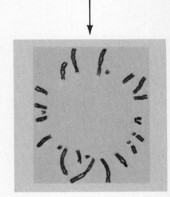

Metaphase

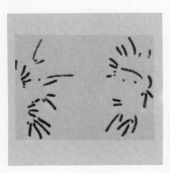

Anaphase

Figure 1.9 Mitosis in cells of the desert locust *Shistocerca gregaria* as viewed through the light microscope (600–850×). The metaphase view is from one pole of the spindle. Note the similarity with the stages illustrated in Figure 1.8.

that they can be recognized as discrete entities rather than as a disorganized jumble of thread (see Figures 1.8 and 1.9). As time goes on, the chromosomes become still shorter and thicker and thus easier to scrutinize. Figure 1.10 shows the chromosomes in a human cell during late prophase. Several features of chromosome structure can readily be observed. First, the chromosomes have replicated; they look double. Each chromosome consists of halves lying close together side by side, and each half of the chromosome is called a **chromatid**. Because the two chromatids in each chromosome are genetically identical, the partner chromatids are often called **sister chromatids**.

A second point to note is that, although each chromosome has replicated almost everywhere along its length, there is one specific point or region on the chromosome that still appears single, like the crosspoint of an X. This special point is called the **centromere**, and it will become visibly doubled later in mitosis. Because they are still both attached to a single centromere, sister chromatids are technically considered to be parts of a single chromosome during prophase. Only later in mitosis, when the centromere itself finally splits apart, will each sister chromatid be considered a chromosome in its own right. Thus, during prophase, a human cell has 92 chromatids but 46 chromosomes.

A third point to note is that the chromosomes in a cell do not all look the same. Some are relatively long, some short, and some intermediate in size. Some have their centromere in the middle, others near an end, and still others practically at the tip. But a regularity or pattern can be discerned. Each chromosome has a match, a sort of partner, another chromosome that looks just like it somewhere else in the same nucleus. It may require some hunting to find the chromosomes that match in size and appearance because the chromosomes are in no particular order in the nucleus; their arrangement appears haphazard. The pairs of chromosomes that match in size and appearance are called **homologous chromosomes**, or **homologues** for short. Chromosomes that do not match are called **nonhomologues**.

The 46 chromosomes in a cell in a human female are shown in Figure 1.11*a*. When the individual chromosomes are cut out of the photograph and arranged in pairs by size, as shown in Figure 1.11*b*, the representation is called a **karyotype** of the individual. Note in the karyotype in Figure 1.11*b* that there are 23 pairs of homologous chromosomes. Although the homologous pairs of chromosomes in some organisms are sufficiently distinctive to be recognized quite easily (see Figure 1.9 for one example), the homologous pairs of chromosomes in human cells are difficult to identify because many chromosomes are too similar in size and centromere position to be readily distinguished.

As will be discussed in the next chapter, special staining procedures have been devised to reveal which human chromosomes are homologues. It should be pointed out here that in the human male, in contrast to the female, one pair of chromosomes does not match in size. The chromosomes in this nonmatching pair are involved in the determination of sex, and they will be discussed in later chapters. Yet even though the sex chromosomes are of unequal size, they are often considered to be homologous.

Prophase is followed by **metaphase**. The beginning of metaphase is signaled

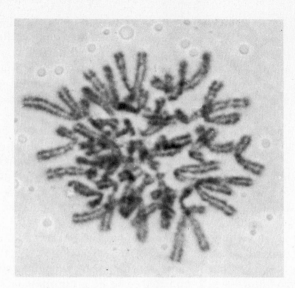

Figure 1.10 Micrograph of chromosomes in a human cell during late prophase of mitosis. As can be seen clearly in a number of chromosomes near the periphery, each chromosome at this stage consists of two chromatids lying side by side and connected to a common centromere.

by the breakdown and rapid dissolution of the nuclear envelope. The **spindle** becomes especially prominent in metaphase (see Figure 1.8), although it actually began to be formed during prophase. It is composed of fibers, mainly microtubules, that arch over and through the region occupied by the chromosomes. The spindle has roughly the shape of a football, with the bulge in the middle engulfing

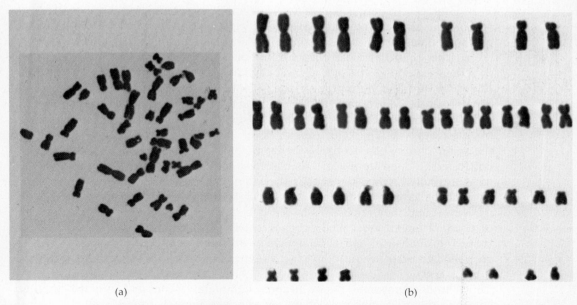

(a) (b)

Figure 1.11 Metaphase spread (a) and karyotype (b) of the 46 chromosomes found in the somatic cells of a normal female.

the region occupied by the chromosomes. Some of the spindle fibers become attached to the centromeres of the chromosomes, and a tugging of the fibers on the centromeres begins. Each centromere is pulled or pushed into position approximately in the middle of the spindle. Imagine a plane cutting crosswise through the spindle at an equal distance from the ends of the spindle. This imaginary plane is called the **equatorial plate**, or the **metaphase plate**, and the centromeres of the chromosomes are brought to lie on this plane. The metaphase sketch in Figure 1.8 is an edge-on view of the equatorial plate because the imaginary plane runs through the centromeres. The metaphase photograph in Figure 1.9 is a view perpendicular to the equatorial plate as seen from one end of the spindle. Although the two views look rather different, they are just different aspects of the same three-dimensional structure. When the centromeres have been aligned on the equatorial plate, the cell is ready for the actual separation of the sister chromatids.

Anaphase commences when the centromeres of the chromosomes become visibly double and begin to separate (see Figures 1.8 and 1.9). At the moment the centromere splits, each chromosome is considered to have divided; each half (formerly called a *chromatid*) is now considered to be a chromosome in its own right. Immediately after the centromeres have split in anaphase, therefore, a human cell contains 92 chromosomes.

When a centromere has split, its halves begin to be pulled toward opposite poles of the spindle, toward the pointed ends. This moving apart happens because the spindle fibers attached to the moving centromeres shorten during anaphase. In addition, the whole cell seems to elongate slightly along the direction of the spindle.

Each centromere is part of a chromosome, of course, and as the centromere moves, the arms of the chromosome are pulled passively along. (Very rare cells can sometimes be found that contain an abnormal chromosome with no centromere. This centromereless chromosome does not move like the others; it tends to remain behind as the others are pulled away, and it is often pushed by other forces toward the periphery of the spindle.) Sister chromatids first begin to separate at the centromere, and the tugging of the spindle fiber on the centromere is initially counteracted by the apparent stickiness of the chromatids and sometimes by the winding of the arms of sister chromatids around each other. When the sister chromatids do finally become completely separated, they are rapidly pulled to the opposite poles. The chromosomes assume characteristic shapes as they are pulled along, the shapes depending on the position of the centromere. Chromosomes with their centromere near the middle are called **metacentric** chromosomes, and they appear V-shaped during anaphase; chromosomes with arms of unequal length are called **submetacentric** chromosomes, and they appear J-shaped at anaphase; and chromosomes with their centromere very near the tip are called **acrocentric** chromosomes, and they appear rod-shaped during anaphase (Figure 1.12).

The chromosomes finally become completely separated into two groups, with one group at each pole of the spindle (see Figures 1.8 and 1.9). In normal human cells, each group contains exactly 46 chromosomes. Barring rare acci-

Type of chromosome	Appearance at metaphase	Appearance at anaphase

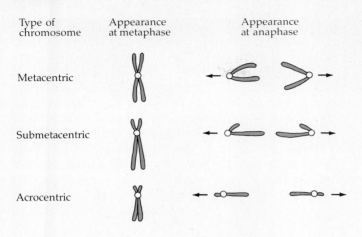

Figure 1.12 Definition of types of chromosomes (metacentric, submetacentric, and acrocentric) in terms of centromere position and appearance at metaphase and anaphase of mitosis.

dents, such as the addition or loss of a chromosome, each group consists of an exact duplicate of the chromosomes that were present in the cell before it began mitosis; that is, the group of chromosomes at one pole is identical with the group at the other pole.

The next stage of mitosis, **telophase**, is marked by the reappearance of a nuclear envelope around each group of chromosomes, the breakdown and disappearance of the spindle, and the division of the cytoplasm (see Figure 1.8). In animals, the cell literally pinches itself in two (Figure 1.13). A furrow due to the action of microfilaments develops around the circumference of the cell along the edge of the equatorial plate, which now separates the two groups of chromosomes. This furrow pinches in like a noose tightening around a balloon. Finally the two daughter cells are completely separated. (In plants, a cell wall grows, forming a partition between the daughter cells.)

Cytoplasmic division divides the contents of the cytoplasm—mitochondria, ribosomes, and so on—into roughly equal parts. Whereas the division of the chromosomes is highly exact, the division of the cytoplasmic contents is more irregular. It does not seem to matter whether one daughter cell gets a few more mitochondria or ribosomes than the other. In any case, the process of mitosis is now complete. Where before there was one cell, now there are two, and the two daughter cells are genetically identical.

After telophase, each daughter nucleus fades back into an interphaselike appearance, and the mitotic cycle may begin anew. Each telophase nucleus has received one complete diploid set of chromosomes, and, if the cell were to divide again, it would be important to replicate the chromosomes before proceeding into the next prophase.

At this point, it might be helpful to review the major aspects of chromosome behavior during the various stages of mitosis:

Interphase: Chromosomes replicate.

Prophase: Chromosomes first become visible through light microscope.

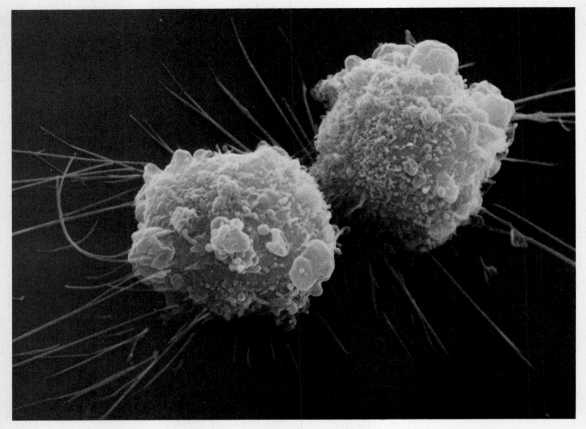

Figure 1.13 A mouse cell late in telophase nearing the completion of cell division (approx. 4000×). The fibers extending out from the cells are "retraction fibers" along which each daughter cell will flatten out after mitosis is completed.

> **Metaphase:** Chromosomes align with centromeres on equatorial plate.
>
> **Anaphase:** Centromeres split and chromosomes move to opposite poles.
>
> **Telophase:** Chromosomes resume interphase appearance; cytoplasm divides; nuclear envelope reappears.

From a genetic point of view, the most important feature of mitosis is that the daughter cells receive identical complements of chromosomes. To paraphrase our earlier brief summary statement of mitosis: the chromosomes replicate, and the genetically identical sister chromatids pull apart as the cell pinches into two.

CAPTURING CHROMOSOMES IN METAPHASE

Chromosomes are easiest to examine and count in metaphase (see Figure 1.11). The problem, though, is that metaphase is fairly short, so not many cells are to be found in metaphase at any one time. The lengths of time spent by cells in the different stages of mitosis depend on the type of cell and the organism it comes from, as well as on such other circumstances as temperature and age.

In human tissues, the situation for chromosome study would seem to be very unfavorable. Certain white blood cells cultured in the laboratory divide about every 18 hours; 17 of these hours are spent in interphase, and metaphase takes just a few minutes. Luckily, a chemical trick can be used to capture chromosomes at metaphase. The trick is to treat cells with the drug **colchicine** (or with a less toxic chemical derivative of it). Colchicine is a spindle poison. By binding chemically with the molecules that make up the spindle, it prevents the formation of the spindle during mitosis. Cells treated with colchicine cannot pass through anaphase. Therefore, a culture of cells treated with colchicine will accumulate cells in metaphase because, no matter where the cell was in the mitotic cycle when the colchicine was added, sooner or later it must reach metaphase. Since a spindle is required to continue into anaphase, cells that lack a spindle simply stop at metaphase. Colchicine is used routinely for chromosome study in virtually all human genetics laboratories.

GENETIC INFORMATION IN THE CHROMOSOMES

Figure 1.14 depicts a highly schematic view of the organization of DNA in chromosomes. The figure is greatly oversimplified because an actual molecule of DNA is extremely long—much longer than a chromosome. For example, the length of the longest human chromosome at metaphase is about 10 μm (10^{-3} cm); if the DNA in this chromosome were fully extended, its length would be about 8 cm; that is, the extended length of the DNA is some 8000 times the length of the chromosome itself.

Details of how DNA is packaged in the chromosome are not understood. It is known that the DNA molecule becomes associated with certain types of proteins called **histones**, which form a sort of beads-on-a-string structure, the beads being known as **nucleosomes** (see Figure 1.15b). When the spacer DNA between the nucleosomes becomes wrapped up in the beads, a nucleosome fiber 100Å in diameter is formed (Figure 1.15c). (An *angstrom* unit, Å, is a unit of length equal to 0.0001 μm.) The nucleosome fiber undergoes further coiling to produce a **chromatin fiber** approximately 300–500Å in diameter (Figure 1.15d). This chromatin fiber undergoes still higher levels of coiling (Figure 1.15e and f) to produce the chromatid seen at metaphase. Although some of the details of the coiling in Figure 1.15 are speculative, the complex coils and gyrations of the fundamental chromatin fiber are evident in Figure 1.16, which is an electron micrograph of one of the smallest human chromosomes at metaphase.

The genetic information in a chromosome resides in its DNA. As noted earlier, each DNA molecule consists of two intertwined strands. These strands carry genetic information in the form of a chemical "alphabet," so each strand of DNA may be thought of as a tiny ticker tape with genetic instructions printed on it. The most important unit of genetic information is the gene, which, to use the ticker-tape analogy, corresponds to a segment of the ticker tape. A typical gene is thought to consist of several thousand chemical "letters." Most genes carry the information to direct the manufacture of a particular type of protein in

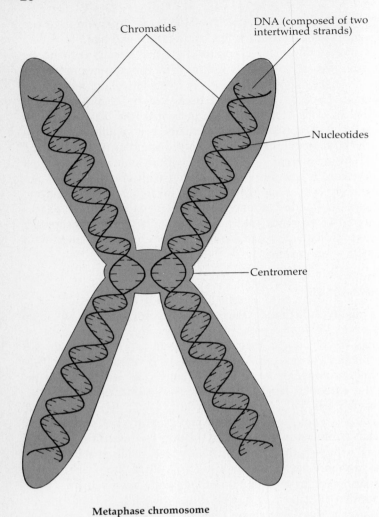

Chromatids

DNA (composed of two intertwined strands)

Nucleotides

Centromere

Metaphase chromosome

Figure 1.14 Highly schematic version of the molecular structure of a eukaryotic metaphase chromosome, emphasizing that each chromatid contains a single long molecule of DNA, which, in turn, is composed of two intertwined strands of nucleotides linked end to end.

the cell (such as hemoglobin), and the protein, in turn, carries out some important cellular function. (Exceptional genes that do not code for proteins include, for example, genes that code for the RNA constituents of ribosomes.) Thus, cellular functions are ultimately directed by genes, most of which exert their influence by dictating the makeup of proteins by means of a process to be described in Chapter 9. Each gene has associated with it in the DNA some information about its control, about whether or not its particular protein should be made in a particular cell at a particular time.

The alphabet of DNA consists of only four chemical letters, called **nucleotides**. Nucleotides are the chemical constituents of DNA, and millions of them are chemically linked together, end to end, to form each strand in a molecule of DNA (the ticker tapes). The total number of nucleotides in a human cell is about 6 billion. That is as many letters as would appear in 10,000 books the size of this

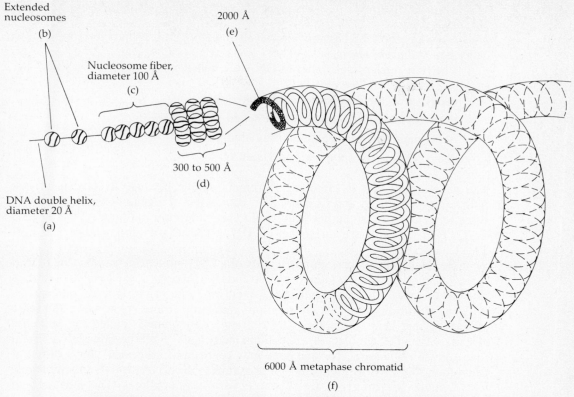

Extended
nucleosomes

(b)

2000 Å

(e)

Nucleosome fiber,
diameter 100 Å

(c)

300 to 500 Å

(d)

DNA double helix,
diameter 20 Å

(a)

6000 Å metaphase chromatid

(f)

Figure 1.15 One model of chromosome structure. (*a*) Extended DNA double helix. (*b*) DNA complexes with histones to form extended nucleosomes like beads on a string. (*c*) DNA spacer between nucleosomes becomes wrapped up in nucleosomes to form nucleosome fiber. (*d*)–(*f*) Higher levels of chromosome structure brought about by coiling of nucleosome fiber. Although parts (*a*)through (*d*) are well established, parts (*e*) and (*f*) are somewhat speculative.

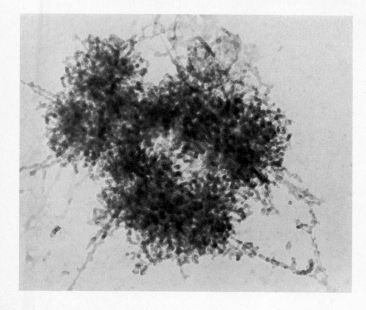

Figure 1.16 Electron micrograph of human E-group chromosome showing extensively coiled and folded fiber.

one. As will be noted later, though, not all the DNA in human cells is thought to carry genetic information.

The four nucleotides found in DNA are called

deoxyadenosine phosphate (abbreviated **A**)

deoxythymidine phosphate (abbreviated **T**)

deoxyguanosine phosphate (abbreviated **G**)

deoxycytidine phosphate (abbreviated **C**)

The genetic alphabet is therefore extremely simple because it consists of just four letters—A, T, G, and C. Moreover, each word in a gene consists of exactly three letters. Thus, an actual gene could carry the information CAT ACT TAG GAG, but, as there are no spaces in DNA, the information would read CATACTTAGGAG.

Because a gene corresponds to part of a DNA molecule, and because DNA molecules consist of two intertwined strands, each gene consists of two strands (the ticker tapes). For each gene, however, only one of the two strands carries the genetic information; this information-containing strand is known as the **sense** strand for that gene. The other DNA strand in the gene is called the **antisense** strand. A long DNA molecule will contain many genes, and the same DNA strand that represents "sense" for one gene may represent "antisense" for a different gene down the way. In Figure 1.17, for example, the DNA strands in a molecule that contains several genes are shown as parallel heavy lines. The sense strand for genes 1, 2, and 4 is the top strand (and the antisense strand for these genes is the bottom one), whereas, for gene 3, the situation is the reverse.

On the right of Figure 1.17, corresponding small segments of the DNA strands have been enlarged for closer inspection of their nucleotide sequences. There is a precise relationship between the corresponding nucleotide sequences, although the relationship may not be immediately obvious. However, the relationship between the sense strand and the antisense strand is quite simple: Wherever one strand carries an A, the other carries a T; and wherever one strand carries a G, the other carries a C.

It is an amusing coincidence that the genetic information in the sense strand is encoded in the antisense strand in a manner that spies used hundreds of years ago to conceal secret messages, and some important characteristics of DNA and its replication can be understood in terms of this spy analogy without the need for chemical details. The code (actually a type of cipher) is known as an alphabetic substitution. In alphabetic substitution, each letter of a sense message (the *plaintext,* as code-breakers call it) is rendered in the enciphered antisense message (the *ciphertext*) by a different letter of the same alphabet. Thus, for example, if we choose a simple cipher in which plaintext C is replaced with ciphertext G, plaintext A is replaced with T, and T is replaced with A, then the plaintext message CAT would be enciphered as GTA. Recall now that the genetic information in the sense strand is in an alphabet of just four letters—A, T, G, and C. The genetic cipher actually used in DNA is the following:

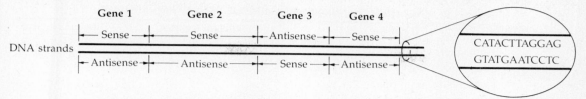

Figure 1.17 Diagram of DNA (the strands are shown uncoiled for ease of representation) with an expanded region at the right showing the nucleotide pairs in the strands (A with T and G with C). Genes correspond to segments of DNA, but the actual genetic information resides in only one strand (the sense strand). As illustrated, a long molecule of DNA contains many genes, and the DNA strand that is the sense strand for one gene can be the antisense strand for another gene along the way.

Replace A in sense strand (plaintext) with T in antisense strand (ciphertext).

Replace T in sense strand (plaintext) with A in antisense strand (ciphertext).

Replace G in sense strand (plaintext) with C in antisense strand (ciphertext).

Replace C in sense strand (plaintext) with G in antisense strand (ciphertext).

Thus, if the sense strand carries CAT, the antisense strand across the way will carry GTA. The longer sequence CATACTTAGGAG in the sense strand is present in the other strand as GTATGAATCCTC (see Figure 1.17). When the sense and antisense strands are close together so that corresponding nucleotides are paired, as in Figure 1.17, an A in one strand will always be paired with a T in the other strand across the way, and a G in one strand will always be paired with a C in the other strand.

One important feature of the genetic "cipher" is that the encipherment of the ciphertext gives back the plaintext. Thus, for example, the ciphertext for CAT is GTA, and the ciphertext for GTA is CAT; the plaintext message comes back again. This feature of the genetic cipher is important because, when the DNA strands separate during DNA replication, the new strand added to the sense strand will be identical to the old antisense strand, and the new strand added to the antisense strand will be identical to the old sense strand. In this manner, genetic information in the DNA is faithfully replicated generation after generation.

In summary, a gene corresponds to a segment of a DNA molecule. The genetic information in the gene is contained in the sequence of nucleotides (A, T, G, and C) in the sense strand. The antisense strand contains the genetic information in the form of a cipher in which A is replaced with T, T with A, G with C, and C with G. (The underlying reasons for this relationship between the nucleotide sequences of the sense and antisense strands will be discussed in Chapter 8.)

Table 1.2   CHROMOSOME NUMBERS IN VARIOUS PLANTS AND ANIMALS

Organism	Diploid Number	Haploid Number
Human *(Homo sapiens)*	46	23
Chimpanzee *(Pan troglodytes)*	48	24
Gorilla *(Gorilla gorilla)*	48	24
Dog *(Canis familiaris)*	78	39
Cat *(Felis domesticus)*	38	19
House mouse *(Mus musculus)*	40	20
Carp *(Cyprinus carpio)*	104	52
Fruit fly *(Drosophilia melanogaster)*	8	4
Wheat *(Triticum aestivum)*	42	21
Corn *(Zea mays)*	20	10
Tomato *(Lycopersicon esculentum)*	24	12
Tobacco *(Nicotiana tabacum)*	48	24
Garden pea *(Pisum sativum)*	14	7
Pink bread mold *(Neurospora crassa)*	14	7
Yeast *(Saccharomyces cerevisiae)*	34	17

A cell contains as many chromosomes as its nucleus contains molecules of DNA. The number of chromosomes varies from species to species; some examples are given in Table 1.2. The diploid number of chromosomes is the number found in all cells of the body except the gametes; these nongametic cells are called **somatic cells**. The haploid number of chromosomes is the number found in gametes (sperm or eggs in the case of animals).

Recall that in somatic cells, chromosomes actually occur in homologous pairs. Each chromosome and its homologue carry genetic information corresponding to the same genes, but the DNA molecules are not necessarily identical in every detail. The DNA in one chromosome may contain errors or "misprints" called **mutations.** The information that may be garbled in one chromosome by a mutation is often supplied in usable form by the homologue, because the odds are that both chromosomes will not have mutations in exactly the same genes. Mutations of the same gene in both homologous chromosomes do occur, though, and the result may be a hereditary disease in the person who carries these chromosomes. In some cases, moreover, a mutation in only one homologue of a pair may be sufficient to cause a hereditary disease. (One exception to the rule that a chromosome and its homologue carry genetic information for corresponding genes should be mentioned here. The pair of chromosomes that determine sex— the X and Y chromosomes—carry different genes. These chromosomes will be discussed in the next chapter and in detail in Chapter 5.)

When chromosomes are viewed through the light microscope, as in Figure 1.10, none of the underlying chemical details regarding DNA can be seen. Such details lie well below the limit of resolution of the light microscope. Nevertheless, a great deal about genetics can be learned from the study of chromosomes. In the next chapter, we will examine human chromosomes more closely and discuss the important process of cell division—meiosis—in which gametes are formed.

SUMMARY

1. **Human genetics** is the study of the inherited characteristics of human beings. Knowledge of human genetics is of practical importance because many families have individuals affected with conditions that are, at least in part, due to genetic factors. Knowledge of human genetics is also subject to misinterpretation or misuse, as evidenced by the eugenics movement early in this century.

2. Hereditary factors, called **genes**, are composed of a chemical **deoxyribonucleic acid (DNA)** found in cells. Much of genetics is therefore concerned with cells. Cells are the fundamental units of living things, the smallest units of life.

3. Two fundamentally different types of cells are to be distinguished: **prokaryotic cells**, which are relatively primitive cells lacking a nucleus, and **eukaryotic cells**, which have a nucleus. Examples of prokaryotes are bacteria and blue-green algae. Eukaryotic cells are those of plants, animals. and humans.

4. Eukaryotic cells contain internal structures called **organelles**. The most important of these in genetics is the **nucleus**, which contains the DNA in the form of **chromatin**. The nonnuclear part of a eukaryotic cell is called the **cytoplasm**, and it contains such organelles as the **ribosomes**, which are the sites of protein synthesis.

5. **Mitosis** is a type of cellular reproduction in which one cell divides to produce two genetically identical daughter cells. For convenience, the process of mitosis is considered to consist of five stages: **interphase** (DNA replicates and chromatin begins to coil into discrete chromosomes), **prophase** (chromosomes first become visible in the light microscope as pairs of **sister chromatids** attached to a common **centromere**), **metaphase** (centromeres of chromosomes align on a cell's equatorial plate), **anaphase** (centromeres split and sister chromatids proceed to opposite poles of the spindle), and **telophase** (chromosomes regain their interphase appearance and the cytoplasm divides).

6. A **chromosome** consists of a tightly coiled fiber of chromatin, which, in turn, consists of a long, coiled molecule of DNA along with certain other constituents, principally **histones**.

7. A molecule of DNA consists of two intertwined strands, each composed of a long sequence of **nucleotides** linked end to end. Four nucleotides are found in DNA; these are denoted by the symbols A, T, G, and C. A gene corresponds to a segment of such a DNA molecule. The genetic information in the gene consists of the sequence of nucleotides along one of the DNA strands (the **sense strand**); the other DNA strand for the gene is called the **antisense strand**. The relationship between the nucleotide sequences in the sense and antisense strands is as follows: Wherever one strand carries an A, the other carries a T; and wherever one strand carries a G, the other carries a C.

8. Most genes exert their influence on cells by providing instructions for the cell to manufacture a particular type of protein, such as hemoglobin.

Genes may contain errors called **mutations**, which can lead to the manufacture of defective proteins and thus a genetically caused disorder. However, since **homologous chromosomes** carry genetic information for corresponding genes, a mutation in a gene in one chromosome is often compensated for by a normal gene in the homologue.

KEY WORDS

Anaphase	Equatorial plate	Prokaryote
Centromere	Gamete	Prophase
Chromatid	Haploid	Replication
Chromosome	Homologous chromosomes	Ribosomes
Cytoplasm	Interphase	Sister chromatid
Diploid	Metaphase	Spindle
DNA	Mitosis	Telophase
Eugenics	Nucleotides (A, T, G, C)	Zygote
Eukaryote	Nucleus	

PROBLEMS

1.1. Why did the eugenics movement favor compulsory sterilization and the restriction of immigration of certain ethnic groups?

1.2. What are the principal differences between prokaryotic and eukaryotic cells?

1.3. Define the functions of the following cellular constituents: DNA, ribosomes, mitochondria, chloroplasts, ATP, spindle, cell membrane.

1.4. What is the volume of a human egg considered as a sphere of radius 65μm? (The volume of a sphere is given by the formula $V = \frac{4}{3}\pi r^3$, where r is the radius.)

1.5. What is the volume of a human sperm head considered as a cylinder 1 μm in radius and 4 μm in length? (The volume of a cylinder is given by the formula $V = \pi r^2 l$, where r is the radius and l is the length.)

1.6. What is the difference between a gamete and a zygote?

1.7. Which cells in a normal human do *not* have 46 chromosomes?

1.8. A triploid cell has three complete haploid sets of chromosomes. How many chromosomes would be found in a triploid human cell?

1.9. If a human cell undergoes mitosis in the brief presence of colchicine so that one anaphase is eliminated, how many chromosomes will the resulting cell have?

1.10. The Guinea baboon, *Papio papio,* has a diploid complement of 42 chromosomes. How many chromosomes would be found in a normal sperm? A normal egg?

1.11. As will be discussed in Chapter 4, it is possible to grow mammalian somatic cells (i.e., diploid body cells) in laboratory culture and even to fuse cells from two different species. The gibbon, *Hylobates lar,* has a haploid complement of 22 chromosomes. In somatic cell fusion involving cells of the gibbon and the baboon (see problem *1-10*), how many chromosomes would be found in hybrid cells?

1.12. Why are sister chromatids genetically identical?

1.13. In prophase of mitosis, how many *chromatids* are present in a human cell? How many *chromosomes*?

1.14. In a cell undergoing mitosis, how could you tell the difference between late metaphase and early anaphase?

1.15. In a cell irradiated with x-rays, a normally submetacentric chromosome is missing and in its place is a somewhat shorter metacentric chromosome. How could this be explained?

1.16. One strand in a DNA molecule has the nucleotide sequence ATGCCGTGCA. What is the nucleotide sequence of its partner strand?

FURTHER READING

Alberts, B., D. Bray, J. Lewis, M. Raff, K. Roberts, and J. D. Watson. 1983. Molecular Biology of the Cell. Garland, New York. A comprehensive, up-to-date textbook of cell biology.

Diener, T. O. 1980 (January). Viroids. Scientific American 244:66–73. The smallest known agents of disease—essentially ''naked'' strands of RNA.

Dustin, P. O. 1980 (August). Microtubules. Scientific American 243:66–76. On the manifold functions of these proteinaceous filaments.

Dyson, R. D. 1978. 2d ed. Essentials of Cell Biology. Allyn and Bacon, Boston. A good overview of the structure and function of cells and their organelles.

Grivell, L. A. 1983 (March). Mitochondrial DNA. Scientific American 248:78–89. This wonderful organelle has its own genetic system.

Hayflick, L. 1980 (January). The cell biology of human aging. Scientific American 242:58–65. What causes cells to grow old?

Lake, J. A. 1981 (August). The ribosome. Scientific American 245: 84–97. A careful examination of this marvelous little factory.

———. 1983. Evolving ribosome structure: Domains in archaebacteria, eubacteria, and eukaryotes. Cell 33: 318–19. Excellent, brief look at ribosome structure.

Lodish, H. F., and J. E. Rothman. 1979 (January). The assembly of cell membranes. Scientific American 240:48–63. Discusses the structure and growth of the film that holds cells together.

Ludmerer, K. M. 1971. Genetics and American Society: A Historical Appraisal. Johns Hopkins University Press, Baltimore. Still the best available account of the eugenics movement in the United States.

Milunsky, A. 1979. Know Your Genes. Avon Books, New York. Despite its cutesy title, this is a good introduction for beginners.

Porter, K. R., and J. B. Tucker. 1980 (March). The ground substance of the living cell. Scientific American 244: 56–57. Careful examination of the cell reveals a gossamer lattice of filaments.

Therman, E. 1980. Human Chromosomes: Structure, Behavior, Effects. Springer-Verlag, New York. Includes an excellent account of mitosis.

Woese, C. R. 1980 (June). Archaebacteria. Scientific American 244:98–122. Discusses a form of life that is neither prokaryote nor eukaryote but constitutes a kingdom of its own.

chapter 2

Chromosomes and Gamete Formation

Approximately 1 out of every 155 live-born children has some major chromosomal abnormality. As will be discussed in detail in Chapter 6, some of these children have physical abnormalities that can be recognized at birth; others are apparently normal at birth but suffer later impairment of physical or mental development; still others are physically and mentally normal, but their chromosome abnormality leads to a high risk of physical or mental abnormalities among their children.

This 0.6 percent (1/155) of live-born children with abnormal chromosomes are sufficiently normal in their development to be born alive. About 15 percent of all diagnosed pregnancies end in spontaneous abortion, and about half of these are associated with major chromosomal abnormalities of the fetus.

One subject of this chapter is the number and appearance of the chromosomes in cells of normal people. There is some variation in the appearance of the chromosomes in normal individuals; the nature and extent of this variation will be noted later. Bear in mind that the immediate discussion will necessarily be about the "typical" set of chromosomes. The descriptions of chromosomes will apply to people of all ages. For the most part, they will apply to the chromosomes in the cells of both sexes, with the exception of the X and Y chromosomes, which determine sex. The descriptions will also apply to the chromosomes of people of all colors, sizes, races, nationalities, and so on; that is to say, the kinds of chromosomes in the cells of all normal human beings are essentially the same, with the exception of certain normal variants to be noted later.

VIEWING HUMAN CHROMOSOMES

First it should be noted how chromosomes are prepared for examination and from which cells they are derived. Chromosomes are best examined in metaphase of mitosis or in the corresponding stage in meiosis—the special kind of cell division that gives rise to eggs and sperm. Cells undergoing mitosis are usually the ones examined. Contrary to what one might expect, however, the cells that are examined are not the ones that are rapidly dividing in the body. Cells of the bone marrow, lymph nodes, spleen, and other rapidly dividing tissues are generally difficult to obtain; getting samples of them requires procedures that can be painful. In some cases, even surgery is required. So the usual procedure is for the physician to take cells that are easy and painless to get. Normally these are nondividing cells, but they can be stimulated to divide in a test tube or culture flask in the laboratory.

A typical procedure in the study of human chromosomes is outlined in Figure 2.1. A small amount of blood is taken from the patient. (In young babies, blood is usually obtained from a heel prick.) The blood is added to a solution containing a chemical that prevents it from coagulating, and a small amount of this solution is then added to a rich nutrient broth containing an ingredient extracted from the juice of the red kidney bean. This substance stimulates certain cells to undergo mitosis. In the sample of blood are red blood cells and several kinds of white blood cells. One kind of white blood cell, the small lymphocyte, is the type stimulated to undergo mitosis.

The cultures that contain the small lymphocytes in nutrient broth are incubated for about three days, during which time they divide repeatedly. Then colchicine is added to stop the cells in metaphase, and about $1\frac{1}{2}$ hour later the cells are suspended in a very weak salt solution (one with a lower salt concentration than that inside the cells) and allowed to imbibe water for a few minutes to swell them, to help disperse the chromosomes, and to aid in disentangling the arms of the sister chromatids. (Figure 2.2 demonstrates the importance of colchicine and the low-salt solution in obtaining satisfactory preparations of human chromosomes.) The cells are fixed by adding an acid-alcohol mixture to preserve

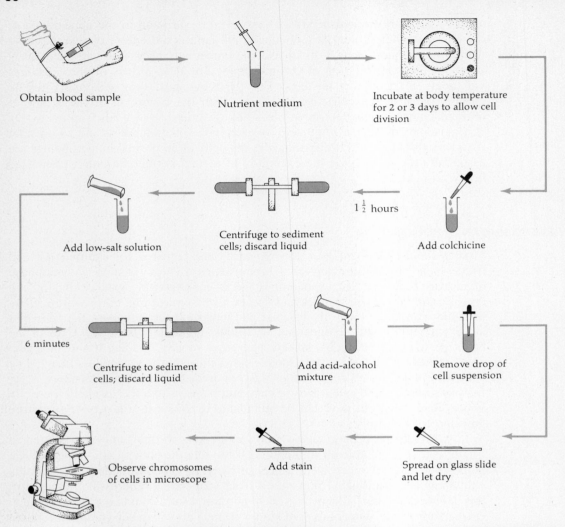

Obtain blood sample

Nutrient medium

Incubate at body temperature
for 2 or 3 days to allow cell
division

Add low-salt solution

Centrifuge to sediment
cells; discard liquid

$1\frac{1}{2}$ hours

Add colchicine

6 minutes

Centrifuge to sediment
cells; discard liquid

Add acid-alcohol
mixture

Remove drop of
cell suspension

Observe chromosomes
of cells in microscope

Add stain

Spread on glass slide
and let dry

Figure 2.1 Outline of experimental procedure for viewing human chromosomes in metaphase of mitosis. Note especially the addition of colchicine (to prevent formation of the spindle) and the brief soaking in a low-salt solution (to swell the cells and disperse the chromosomes).

them and retain their internal structures and then spread out on a small glass plate or slide and allowed to dry. This drying separates the chromosomes in a cell from one another so they can be observed individually. Then the chromosomes are stained with a dye, whereupon the slides are washed, dried, and examined under a light microscope.

Sometimes it is important to examine cells from several tissues in addition to white blood cells. Studying cells from several tissues might be necessary in some cases to diagnose or to check whether a patient is a chromosomal mosaic.

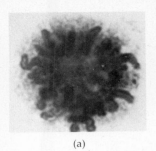
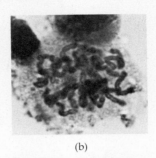
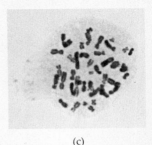

(a)   (b)   (c)

Figure 2.2 Human cells in metaphase prepared as in Figure 2.1 but (a) without colchicine or soaking in low-salt solution, (b) with colchicine but without soaking in low-salt solution, and (c) with both colchicine and soaking in low-salt solution. In (a) the metaphase chromosomes are in close proximity to one another, held by the fibers of the spindle. In (b) the spindle has been disrupted by colchicine, but the chromosomes are still jumbled and difficult to analyze. The chromosomes in the swollen cell (c) are much more spread out and easier to examine.

A **mosaic** is a person who has two or more genetically different types of cells or tissues; more specifically, a **chromosomal mosaic** is an individual who has one chromosome constitution in some cells and a different chromosome constitution in other cells. The vast majority of people have exactly the same chromosome constitution in virtually all their cells, so the constitution of small lymphocytes is the same as that of brain cells, nerve cells, muscle cells, and others. In chromosomal mosaics, this is not the case. Some of the patient's cells may be normal, whereas others may have, for example, a missing chromosome or an extra one. The more cell types that can be examined, the more likely it is that the abnormal cells in a mosaic individual will be detected.

In summary, the chromosomes of small lymphocytes in metaphase of mitosis are the ones most easily obtained and usually examined.

IDENTIFICATION OF HUMAN CHROMOSOMES

Figure 2.3 shows the chromosome constitution of a normal female (Figure 2.3a and b) and a normal male (Figure 2.3c and d). An unarranged photograph of the chromosomes in a nucleus, as in Figure 2.3a and c, is called a **metaphase spread**; when the chromosomes in a metaphase spread are cut out and rearranged as in Figure 2.3b and d, the arrangement is called a **karyotype**.

Figure 2.3 shows that normal cells have 46 chromosomes comprising 23 homologous pairs. The homologous chromosomes match in size and shape, except for the sex chromosomes. There are two **sex chromosomes** in both males and females. In females, the two sex chromosomes match in shape and size and are truly homologous in the sense that they carry corresponding genes; this sex chromosome is designated the **X chromosome** (see Figure 2.3b). In males, the two sex chromosomes do not match in shape and size, nor do they carry corresponding genes; one of the sex chromosomes is an X (the same X chromosome as in females), and the other sex chromosome is designated the **Y chromosome** (see Figure 2.3d). In addition to the sex chromosomes—XX in females and XY

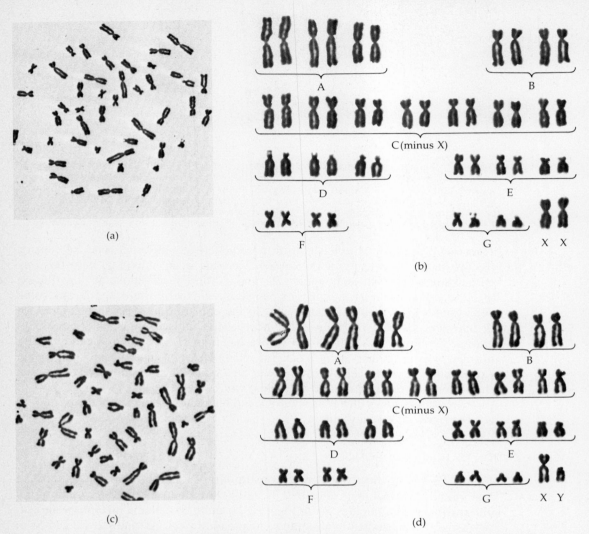

Figure 2.3 Metaphase spread (*a*) and karyotype (*b*) of a normal female; metaphase spread (*c*) and karyotype (*d*) of a normal male. Note that the only difference in the karyotypes is in the sex chromosomes: Females have two X chromosomes, and males have one X chromosome and one Y chromosome. (Identification of the X chromosome among the C-group chromosomes is somewhat arbitrary without special staining procedures.)

in males—the cells of both sexes have 44 other chromosomes that are called **autosomes** to distinguish them from the sex chromosomes; these 44 autosomes represent 22 homologous pairs. The 44 autosomes together with the 2 sex chromosomes provide the grand total of 46 chromosomes in the cells of normal individuals of both sexes.

An otherwise well-informed science reporter has claimed that the X chromosome is designated ''X'' because it is shaped rather like the letter X when viewed under the light microscope, and the Y is shaped rather like the letter Y. The true story behind the nomenclature of the X and Y chromosomes is much more interesting. Most animals have sex chromosomes. The first sex chromosome was recognized in male grasshoppers around 1890 as a chromosome that did not have a homologue; this strange chromosome was called the "X factor." Those were the days when it was fashionable to refer to things that were not understood as "X this" or "X that," which was an easy way to cloak the subjects in mystery and romance (the term "x-ray" is another legacy of the period). About 10 years later it was suggested that this "X factor" was the chromosome that determined sex, and within a few years it had become obvious that in many insects the males are XY. The Y chromosome was so designated simply because it is a sort of partner of the X, and X's partner in the roman alphabet is the letter Y. Although in mammals and many insects the XY sex is male and the XX female, it should be noted in passing that the XY sex is not always the male. In birds, moths, butterflies, and related species, the sex-chromosome situation is reversed—males correspond to XX and females to XY.

Throughout the history of genetics, the designations X and Y for the sex chromosomes have stuck. In most organisms, including humans, the autosomes are designated by numbers. According to agreements reached at international conferences on human chromosome nomenclature, the 22 pairs of autosomes are numbered by size from longest to shortest. A problem with this numbering system is that the lengths of the chromosomes are sufficiently variable from cell to cell that one cannot always be sure which chromosome is which. One can decide with some confidence whether a chromosome is long, medium, or short, however; and one can tell from the lengths of the arms whether a chromosome has its centromere near the middle (making it a **metacentric** chromosome), near the tip (an **acrocentric** chromosome), or somewhere between (a **submetacentric** chromosome). Based on overall length and centromere position, the chromosomes (including the sex chromosomes) can be sorted readily into seven groups, which are assigned the letters A through G (see Figure 2.3b and d).

The criteria by which the chromosomes are assigned to groups A through G are listed in Table 2.1. Chromosome complements are customarily arranged and presented as karyotypes of the sort depicted in Figure 2.3b and d. When a karyotype is produced, the chromosomes are individually cut out of a photograph

Table 2.1   DEFINITION OF HUMAN CHROMOSOME GROUPS A THROUGH G

Group	Chromosome numbers	Distinguishing characteristics
A	1 through 3	Large metacentrics
B	4 and 5	Large submetacentrics
C	6 through 12 plus X	Medium submetacentrics
D	13 through 15	Medium acrocentrics
E	16 through 18	Medium submetacentrics, but smaller than C group
F	19 and 20	Small metacentrics
G	21 and 22 plus Y	Small acrocentrics

of a metaphase spread. The chromosomes are then separated into groups A through G, the homologous pairs of chromosomes in each group are identified, the chromosomes are arranged in pairs, and the pairs are numbered from longest to shortest (except for chromosomes 21 and 22). Karyotyping hundreds or thousands of individuals can be extremely tedious, of course, but various computer-assisted methods are currently being devised and perfected. Although the assignment of chromosomes to groups is relatively easy, distinguishing among the chromosomes within a group (and especially trying to identify the homologous pairs) has often been pure guesswork. The exercise has been judged ''more amiable than realistic.''

Fortunately, methods have been devised for positive identification of each chromosome. One aid to chromosome identification is called **autoradiography**. In autoradiography, a chromosome is induced to take a picture of itself. Nucleotides incorporated by the cell into the DNA of chromosomes during chromosome replication can be added to the culture medium in radioactive form. These radioactive nucleotides are unstable; they will eject electrons. An electron that hits a photographic film will cause a spot to appear, leaving a record of its presence. Any part of a chromosome that contains the radioactive nucleotide will produce spots over the radioactive segment when overlaid with unexposed film and left in the dark to expose for several days; a chromosome that contains none of this material will not cause darkening of the film.

Figure 2.4 is an autoradiograph karyotype of a cell from a normal female.

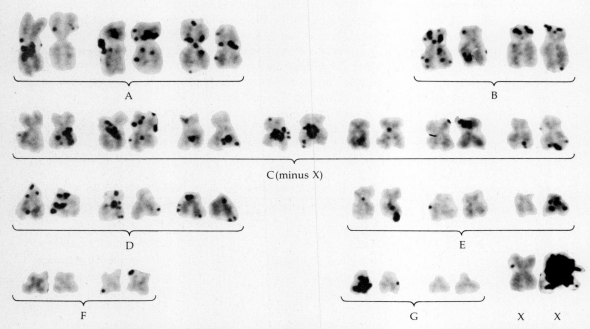

Figure 2.4   Autoradiograph of chromosomes from a normal female. Note the extremely heavy labeling of one of the X chromosomes, which is the inactive X to be discussed in Chapter 5.

Note that one of the X chromosomes is very heavily labeled with radioactivity, which is characteristic of one of the X chromosomes in normal females. The basis of this difference between the X's will be discussed further in Chapter 5.

The most elegant procedures for identifying chromosomes use special stains that bind preferentially to certain segments of the DNA in the chromosomes because these regions are particularly rich or poor in some chemical constituents. Such stains produce a visible pattern of **bands** or crosswise striations on the chromosomes. Some stains, for example **quinacrine**, fluoresce brightly when examined under certain wavelengths of light. Chromosome 21 fluoresces more brightly than chromosome 22 when stained with quinacrine, for example; and the Y chromosome fluoresces brilliantly (Figure 2.5). The fluorescent stains are thought to interact primarily with regions of DNA that are relatively rich in A-T nucleotide pairs (and correspondingly poor in G-C nucleotide pairs); bands in chromosomes produced by the stains are called **Q bands.**

At present, the most common chromosome-banding method uses the **Giemsa** stain, a mixture of two dyes. When properly used, Giemsa produces the spectacular pattern of bands known as **G bands** shown in Figure 2.6. Note in the figure that each chromosome has a unique pattern of bands and that homologous chromosomes can readily be identified by their identical banding patterns. Altogether, Giemsa produces about 300 G bands in the haploid set of human chromosomes. (Recall that the haploid set consists of 23 chromosomes, one chromosome from each of the 23 homologous pairs.) Since humans are widely believed to have between 30,000 and 90,000 genes, each G band, on the average, represents 100 to 300 genes.

In addition to Q bands and G bands, several other banding procedures have

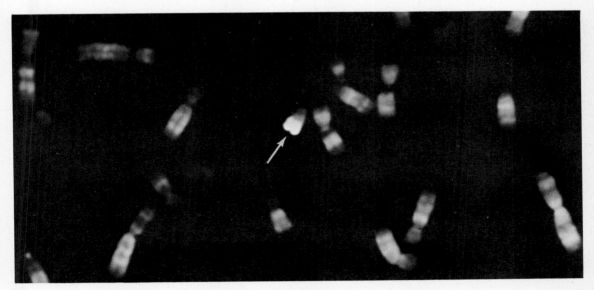

Figure 2.5 Quinacrine-induced fluorescence of human chromosomes. Note the particularly bright fluorescence of the long arm of the Y chromosome (arrow).

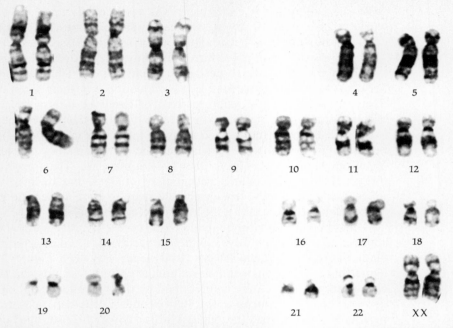

1   2   3   4   5

6   7   8   9   10   11   12

13   14   15   16   17   18

19   20   21   22   XX

Figure 2.6 Karyotype of a human female showing the spectacular banding pattern observed with the Giemsa staining technique.

been developed. One procedure highlights the centromeric region of the chromosomes (producing **C bands**), for example; another reverses the Giemsa effect (providing **R bands**)—it stains the regions Giemsa leaves unstained and vice versa. Figure 2.7 is a composite of the various banding patterns observed in human metaphase chromosomes.

NORMAL VARIATION IN HUMAN CHROMOSOMES

Some chromosomes in the human set do not look the same in every person; that is, minor variants of certain chromosomes do occur in the human population. Examples of such variants are shown in Figure 2.8. Chromosome 1 at the left of Figure 2.8a has a long arm that is noticeably longer than that of its homologous counterpart. Figure 2.8b shows chromosome 21 (right) with giant satellites. (A **satellite** is a little blob of chromosomal material attached to the short arm of a chromosome by a slender filament. Note in Figure 2.7 that the chromosomes in groups D and G normally have satellites.) Figure 2.8c and d illustrates variation in the length of the long arm of the Y chromosome; the Y chromosome in Figure 2.8c is shorter than chromosome 22 (normally the Y is slightly longer), whereas the Y in Figure 2.8d is substantially longer than chromosome 22.

The prevalence of the sort of chromosome variation shown in Figure 2.8 has been examined in a number of studies, of which the results of a Boston study of newborn males are typical. Among the 13,751 Boston newborns who were

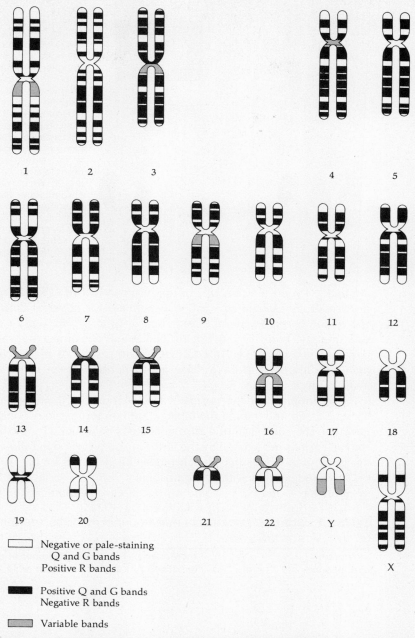

Negative or pale-staining
Q and G bands
Positive R bands

Positive Q and G bands
Negative R bands

Variable bands

Figure 2.7 Banding pattern of human chromosomes obtained as a composite of various special staining procedures. G bands are Giemsa bands (see Figure 2.6); R bands are reverse Giemsa bands (R bands are heavy where G bands are light and light where G bands are heavy); and Q bands are quinacrine fluorescent bands (see Figure 2.5). Variable bands are ones that do not always show up.

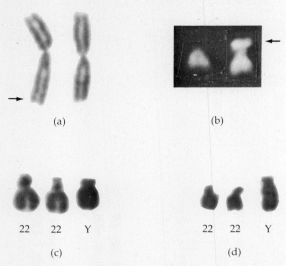

Figure 2.8 Some variants found in the normal chromosome complement. (*a*) An "uncoiler" chromosome 1 on the left and its normal homologue on the right; note that the long arm of the uncoiler chromosome (arrow) is about 15 percent longer than that of its homologue. (*b*) A normal chromosome 21 (stained with quinacrine) and a chromosome 21 with giant satellites (arrow). (*c*) A short Y chromosome beside a pair of chromosome 22 from the same cell for comparison; the short Y is the same length or shorter than chromosome 22—in contrast to a normal Y, which is slightly longer than chromosome 22. (*d*) A long Y chromosome beside a pair of chromosome 22 from the same cell for comparison; the long Y is much longer than chromosome 22. Each pair or set of chromosomes has been enlarged to a different scale.

studied, 84 (i.e., 0.6 percent, or approximately 1 in 155) had major chromosome abnormalities associated with severe mental or physical defects. In addition to the major chromosome abnormalities, a total of 462 (i.e., 3.36 percent, or approximately 1 in 30) individuals had minor chromosome variants of the type depicted in Figure 2.8. In contrast to the results of major chromosome abnormalities, infants who have these minor variants seem to be completely normal and healthy.

A summary of the normal variants found in the Boston study is given in Table 2.2. To understand the table, a note on terminology is necessary. In human genetics, the short arm of a chromosome is designated by the letter *p* (for *petite*); the long arm is designated by the letter *q*. A plus (+) sign following a chromosome designation indicates that the chromosome or arm is longer than normal; a minus (−) sign indicates that the chromosome or arm is shorter than normal. Thus, the symbols Dp+ and Gp+ would represent chromosomes with large satellites (i.e.,

Table 2.2 NORMAL VARIANTS IN HUMAN CHROMOSOMES FOUND IN A LARGE BOSTON STUDY

Variant chromosome	Arm	Increased (+) or diminished (−) size	Number of cases (out of 13,751)	Percent
D	p	+	51	0.37
G	p	+	51	0.37
Y	q	+	240	1.75
Y	q	−	52	0.38
Other			68	0.50
Total			462	3.36

Source: Data from S. Walzer and P.S. Gerald, 1977, a chromosome survey of 13,751 consecutive male newborns, in E.B. Hook an I.H. Porter (eds.), Population Cytogenetics: Studies in Humans, Academic Press, New York, pp. 45–61.

enlarged short arms). Similarly, Yq+ represents a Y chromosome of the type shown in Figure 2.8*d*—one with an unusually long (+) long arm (q). Likewise, Yq− represents a Y chromosome of the type shown in Figure 2.8*c*—one with an unusually short (−) long arm (q).

As is evident in Table 2.2, a significant fraction of normal individuals has minor chromosomal variants, typically enlarged satellites of D or G chromosomes or variation in Yq. Not shown in Table 2.2 is yet another normal variant—a metacentric (instead of submetacentric) C-group chromosome, which banding methods reveal to be chromosome 9. The incidence of the metacentric C is correlated with race: the variant occurs in approximately 1 out of 1350 Caucasians but in approximately 1 out of 75 non-Caucasians. There is thus a nearly twentyfold greater incidence of the metacentric C in non-Caucasians.

Considering that chromosomes carry the genes, it may seem strange that certain chromosomes or parts of chromosomes can vary in size without having dramatic or even detectable effects on the individual. Several explanations for this can be offered. Perhaps the differences in length do not reflect differences in the amount of genetic information in the chromosomes as much as they reflect differences in the amount of coiling of the chromosomes: the more tightly coiled a chromosome is, the shorter it will appear. Or it may be that the variation in length occurs in parts of chromosomes that are relatively inert genetically—that is, in parts that do not carry much genetic information to begin with. It is known, for example, that the factors on the Y chromosome that determine maleness are located on the short arm of the Y, whereas the normal variation observed in the overall length of the Y is due exclusively to differences in the length of the long arm. Perhaps the long arm of the Y (and certain other chromosomal regions as well) carries so little or such a special kind of genetic information that the variation in size makes little or no difference.

In any case, the normal types of chromosomal variation summarized in Table 2.2 show that normal individuals can have chromosomal (i.e., genetic) differences. Chromosomal variation is just the tip of the iceberg of genetic variation, however. As will be discussed in Chapter 14, there is so much subtle variation in the DNA among individuals—variation not visible in the microscope—that no two individuals (with the exception of identical twins) are ever genetically alike.

SEXUAL AND ASEXUAL INHERITANCE

Now we must consider the special kind of cell division that is involved in the formation of gametes. This gamete-forming type of division is called **meiosis**, and it is the basis of **sexual inheritance**—that is, inheritance mediated through the union of gametes and involving two sexes. Mitosis, discussed in the preceding chapter, is the cellular basis of **asexual inheritance**—that is, inheritance from cell to cell without the involvement of gametes.

The process of sexual inheritance is more subtle than that of asexual inheritance for several reasons. With sexual inheritance, each child has two parents, and both contribute equally to the child's genetic constitution. Despite the parents' equal contributions, however, the children do not tend to resemble one parent or the other in particular, nor do the children look exactly intermediate or like a blend of the parents. Although the resemblance between parent and child is sometimes especially striking, most children exhibit a mixture of traits or features. Some of these traits can be recognized (more or less) in the parents or other relatives; others are unique to the child. Finally, with sexual inheritance, the offspring produced are genetically different from one another; with asexual inheritance, by contrast, the daughter cells are genetically identical. The genetic diversity of offspring produced through sexual inheritance makes it seem as if sexual inheritance is completely unpredictable, almost whimsical.

But there is regularity in sexual inheritance nevertheless; the rules of sexual inheritance are known. These rules are statistical, involving probabilities. We can say with complete confidence that very nearly half of all children born in the world today will be female, but we cannot say without special tests what the sex of any particular unborn child will be. One might think at first that statistical rules are of no value because they cannot be used to predict particular events, but this would be wrong. For potential parents to know in advance that they have a high risk of producing an abnormal child—say, 25 or 50 percent—is to forewarn them, even though any particular child they have may be normal. Often the high risk alone may be sufficient to deter them from having children; this decision prevents a terrible anxiety during each pregnancy and it relieves society of the social and medical costs of severely malformed children. The value of risk probabilities in this context is similar to that of gruesome highway statistics to deter drunk drivers.

THE STRATEGY OF MEIOSIS

The statistical rules of inheritance have their origin in the process of meiosis. This meiotic type of cell division has evolved to satisfy simultaneously two conflicting requirements. On the one hand, the genetic constitution of the offspring, not only humans but all species, should be similar to that of the parents. Some similarity is desirable because the very survival of the parents demonstrates that their genetic constitutions are successful, and it is generally unwise to break up winning combinations of genes. On the other hand, the genetic constitution of the offspring should not be *too* similar to the parents. Some differences are desirable because children inevitably grow up in an environment that is different in many ways from the one their parents grew up in. The survival of the parents shows that the genes carried by them are successful, but successful in the *old* environment. Sexual reproduction must equip at least some of the offspring with the ability to cope with a new and unpredictable environment and thereby allow the species to evolve, while at the same time preserving to some extent the harmoniously working genes in the parents. Meiosis satisfies both these requirements by allowing only half the genes from each parent to be passed on to each child, but the nature of the process is also to reshuffle the genes in such a way that each child will have a combination of genes unique to it alone.

In humans, the meiotic process occurs only in certain cells of the ovaries and testes. These cells are the **oocytes** of the ovaries and the **spermatocytes** of the testes. Since the processes are so different, it is instructive to contrast meiosis with mitosis. Recall that the most important event in early mitosis is the replication of each chromosome; the two sister chromatids are then separated by being pulled in opposite directions into what will become the nuclei of two daughter cells. The two daughter cells produced by mitosis each have a full diploid complement of 46 chromosomes. Meiosis is very different; when complete, it results in four daughter cells, not two, and each daughter cell has a haploid set of 23 chromosomes—one representative from each of the homologous pairs. The daughter cells produced by meiosis develop into the egg or the sperm, and each carries 23 chromosomes; when the egg and sperm come together in fertilization, their fusion restores the normal diploid complement of human chromosomes: 23 pairs of homologues, 46 chromosomes altogether.

THE MECHANICS OF MEIOSIS: PROPHASE I

Meiosis, like mitosis, begins with an interphase. This is called **interphase I**. From the standpoint of genetics, the most important event that occurs during interphase I is the replication of the DNA in the chromosomes.

After DNA replication is accomplished and the cell has mobilized for division in various other ways, the cell enters **prophase I**, marked by the first appearance of the long, threadlike chromosomes convoluted throughout the nucleus. Figure 2.9 illustrates the appearance of the chromosomes at various times during prophase I, and the accompanying photographs are from testes cells of the grasshopper *Chorthippus parallelus*. The sketches in Figure 2.9 depict prophase I in an organism that has a diploid number of four chromosomes (two homologous pairs, one shown as metacentric and the other as submetacentric); the shaded chromosomes represent those that were contributed to the individual through the sperm, and the unshaded ones came from the egg. As seen in Figure 2.9, prophase I is a very long stage comprising several noteworthy events that decisively distinguish it from anything seen in mitosis. When first seen in prophase I (see Figure 2.9*a*), the chromosomes are single threads, unlike their appearance in prophase of mitosis when they are clearly already doubled, with their two sister chromatids intimately intertwined. Even though the DNA has been replicated in interphase I, the chromosomes themselves in prophase I of meiosis have not yet become visibly double. Throughout all of prophase I, the nucleus increases slowly in volume as the chromosomes undergo coiling so that the chromosomes appear to become progressively shorter and thicker.

An event unique to meiosis and of enormous importance occurs when the chromosomes are still quite extended. The homologous chromosomes pair up. The homologues find each other somehow and come to lie side by side, gene for gene, like zippers coming together tooth for tooth. The forces that bring about this alignment are unknown. The homologous chromosomes pair first at a few regions near the tips (see Figure 2.9*b*); the pairing then progresses zipperlike from these few points. This pairing of homologous chromosomes is called **synapsis**. When synapsis of each of the homologous pairs of chromosomes is com-

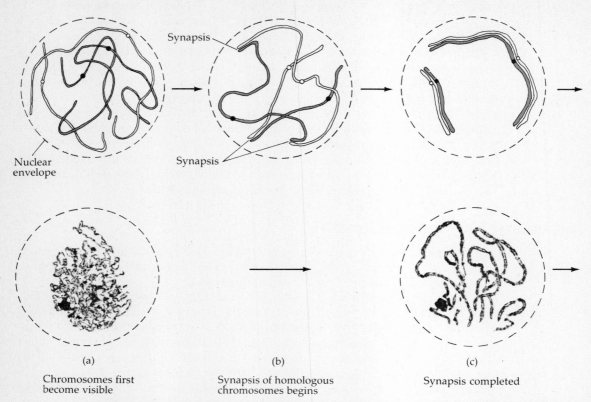

Synapsis

Nuclear
envelope

Synapsis

Synapsis

(a)   (b)   (c)

Chromosomes first   Synapsis of homologous   Synapsis completed
become visible   chromosomes begins

Figure 2.9   Principal events in prophase I of meiosis in a hypothetical cell that has two pairs of homologous chromosomes (top). Corresponding stages in testes cells of the grasshopper *Chorthippus parallelus* as viewed in the light microscope (bottom). Note particularly the pairing (synapsis) of homologous chromosomes in (*b*) and the appearance of chiasmata (crosslike structures) in (*e*). (Photos 600–900×).

pleted, the nucleus will contain as many structures as there are chromosomes in the haploid set (two in Figure 2.9*c*). In human cells, the nucleus in the stage corresponding to Figure 2.9*c* has 23 structures visible in it, each consisting of two homologues intimately paired along their length. One exception in human cells should be noted, however. In males, the X and Y chromosomes do not synapse along their entire length but only tip to tip, with the tip of the short arm of the X pairing with the tip of the short arm of the Y. Throughout the period when synapsis occurs, the chromosomes continue to become shorter and thicker (compare Figure 2.9*c* with 2.9*d*).

The next noteworthy event occurs after synapsis is completed. The chromosomes finally become visibly double (see Figure 2.9*e*). Each chromosome is now seen to consist of a pair of sister chromatids attached to a single centromere. Thus, the synapsed homologues consist of two centromeres side by side, with each centromere having attaching two sister chromatids. The whole unit is called

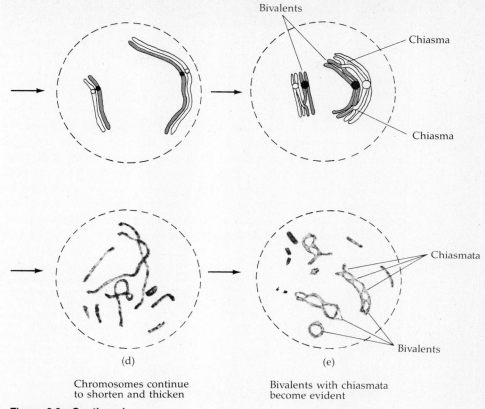

(d)

Chromosomes continue
to shorten and thicken

(e)

Bivalents with chiasmata
become evident

Figure 2.9 Continued

a **bivalent**, and the chromatids attached to different centromeres are called **nonsister chromatids**. As illustrated in Figure 2.10, each bivalent consists of four chromatids (i.e., two pairs of sister chromatids) attached pairwise to two centromeres and synapsed lengthwise. The important distinction between sister and nonsister chromatids is that sister chromatids, being replicas of the same chromosome, must be genetically identical; nonsister chromatids, being replicas of homologous chromosomes, may be genetically different. In human cells in the stage corresponding to Figure 2.9e, the number of chromatids is 92, the number of centromeres is 46, and the number of bivalents is 23.

Within the bivalents are certain crosslike structures called **chiasmata** (the singular is **chiasma**), as illustrated in Figures 2.9e and 2.10. Chiasmata are the visible manifestations of a process called **crossing-over**, during which nonsister chromatids undergo an exchange of parts. The genetic importance of crossing-over is best made clear by analogy. The analogy is not perfect because the bivalent is unique—unlike anything else in the world. Nevertheless, for purposes of discussion, imagine two lengths of ticker tape that both carry letters and words

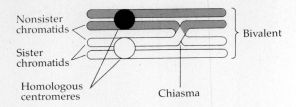

Figure 2.10 Expanded view of a bivalent late in prophase I, corresponding to Figure 2.9e. Note that the chiasma is indicative of crossing-over between nonsister chromatids.

designed to convey the same meaning. The tapes are assumed to be identical except where misprints occur (Figure 2.11a). The ticker tapes correspond to homologous chromosomes, and each one carries a single molecule of DNA, as described in Chapter 1. Arrange these tapes so that they are aligned letter to letter. This arrangement represents synapsis. Now let each tape be duplicated, but be sure that each copy is still aligned letter for letter with its original. Select some point along the length of the tape and staple the copy and the original together. Do this as well to the other copy and its original, being careful to insert

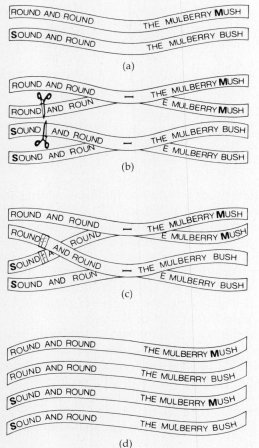

(a)

(b)

(c)

(d)

Figure 2.11 A surprisingly good analogy can be drawn between genetic information in chromosomes and letters on a ticker tape. (a) Two ticker tapes are paired as in synapsis; note that each tape bears a misprint (boldface letter). (b) The tapes have become visibly double, and breaks (cuts by scissors) occur at corresponding positions in nonsister chromatids. (c) The breaks are rejoined using transparent tape, forming a chiasmalike cross. (d) The resulting products; note that two products (the ones not involved in the splice) are like the original tapes; of the two products involved in the splice, one bears both misprints and one is entirely free of misprints.

this second staple in the same relative position as the first. These staples represent the centromeres. This duplication and stapling produce two pairs of tapes, each pair attached to one centromere, and the tapes are aligned lengthwise. The whole structure represents a bivalent.

Now the important part occurs. Proceed along the tapes until you come to some position—any position. Cut two of the tapes at exactly the same position, being careful that the two strands that are cut are attached to different centromeres (i.e., are nonsister chromatids), as shown in Figure 2.11*b*. Now interchange the severed ends and repair the broken strands with transparent tape (see Figure 2.11*c*). Be careful not to move the ends physically; leave them where they are, but take two small pieces of tape and bridge the gaps. What you have now accomplished is a splice. Two of the strands in the bivalent, the uncut ones, are identical to their parent strands. The other two strands are spliced; each is identical along part of its length with one of the parent strands, and along the rest of its length it is identical with the other parent strand (see Figure 2.11*d*).

The correspondence between this paper-and-tape model and an actual chiasma is shown in Figure 2.12. Crossing-over, represented by the chiasma, corresponds to the splice between nonsister chromatids. The splicing does not involve scissors and tape, of course, and the actual mechanism of the process is not fully understood, but the currently accepted view of the process is illustrated in Figure 2.13, which represents the two intertwined DNA strands in each chromatid as parallel lines. (Recall from Chapter 1 that a complete DNA molecule

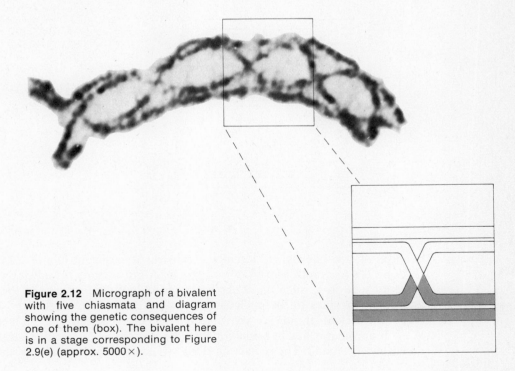

Figure 2.12 Micrograph of a bivalent with five chiasmata and diagram showing the genetic consequences of one of them (box). The bivalent here is in a stage corresponding to Figure 2.9(e) (approx. 5000×).

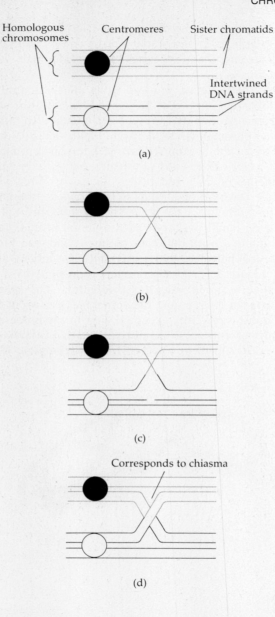

Homologous chromosomes   Centromeres   Sister chromatids

Intertwined DNA strands

(a)

(b)

(c)

Corresponds to chiasma

(d)

Figure 2.13 Current view of crossing-over at the DNA level. The DNA in each chromatid is shown as two parallel lines. (*a*) Synapsed chromosomes with a break or nick in one DNA strand of each of two nonsister chromatids. (*b*) The broken DNA strands are reconnected between the chromatids. (*c*) Remaining DNA strand in each chromatid is broken by enzymes in the nucleus. (*d*) Broken DNA strands are reconnected crosswise as before, resulting in crossing-over between two nonsister chromatids.

consists of two intertwined strands as illustrated earlier in Figure 1.14.) In part (*a*), one DNA strand in each of two nonsister chromatids is shown as having a break or nick in it; such nicks are produced by certain enzymes in the nucleus, and other enzymes in the nucleus can connect broken DNA strands. In Figure 2.13*b*, the nicked strands are shown as having been repaired, but the DNA strand in each chromatid has become connected with the corresponding strand in the nonsister chromatid. The remaining DNA strand in each chromatid is then broken by enzymes (part *c*), and these are cross-connected as before (part *d*). The resulting bivalent has a pair of spliced DNA strands, much as in Figure 2.11. The chromosome strands (chromatids) that are actually involved in the splice are called **crossover strands** or **crossover chromosomes**, and they are said to have **recombined** or to be **recombinant chromosomes** because of the intermixture of genetic information between them. The genetic consequences of crossing-over will be discussed further in Chapter 4.

Chiasmata may have a physical function in the bivalents. Shortly after the chromosomes first become visibly double in prophase I, the nonsister chromatids in the bivalent appear to repel one another. Each pair of sister chromatids remains tightly associated, but they move apart from the other pair of sister chromatids. (See the photograph in Figure 2.12 for an example.) This repulsion is often most marked in the regions around the centromeres, because the physical moving apart of the centromeres is easiest to see. In any case, the homologous chromosomes would probably dissociate completely if it were not for the chiasmata holding them together. Most bivalents have at least one chiasma. The shortest human bivalents (involving chromosomes of the G group) have one or two chiasmata; the longest ones (involving chromosomes of group A) have four, five, or six. The number varies from cell to cell and from bivalent to bivalent, but in males the total number of chiasmata in all the bivalents taken together averages about 54, or 2.3 per bivalent (Figure 2.14). All bivalents usually have at least one chiasma, with the exception of the XY chromosome pair in males. No crossing-over between the X and Y occurs, and no chiasmata are observed. The two X chromosomes in females, on the other hand, behave in this respect like the autosomes, forming typical bivalents and exhibiting chiasmata.

The crucial sequence of events in prophase I is marked by the first appearance of the chromosomes, the pairing of homologues, the visible doubling of the chromosomes to make the bivalents four-stranded structures with chiasmata, and the apparent repulsion of homologous chromosomes in the bivalents. As the repulsion of the homologous chromosomes continues, the chiasmata sometimes

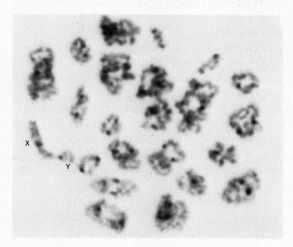

Figure 2.14 Bivalents in prophase I in a human male showing numerous chiasmata. Note the tip-to-tip association of the X and Y chromosomes.

seem to "slip" toward the tips of the chromosomes in much the way that a loose knot between two ropes can be slipped toward and even off the ends by pulling the ropes apart. The bivalents continue to shorten and thicken through more intense coiling toward the end of prophase I. The spindle begins to form, and the bivalents tend to migrate to the periphery of the nucleus near the nuclear envelope. Shortly thereafter, the nuclear envelope breaks down, and this breakdown marks the end of prophase I.

THE MECHANICS OF MEIOSIS: LATER EVENTS

Figure 2.15 summarizes the events that follow prophase I. The photo inserts are from testes cells of the grasshopper *Chorthippus parallelus*. In **metaphase I**, the spindle is completed and the bivalents move so that they come to lie on the equatorial plate (see Figure 2.15a). The orientation of the bivalents on the plate is usually quite precise. Each bivalent lines up so that its two centromeres lie on opposite sides of the plate, pointing toward the poles of the spindle.

Anaphase I commences when the two centromeres of each bivalent actually begin to separate as they are pulled in opposite directions (see Figure 2.15b). The pair of chromatids associated with each centromere follows along, and the chiasmata that held the bivalents together are finally disengaged as they slip off the ends. Note the results of crossing-over in Figure 2.15b; the recombinant chromatids are those that carry partly paternal (shaded) and partly maternal (unshaded) material. As the centromeres proceed toward the poles in anaphase I, the chromatids seem to lose their attraction to each other and the arms are free to flop around somewhat, but the chromatids remain attached to the same centromere. This is the most important feature of anaphase I: The two centromeres of each bivalent are separated from each other, but the centromeres themselves do not split. Anaphase I is thus very different from anaphase of mitosis, in which the centromeres do split. The separation of homologous centromeres in anaphase I results in a halving of the chromosome number. Recall that the number of chromosomes in a cell is determined by counting the number of centromeres. Since homologous centromeres pull apart during anaphase I, each daughter cell receives the haploid number of centromeres and hence the haploid number of chromosomes. (Note, however, that the number of chromatids in each daughter cell is still the diploid number because each centromere carries two chromatids.)

The end of anaphase I coincides with the arrival of the chromosomes at the poles of the spindle. Only one homologue of each bivalent has gone to each pole, and in human cells there will be 23 nonhomologous chromosomes at each pole. In **telophase I**, the chromosomes uncoil slightly and in some species a nuclear envelope begins to reform around each of the two groups of chromosomes. In many species (including humans), the cytoplasm divides during telophase I (see Figure 2.15c). In human males, the cytoplasm is divided roughly equally so that two daughter cells of equal size are formed. Each of these cells undergoes the

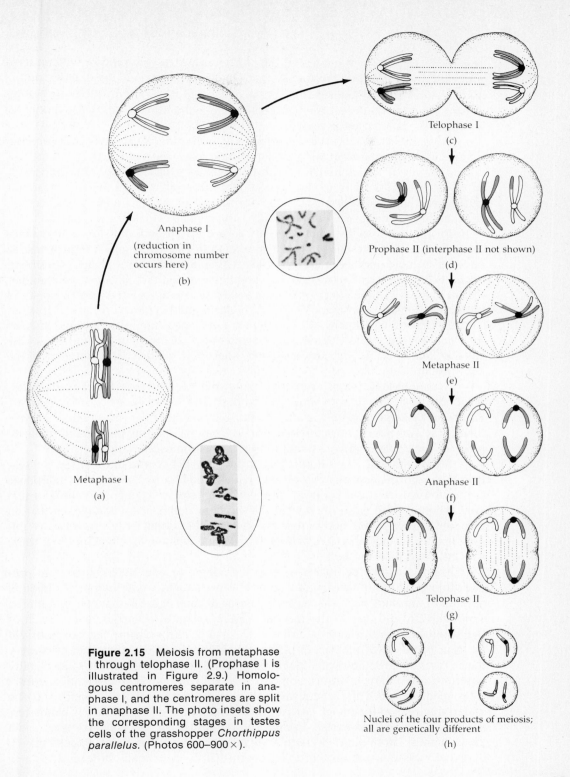

Anaphase I

(reduction in chromosome number occurs here)

(b)

Telophase I

(c)

Prophase II (interphase II not shown)

(d)

Metaphase II

(e)

Anaphase II

(f)

Telophase II

(g)

Metaphase I

(a)

Nuclei of the four products of meiosis; all are genetically different

(h)

Figure 2.15 Meiosis from metaphase I through telophase II. (Prophase I is illustrated in Figure 2.9.) Homologous centromeres separate in anaphase I, and the centromeres are split in anaphase II. The photo insets show the corresponding stages in testes cells of the grasshopper *Chorthippus parallelus*. (Photos 600–900×).

steps that follow. In human females, the division of the cytoplasm is grossly unequal. The cytoplasm divides so that one of the daughter cells receives almost all of it. This cell is destined to give rise to the egg. The other cell formed by the division is very small and lies just outside the membrane of the larger one. This small cell is called a **polar body**.

After the division of the cytoplasm, the nuclei in the daughter cells undergo a very abbreviated and incomplete interphase called **interphase II**. The chromosomes do not uncoil completely, and the nuclear envelope is not fully reformed. Most important, there is no synthesis of DNA or replication of the chromosomes during interphase II. For all practical purposes, human cells undergoing meiosis proceed almost directly from telophase I to prophase II, the next stage.

Prophase II begins what amounts to a conventional division of the chromosomes, almost like mitosis (see Figure 2.15d). However, nuclei in prophase II have the haploid number of chromosomes (23 in humans), not the diploid number. As noted earlier, the first meiotic division has reduced the chromosome number. Recall that each of the chromosomes in prophase II is associated with two distinct chromatids. These chromatids coil tightly again and become shorter and thicker. Toward the end of prophase II, the nuclear envelope disappears (in those species in which it has formed), the spindle begins to be set up again, and at **metaphase II** (see Figure 2.15e) the chromosomes line up with their centromeres on the cell's equatorial plate.

In **anaphase II**, the centromeres finally split into two parts (see Figure 2.15f). Anaphase II is thus a conventional anaphase of the sort that occurs in mitosis. After the centromeres split, each chromosome has its own independent centromere. The two new centromeres that arise from the splitting proceed to opposite poles of the spindle, dragging the arms of the chromosomes along.

In **telophase II**, the chromosomes uncoil and become diffuse, the nuclear envelope reforms, and the cytoplasm divides (see Figure 2.15g). As in telophase I, the division of the cytoplasm in males produces two daughter cells of equal size, whereas in females the division of the large cell is again unequal so that an egg and another polar body are formed. (The first polar body occasionally undergoes the second meiotic division also, so that at the end of the meiosis, there may be three polar bodies.)

Meiosis is now complete. The spermatocyte or oocyte has divided to produce four products of meiosis, and each of these daughter cells carries the haploid number of chromosomes. The egg carries one of each of the autosomes, numbered 1 through 22, and one X chromosome. Each sperm also carries one of each of the autosomes, and it carries either an X or a Y chromosome but not both. Of the four products of meiosis in males, two carry the X and the other two carry the Y. As is apparent in Figure 2.15h, each of the products of meiosis is genetically different; they carry various combinations of paternal (shaded) and maternal (unshaded) genes. This genetic diversity among the products of meiosis is the underlying reason for the genetic diversity among the offspring that is characteristic of sexual reproduction. Earlier we mentioned that a typical human male expels about 350 million sperm in a single ejaculate; the meiotic process makes it virtually certain that each of these sperm will be genetically different!

SUMMARY OF MEIOSIS

At this point it may be useful to summarize the principal characteristics of chromosome behavior during meiosis.

Interphase I: DNA synthesis and chromosome replication occur.

Prophase I: Chromosomes become visible as single strands; synapsis occurs; chromosomes become visibly double, producing bivalents within which chiasmata (the physical manifestations of crossing-over) are evident.

Metaphase I: Bivalents align on the equatorial plate.

Anaphase I: Homologous centromeres pull apart, causing a reduction in chromosome number.

Telophase I: Cytoplasm divides.

Interphase II: Chromosomes uncoil slightly; there is no DNA synthesis or chromosome replication.

Prophase II: Chromosomes again coil.

Metaphase II: Chromosomes align on the equatorial plate.

Anaphase II: Centromeres split and chromosomes pull apart.

Telophase II: Cytoplasm divides, completing meiosis.

The stages interphase I through telophase I constitute the **first meiotic division**; the stages interphase II through telophase II constitute the **second meiotic division**.

DEVELOPMENT OF GAMETES

After telophase II, the nuclei undergo what further differentiation is required to form a mature egg or sperm. In males, all the telophase II cells develop into sperm. The cells elongate and the tail of the sperm is formed (Figure 2.16). The nucleus becomes supercondensed in the sperm head, and the bulk of the cytoplasm is stripped away and degraded. What is left in the sperm head is essentially the nucleus. The head of the sperm is a sort of space capsule, and the tail is a propulsion system designed to launch and propel the nucleus toward fertilization.

The stages of meiosis in females are synchronized with events in the life cycle and the menstrual cycle. The earliest stage in meiosis—prophase I—takes place in the ovary of a female when she is still an embryo developing in her mother's uterus. When the embryo is only 4 months old, the deepest layers of the developing ovary contain many cells in the earliest part of prophase I. These cells gradually progress through prophase I. When the embryo is 7 months old, a few cells can be found in late prophase I. Late in prophase I, a curious arrest of meiosis occurs. Instead of proceeding to metaphase I, the bivalents unwind or "decondense" and the nucleus enters an interphaselike state. This state is not really interphase because none of the DNA synthesis that characterizes interphase I occurs. The cells in this interphaselike state are quiescent. In embryos 9 months old and about to be born, the ovary is heavily populated with cells in this quiescent state.

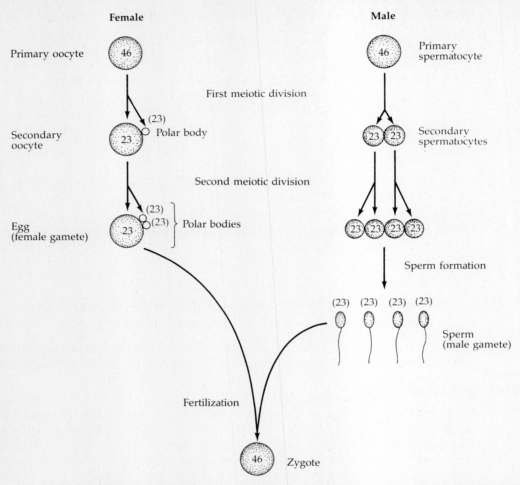

Figure 2.16 Chromosome numbers in various cell types in human reproduction, showing the grossly unequal cytoplasmic division in females leading to one large egg cell and two (sometimes three) tiny polar bodies. The illustration is somewhat oversimplified because in human females the second meiotic division actually takes place in the fallopian tube after fertilization.

Each of the interphaselike cells is surrounded by a large number of helper cells. Each group of helper cells together with the included interphaselike cell is known as a **follicle**. The follicles are quiescent through childhood. At puberty, with the onset of menstruation and its hormonal cycles, one follicle per month reaches maturity and **ovulation** (the shedding of the egg) occurs. (The synchronized rise and fall in the levels of female hormones trigger ovulation, and this is why the artificial interference with hormone levels caused by birth control pills prevents ovulation.) The stimulated follicle increases in size almost 10 times—to 4000 μm or about $\frac{3}{16}$ in—although the oocyte itself, the interphaselike cell, does not increase in size. The follicle bursts open and the oocyte is extruded from the ovary and is caught in the upper part of the fallopian tube, where it begins its journey toward the uterus. As the follicle is ripening, the chromosomes in the oocyte resume meiosis; they coil up again and undergo metaphase I, anaphase I, and telophase I. The tiny polar body is pinched off and sticks to the surface of the oocyte (see Figure 2.16). After the follicle bursts and the egg is shed, the cells go through prophase II and the chromosomes align for metaphase II. Metaphase II, anaphase II, and telophase II occur only after the sperm has penetrated the egg. During telophase II, the egg extrudes another polar body. The fertilized egg at this stage contains two nuclei—its own telophase II nucleus and the nucleus contributed by the sperm. The DNA in both nuclei is replicated, and the two nuclei move toward each other and fuse, producing a full diploid complement of already replicated chromosomes. The fertilized egg immediately enters prophase of mitosis, which occurs while the egg is still progressing slowly down the fallopian tube. About 3 or 4 days after fertilization, the zygote reaches the uterus, and 3 or 4 days later it becomes implanted in the uterine wall.

Earlier we mentioned that the statistical rules of sexual inheritance are the consequences of the process of meiosis. Meiosis determines how the genes will be distributed among gametes, and fertilization determines how these gametes will combine to produce individuals of the next generation. Ordinarily, however, meiosis is not observed directly, nor is fertilization. The genetic consequences of meiosis and fertilization are nevertheless observable in the pattern of inheritance of various traits. These patterns of inheritance—the genetic result of meiosis—are the subject of the next several chapters.

SUMMARY

1. The normal diploid complement of chromosomes in humans consists of 46 chromosomes—44 **autosomes** and 2 **sex chromosomes**. In females, the sex chromosomes are both X chromosomes; in males, one sex chromosome is an X and the other is a Y.

2. The autosomes in humans occur as 22 homologous pairs, which are assigned the numbers 1 through 22 from longest to shortest. Individual chromosomes cannot be identified without special procedures, but each can be assigned to one of seven groups designated by the letters A through G. Each group contains chromosomes that are similar in size and centromere position. The longest chromosomes, which are metacentrics, constitute group A; the shortest ones, which are acrocentrics, constitute group G. The X chromosome is a C-group chromosome; the Y is a G-group chromosome.

3. Individual chromosomes or parts of chromosomes can be identified with special procedures. **Autoradiography** relies on the fact that radioactive nucleotides incorporated in DNA will emit radiation and darken photographic film. Certain fluorescent dyes interact preferentially with regions of DNA rich in A-T nucleotide pairs, so sections of chromosomes that carry such DNA are preferentially stained, producing Q bands.

4. **Giemsa stain**, which produces G bands, gives the highest resolution of human chromosomes and is the most widely used technique for chromosome identification. There are about 300 G bands in the haploid set of chromosomes at metaphase, and each pair of homologous chromosomes has a unique pattern of G bands. Humans are thought to have

between 30,000 and 90,000 genes, so each metaphase G band represents, on the average, 100 to 300 genes.

5. Normal variation among human chromosomes does occur. Approximately 1 individual out of 30 has some minor chromosome variant, such as a D-group or a G-group chromosome with abnormally large **satellites** or a Y chromosome with an atypically sized long arm. These variants usually involve parts of chromosomes thought to carry few (or inactive) genes.

6. **Meiosis** is the basis of sexual inheritance, and it occurs only in the **oocytes** of the ovaries and the **spermatocytes** of the testes. In males, meiosis results in four haploid sperm; in females, it results in a haploid egg and two (sometimes three) polar bodies. Fertilization restores the diploid complement of chromosomes to the zygote.

7. The major features of the first meiotic division are as follows: **interphase I** (DNA and chromosome replication); **prophase I** (synapsis and formation of bivalents containing chiasmata, which are the physical manifestations of crossing-over); **metaphase I** (alignment of bivalents on the equatorial plate); **anaphase I** (separation of homologous centromeres, reducing the chromosome number); **telophase I** (cytoplasmic division). The major events in the second meiotic division are: **interphase II** (abbreviated phase, no DNA or chromosome replication); **prophase II** (chromosomes regain prophase appearance); **metaphase II** (alignment of chromosomes on the equatorial plate); **anaphase II** (centromeres split and separate); and **telophase II** (cytoplasmic division). Telophase II is followed by the formation of sperm or eggs.

8. **Crossing-over** may be thought of as a break and reunion between non-sister chromatids. It results in **recombinant** chromosomes that carry paternal genes along part of their length and maternal genes along the other part. Crossing-over thus allows the formation of chromosomes that carry new combinations of genes.

9. The principal genetic consequence of meiosis is that all gametes will be genetically unique in the sense of carrying different combinations of genes. This diversity of gametic types is the cause of the diversity of offspring formed in **sexual reproduction**.

KEY WORDS

Acrocentric chromosome	Karyotype	Quinacrine
Anaphase (I, II)	Meiosis	R bands
Autosome	Metacentric chromosome	Recombinant chromosome
Bivalent	Metaphase (I, II)	Satellite
Chiasma	Metaphase spread	Sex chromosomes
Chromosome group (A, B, C, D, E, F, G)	Mosaic	Spermatocyte
	Oocyte	Submetacentric chromosome
Crossing-over	Polar body	Synapsis
G bands	Prophase (I, II)	Telophase (I, II)
Giemsa stain	Q bands	X chromosome
Interphase (I, II)		Y chromosome

PROBLEMS

2.1. What is a chromosomal mosaic and how could such a condition be detected?

2.2. In which meiotic anaphase, I or II, do homologous centromeres separate? In which do sister centromeres separate?

2.3. During which meiotic division, I or II, does the X chromosome separate from the Y?

2.4. Of the four products of meiosis in a male, how many carry an X chromosome? How many carry a Y?

2.5. What types of gametes would be formed by a cell in meiosis in which both chromosome 21's went to the same pole at anaphase I, assuming that the second meiotic division proceeded normally?

2.6. Cells in metaphase of mitosis superficially resemble those in metaphase II of meiosis. How could you tell them apart?

2.7. A newborn female has multiple physical abnormalities including an elongated skull, low-set malformed ears, and a small jaw. The infant has 47 chromosomes, the extra one being an E-group chromosome having no G bands in the short arm. Identify the extra chromosome.

2.8. A physically and mentally normal male is found to have 47 chromosomes. The extra chromosome is a G-group chromosome that is found to fluoresce very brightly with quinacrine staining. What is the extra chromosome, and what is the chromosome constitution of the male?

2.9. Identify the illustrated chromosome by its banding pattern.

2.10. Identify the chromosome shown. Is there anything odd about it?

2.11. The chromosome shown here is abnormal. Can you determine from its banding pattern how the chromosome originated?

2.12. The chromosome shown here is abnormal. Can you identify the chromosome and the nature of the abnormality?

2.13. A female carries genes *A* and *B* on one X chromosome and genes *a* and *b* on the other. What X chromosomes would be formed by crossing over between the *A* and *B* genes?

FURTHER READING

Comings, D. E. 1978. Mechanisms of chromosome banding and implications for chromosome structure. Ann. Rev. Genet. 12:25–46. Why chromosome bands are produced.

Grobstein, C. 1979 (June). External human fertilization. Scientific American 240:56–67. A clear discussion of the scientific issues and relevant social concerns.

John, B., and K. R. Lewis. 1965. The Meiotic System. Vol. 6, Pt. Fl. of Protoplasmatologia. Springer-Verlag, New York. A splendid review of meiosis, notable for its extraordinary pictures.

Nora, J. J., and F. C. Fraser. 1981. Medical Genetics, 2d. ed. Lea and Febiger, Philadelphia. A clinically oriented introduction to medical aspects of genetics.

Rhoades, M. M. 1950. Meiosis in maize. J. Heredity 41:57–67. A golden oldie—one of the best descriptive accounts of meiosis.

Therman, E. 1980. Human Chromosomes: Stucture, Behavior, Effects. Springer-Verlag, New York. Includes an excellent account of meiosis.

Tiley, N. A. 1983. Discovering DNA: Meditations on Genetics and a History of the Science. Van Nostrand Reinhold, New York. Brief history of genetics as a highly successful science.

Yunis, J. J. 1976. High resolution of human chromosomes. Science 191:1268–1270. Discusses prophase banding of human chromosomes.

Mendel's Laws and Dominant Inheritance

The mechanics of the meiotic divisions determine the statistical rules of inheritance. However, the rules of inheritance were worked out long before chromosomes had been discovered, before the details of cell division had been described, before chromosomes were known to carry the genes, and even before the overriding importance of the nucleus in inheritance had become clear. These statistical rules of inheritance were discovered by an Augustinian priest, Gregor Mendel, who carried out his most important scientific experiments in the early 1860s in a monastery in what is now Brno, Czechoslovakia, where a monument stands in his honor. (See Figure 3.1 for some memorabilia of Mendel.) Mendel studied the pattern of inheritance of various traits in the garden pea—round versus wrinkled seeds, green versus yellow seeds, and others. He was able to discern certain regularities in their pattern of inheritance. He reported these results in the journal of a local scientific society, which was rather widely disseminated in Europe; he

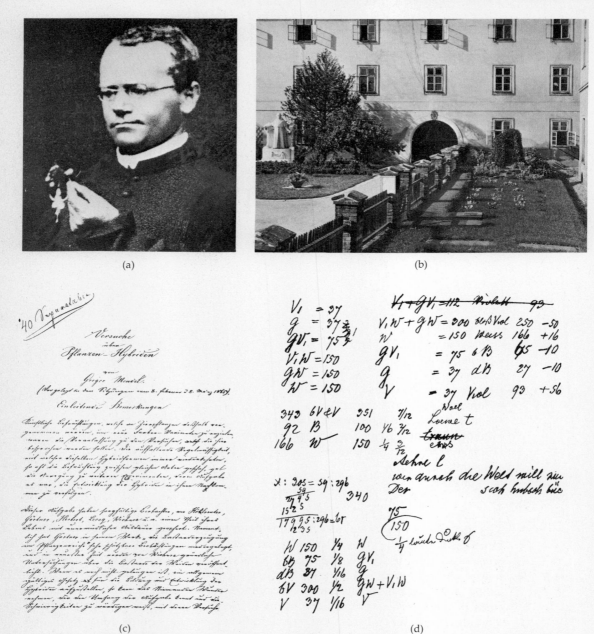

Figure 3.1 (*a*) A portrait of Gregor Mendel, (*b*) part of the monastery garden in which he carried out his pea-breeding experiments, (*c*) handwritten title page of his celebrated paper, and (*d*) a page from one of his experimental notebooks.

also wrote to one of the leading plant breeders of the time, whose name was Nageli, and told him of the results. Nevertheless, the significance of Mendel's findings was completely missed by those who read his paper (if anyone did read it attentively) and by Mendel's correspondents. His elegant work lay unnoticed and unappreciated for 34 years.

Nowadays, there is a natural tendency to think that Mendel was ignored because he was stuck in some backwater monastery and cut off from the world. This is far from the truth. In those days, joining the priesthood was an accepted way for a poor peasant boy like Mendel to receive a higher education. Moreover, the monastery of St. Thomas was at that time the scientific and cultural center of Brno, and Brno was the scientific and cultural center for a large part of southeastern Europe. Many of Mendel's brother monks were scientists of various sorts, so Mendel's interest in experimentation was by no means odd or unusual. Nor was Mendel cut off from the world. For 14 years he taught in the public school system of Brno, and his students recalled him as a kind and gentle man with a lively sense of humor whose teaching talent was in making complex ideas seem simple.

Contemporaries recall Mendel as a man who loved good cigars (up to 20 a day!) as well as good food and wine (St. Thomas was noted for its excellent kitchen and cellar). By no means a quiet drudge puttering in his garden, he was a serious scientist who studied bees, meteorology, and other areas. He was sufficiently well known and respected that in his later years he was made abbot of St. Thomas and undertook an ongoing wrangle with the government over monastery taxation.

Mendel was therefore part of a respected scientific tradition, and isolation was not the primary cause of his neglect. It is more likely that Mendel's work was ignored at the time because no one really understood it. Science often has its own tempo, and certain discoveries must come before others. In Mendel's case, an understanding of statistics was necessary for a proper appreciation of his work. Although a number of prominent scientists in Mendel's time were interested in determining the rules of inheritance, they did not expect these rules to be statistical. Hardly anyone even knew how to think about statistical regularities of any sort, because the intellectual framework of statistics was not laid down until the last half of the nineteenth century and the first half of the twentieth century. A mature appreciation of the unique power and beauty of statistical laws is as difficult to achieve as an appreciation of, say, cubism in art. Both share a perspective on the world that is different from the commonplace, and at first this perspective seems alien, even a little frightening. Appreciation takes time.

Moreover, although the nature of inheritance had caught the attention of some scientists, it was a secondary problem in the biology of the time. The principal interest was in **evolution**—the processes whereby species of organisms become progressively more adapted to their environment. Indeed, on the cold, clear winter evening of February 8, 1865, when Mendel read his paper to the 40 members of the Brno Natural History Society, there was not a single question or discussion pertaining to the paper; instead, the group discussed various issues related to evolution!

Between about 1870 and 1900, the details of cell division and the importance of chromosomes in inheritance gradually became clear. In the late 1870s, the decisive role of the nucleus in cell division and fertilization was demonstrated. Between 1880 and 1900, a whole series of investigators contributed to the understanding of mitosis and meiosis, although the most important discovery is usually considered to be the observation that meiosis reduces the chromosome number by half and that the full chromosome number of a species is restored again only at fertilization. However, the details of meiosis—that chromosomes come in homologous pairs and that bivalents represent homologues paired side by side—were not fully worked out until around 1900.

In the last years of the nineteenth century, interest in the nature of inheritance was rekindled, partly because inheritance was finally perceived to be an important element in evolution and adaptation. A number of people began performing experiments that were, unbeknown to them, very similar to Mendel's. Three men—Hugo de Vries in Holland, Karl Correns in Germany, and Erich von Tschermak in Austria—all working independently, discerned in their own plants the patterns of inheritance that Mendel had seen in his. Only when the three were ready to publish their results did they become aware of Mendel's paper; de Vries, Correns, and Tschermak all published in 1900, and all cited Mendel. With the rediscovery of Mendel's work, it became clear that all living organisms have genes—elements transmitted in inheritance—although in 1900, these were highly abstract entities. In the early 1900s, it was realized that the genes were probably located on the chromosomes because the genes behaved in inheritance as if they were located on the chromosomes. At the same time, it also became evident that Mendel's laws of inheritance have their physical foundation in the process of meiosis.

SOME IMPORTANT TERMINOLOGY

What were these rules of Mendel, these statistical laws? Actually there were three discoveries, covering the phenomena of dominance, segregation, and independent assortment. The discoveries pertaining to segregation and independent assortment are often referred to respectively as *Mendel's first and second laws*. The phenomenon of dominance is really not a generalization about inheritance as much as it is about the way genes act during development. A discussion of dominance will have more meaning if deferred until later; and the second law, that of independent assortment, will be taken up in Chapter 4. The nub of Mendel's theory is in the first law, the *law of segregation*. This law actually rests on several other concepts, which deserve a brief discussion. The first concept is that *heredity traits are determined by genes* (although Mendel used the German word for *factor* since the word *gene* did not exist). The second is that *genes are transmitted from one generation to the next through the gametes* (the egg and sperm in animals). The third concept is that *genes come in pairs*.

The idea that genes come in pairs is the most subtle of Mendel's discoveries. It introduces more nuances that the simple word *gene* can carry. We need two new words. There is a need for a word to specify the place on a chromosome where a gene is located. The word is **locus** (plural **loci**). A locus on a human chromosome may be specified physically by saying, for example: "This locus is so many micrometers from the tip of the short arm of chromosome 11," or "This locus is one of those present in such and such a G band on chromosome 11." A locus can also be specified genetically by saying, for example: "This locus is the position of the gene that causes sickle cell anemia when it is mutated in a certain way." The idea of a locus should convey no impression of the normality or abnormality of the gene carried there. A locus is merely a physical position on a chromosome.

We also need a word to specify the actual gene or genetic information that is present at a locus. The word for this is **allele**. Allele is one of the trickiest words in genetics. It means genes that are alike in one way but different in another. To make this clear, we must first consider what the gene at a locus is like. Recall that a gene is actually a sequence of nucleotides in the DNA at a specific locus on the chromosome, and this sequence of nucleotides is analogous to the sequence of letters on a ticker tape. The two genes that occupy homologous loci are really two instructions, either of which should enable the cell to carry out a particular function. The information is encoded in the sequence of nucleotides in the DNA. The cell has two copies of each gene or instruction; perhaps this has evolved partly to protect the cell against mistakes that may exist in one of them. This results in a benefit similar to the strategy often followed by spies and others for whom a simple misprint can have dire consequences: The same message is sent twice.

Often the genes at homologous loci carry the same sequence. They may both say "round and round the mulberry bush," to use the ticker-tape analogy in Figure 2.11. Sometimes the two copies are different. One may have the normal sequence, "round and round the mulberry bush," and the other can be aberrant in any number of ways. It may have a simple misprint: "round and round the fulberry bush" or "round and pound the mulberry bush." It may have letters missing from the end: "round and round the mul," or from the middle: "round and rd the mulberry bush." It may even have a small addition: "round an roundound the mulberry bush." (Equivalents of each of these kinds of mistakes do occur in genes.) Sometimes both homologous loci carry an abnormal gene.

In any case, the alternative sequences of a gene, including the normal or typical ones and all the variants, are called **alleles**. Alleles are genes that are identical in the sense that they are alternate occupants of the same locus. Of course, only one allele can occupy the locus in a particular chromosome, because the nucleotides in the DNA corresponding to a particular locus must have some particular sequence; this sequence of nucleotides specifies the allele that is present in the chromosome. On the other hand, alleles are genes that are nonidentical in the sense that they carry variant forms of the genetic information of the same gene; these variant forms correspond to different nucleotide sequences or, by the ticker-tape analogy, different sequences of letters.

The distinction between locus and allele is illustrated in Figure 3.2. The locus in the figure corresponds to the position on the chromosome occupied by 12 nucleotides of double-stranded DNA. (Actual loci consist of thousands of nucleotides.) Three alleles are shown: Allele 1 is the normal allele, allele 2 is an allele differing from the normal one only in the first nucleotide position (asterisk),

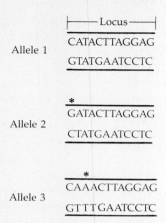

Allele 1

├────── Locus ──────┤

CATACTTAGGAG
GTATGAATCCTC

Allele 2

*
GATACTTAGGAG
CTATGAATCCTC

Allele 3

*
CAAACTTAGGAG
GTTTGAATCCTC

Figure 3.2 A *locus* corresponds to a segment of DNA, illustrated here as parallel lines representing the two strands with A-T and G-C nucleotide pairings between them. An *allele* refers to a particular sequence of nucleotides at a locus. Shown here are three alleles; alleles 2 and 3 differ from allele 1 at the nucleotide positions indicated by asterisks. The word *gene* can mean either "locus" or "allele," depending on the context.

and allele 3 differs from the normal one only in the third nucleotide position. Although a particular locus can carry only one allele, the allele that is present can be any one of those illustrated.

To summarize, the word **gene** is a general catchall word, used fairly loosely to mean the elements that are inherited. The word **locus** means a position on a chromosome. And the word **allele** means a particular form of a gene. Both **locus** and **allele** may be used as synonyms for **gene,** but each word has its own specific shade of meaning. The Mendelian statement "Genes occur in pairs" can now be made more precise. It means "Loci occur in pairs"; that is, for each locus in a chromosome there is a companion locus at the corresponding position in the homologous chromosome. This, of course, follows from the fact that each chromosome in the cell has a homologous chromosome.

As noted in Chapter 2, in humans and many other animals there is an exception to the rule that loci occur in pairs. The exception pertains to the sex chromosomes in males. The X and the Y chromosomes carry different genes. They are homologous chromosomes only in the general sense that they are both sex chromosomes. But during meiosis they pair end to end, not side by side, and no crossing-over occurs between them. The genes carried by these two chromosomes are different. Therefore the loci on the X chromosome are present only once in each cell of a male; the same goes for genes on the Y. The 2 X chromosomes in females and the 22 pairs of autosomes in both sexes do imply that each locus on these chromosomes will be present twice in each cell.

Convenience and brevity often require that the entire sequence of nucleotides or letters in a gene be designated by a single symbol. We may, for example, let the arbitrary symbol *A* stand for the entire normal sequence "round and round the mulberry bush." An aberrant or abnormal sequence of this gene might then be designated by the symbol *a*; for example, *a* might represent the sequence "round and round the fulberry bush." Using upper and lower cases of the same letter is helpful because it is a reminder that *A* and *a* are alleles, either of which may be found at a specific locus. If the locus in question is on one of the autosomes, so that there are two representatives of the locus in every cell, then both loci may carry *A*, both may carry *a*, or one locus may carry *A* and the other *a*. When this locus is described, the genetic constitution of these cells would conveniently be designated as *AA, aa,* or *Aa,* respectively.

This straightforward symbolism does not imply a denial of any other differences between cells of individuals. If one cell of an individual is said to carry *AA* and a cell of a different individual is said to carry *aa,* for example, this draws attention only to the one locus under consideration. The cells may have other important differences as well, but these are intentionally overlooked. This exemplifies the same kind of verbal shorthand used every day when a person says: "Sally has red hair; Mary is a blond." Sally and Mary have innumerable other differences, but their hair color is the only point of interest to the speaker.

One other use of the genetic symbols may possibly cause confusion. The genetic constitution of a *cell* may be *AA, aa,* or *Aa.* But if the cell is a fertilized egg and it is, for example, *Aa,* then the process of mitosis makes it almost certain that each of the cells in the person who develops from this egg will also be *Aa.* One could properly say, for instance, "The fertilized egg that gave rise to Theodore was genetically *Aa*" or "Each of the approximately 10 trillion cells in Theodore's body carries *Aa,*" but both expressions are cumbersome. It is easier to say simply "Theodore carries Aa." It is equally correct to say "Theodore *is Aa.*" The difference in terminology here is as tiny and inconsequential as that in "Theodore's head carries blond hair," "Theodore has blond hair," and "Theodore is blond."

SEGREGATION

Mendel's **law of segregation** states that the homologous loci in an individual separate from one another, or *segregate*, during the formation of eggs and sperm, so that each gamete carries exactly one of the loci, never neither and never both. The law of segregation also states—and this is the statistical part—that in the formation of gametes, half the gametes will carry one of the homologous loci and the other half will carry the other. For Mendel to have inferred this rule from his simple pea-breeding experiments is to his everlasting honor. Once the details of meiosis were worked out, the rule became fairly obvious. An individual cell undergoing meiosis and carrying *Aa* at some pair of homologous loci must produce four telophase II nuclei; two of these nuclei must carry the chromosome with *A* on it and the other two must carry the homologous chromosome with *a* on it. Thus, segregation is a direct consequence of the pairing and separation of homologous chromosomes during meiosis. In human males, each of the telophase II cells goes on to form a mature, functional sperm, so each spermatocyte gives rise to four sperm, two with the *A* allele and two with the *a* allele. In females, only one of the telophase II nuclei remains in the functional egg; but which one of the nuclei becomes the egg nucleus is a matter of chance, so half the time it will carry *A* and half the time *a* (Figure 3.3). We say that the **probability** that a gamete carries *A* is $\frac{1}{2}$ and the probability that it carries *a* is $\frac{1}{2}$. Note that the two probabilities add to 1, which is just the mathematical way of saying that each gamete must carry either *A* or *a* and that these are the only two possibilities.

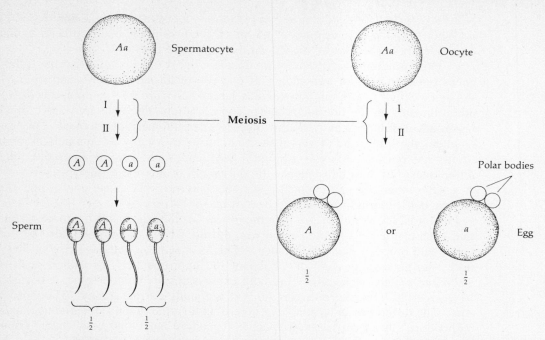

Figure 3.3 Segregation in an *Aa* heterozygote. In males, two *A*-bearing and two *a*-bearing sperm are produced. In females, either an *A*-bearing egg or an *a*-bearing egg is produced, and each of these possibilities is equally likely.

Of course, segregation of *A* from *a* occurs only in *Aa* individuals. Individuals who are genetically *AA* must produce only *A*-bearing gametes; individuals who are *aa* must produce only *a*-bearing gametes. Thus, the probability of an *A*-bearing gamete coming from an *AA* individual is 1, and the probability of an *a*-bearing gamete from *AA* individual is 0. Likewise, the probability of an *A*-bearing gamete from an *aa* individual is 0, and the probability of an *a*-bearing gamete from an *aa* individual is 1.

Ordinarily, the genes themselves are never seen; they are too small to observe except with an electron microscope. Even the chromosomes are far too small to see with the naked eye, and most people never actually see a typical human cell. But this is not important. What is important is that the effects of the genes can be observed in the appearance of the people who carry them. Therefore, the segregation of genes can often be observed merely by examining the kinds of offspring produced by a mating.

Using pea plants, Mendel himself studied, among other traits, the inheritance of round seeds versus wrinkled seeds. In this comparison can be seen one of Mendel's marks of brilliance. He realized that you must always study the inheritance of *differences* between individuals. When every individual is precisely identical in some physical or mental feature, the mode of inheritance of the trait can hardly be determined because it has no pattern of inheritance—other than that all parents and all offspring have the trait. Traits that appear in every individual need not have a genetic basis at all; they might be determined by environmental factors common to all individuals.

Mendel was also careful to focus on only one or a few differences between plants and to ignore the countless other differences he must have seen. He realized as well that the differences to be contrasted must be comparable. For example, in the case of round seeds versus wrinkled seeds, he perceived that trying to determine the inheritance of round seeds versus tall height in the plants would be hopeless because the traits are not comparable. To compare them would be as meaningless as comparing pecans and pineapples; and to study the genetic basis of such a difference would lead to nonsense.

The insight to know exactly what traits are to be compared and contrasted is gained by experience and, sometimes, just good luck. In many cases, the choice of characteristics to be compared is quite straightforward because the comparison is between individuals who are abnormal in some very specific way and those who are normal with respect to this same trait. In other cases, the comparison is not so obvious, and progress in genetics is often made when someone perceives a fruitful way to classify the differences among individuals; this is especially true in human genetics.

There is an amusing and surprisingly accurate analogy to the process of segregation that may eliminate some of the abstractness and mystery associated with the process. The analogy by the famous geneticist J. B. S. Haldane, appears in his book *Heredity and Politics*:

> Nonbiologists often find a certain difficulty in following Mendel's laws, so I am going to use analogy rather freely. In Spain every person has two surnames, one of which is derived from the father and one from the mother. For example, if you are called Ortega y Lopez you derive the name Ortega from your father, it was his father's name; and Lopez from your mother, as it was her father's name. Now I want you to imagine a strange savage nation in which everybody has two surnames; and when a child is born there is a curious ceremony by which he receives one of the surnames from one parent and one from the other, these being drawn at random by a priest, whose business it is. For example, If Mr. Smith-Jones marries Miss Brown-Robinson the children may get the name Smith or Jones from the father and Brown or Robinson from the mother; and will be called Smith-Brown, Smith-Robinson, Jones-Brown, or Jones-Robinson. If the priest draws at random those four types occur with equal frequency. A complication arises from the fact that some people may get the same name from both parents, and be called Smith-Smith, transmitting the name Smith to all their children. If you try to remember that simple scheme you will find no great difficulty in understanding the laws which govern heredity. The things which are analogous to the surnames are called genes.*

For determining the kinds and frequencies of the various types of offspring expected to result from a mating, geneticists often use the sort of squares (called **Punnett squares**, after their inventor) shown in Figure 3.4. Use of the squares is illustrated by means of the Haldane analogy. In Figure 3.4a, Smith-Jones (the

* Haldane, J. B. S. 1938. Heredity and Politics. Norton, New York, pp. 46–47.

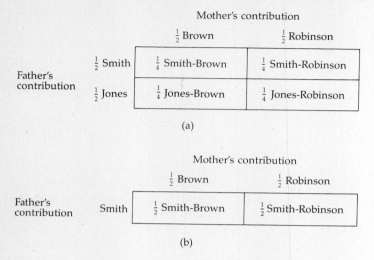

Figure 3.4 Haldane analogy of Mendelian inheritance in which each distinct allele at the locus under consideration corresponds to a distinct surname. The mating in (a) involves four alleles; that in (b) involves three alleles.

father) mates with Brown-Robinson (the mother). The father's contribution to the offspring surname is either Smith or Jones, each with a probability of $\frac{1}{2}$; the mother's contribution is either Brown or Robinson, each with a probabiltiy of $\frac{1}{2}$. The possible offspring surnames are obtained by means of a sort of cross-multiplication of the father's contribution and the mother's contribution; these are shown inside the boxes of the Punnett square. Each of the four offspring surnames has a probability of occurrence of $\frac{1}{4}$ (because $\frac{1}{2} \times \frac{1}{2} = \frac{1}{4}$), which is what Haldane refers to when he says "If the priest draws at random those four types will occur with equal frequency." Segregation of surnames in a mating of Mr. Smith-Smith with Ms. Brown-Robinson is illustrated in Figure 3.4*b*. Because in this case all the offspring must receive the name Smith from the father, only two classes of offspring are possible, and each class has a probability of occurrence of $\frac{1}{2}$ because ($\frac{1}{2} \times 1 = \frac{1}{2}$).

DOMINANCE

Mendel found that seeds carrying two copies of a certain allele were round and that those carrying two copies of another allele were wrinkled. The genetic constitution of the round seeds may be designated as *AA* and that of the wrinkled seeds as *aa*. When plants with these genetic constitutions are crossed, or mated, the offspring (called a **hybrid**) must receive *A* from one parent and *a* from the other. All the hybrid seeds are therefore genetically *Aa*. (A surname analogy for this mating is one in which Mr. Smith-Smith mates with Ms. Brown-Brown, and all the offspring are Smith-Brown.) What do these *Aa* offspring look like? As illustrated in Figure 3.5*a*, Mendel found that they were round—indistinguishable from *AA* seeds. The development of a round seed evidently requires only a single copy of the *A* allele; two copies of the *A* allele do not make a seed perceptibly more round than only one copy. The observation is summarized by saying that the *A* allele is **dominant** to the *a* allele; conversely, the *a* allele is said to be

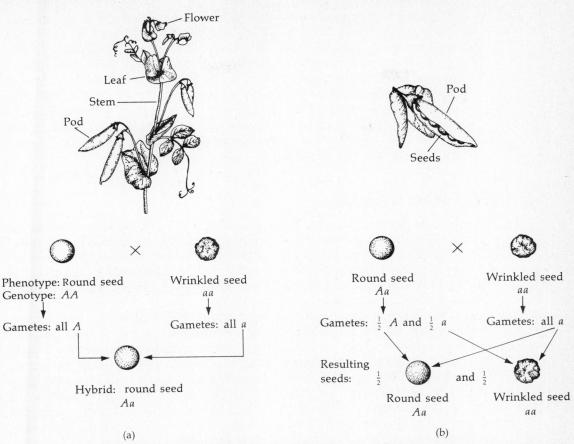

Figure 3.5 Summary of some of Mendel's results in breeding peas with round or wrinkled seeds. (*a*) The mating of a homozygous *AA* plant with a homozygous *aa* plant gives rise to heterozygous *Aa* seeds. The *Aa* seeds are round, like those in the *AA* parent; thus, *A* is dominant to *a*. (*b*) A mating of a heterozygous *Aa* plant with a homozygous *aa* plant. Because of segregation during meiosis in the heterozygous parent, half the resulting seeds will be *Aa* (and therefore round) and half will be *aa* (and therefore wrinkled).

recessive to the *A* allele. The dominance of the allele that causes round seeds is only one example of **dominance**, and more about dominance will be noted later.

Dominance is an example of a rather common occurrence in genetics: individuals who appear to be identical with respect to some trait may not really be identical in the genes that underlie the trait. Because of this, it is of crucial importance to distinguish the genetic constitution of an individual from the physical appearance of the individual. The distinction is made easier to remember because there is a special terminology for it. The **genotype** of an individual is the individual's genetic constitution; and the **phenotype** of an individual is the physical appearance. *AA*, *Aa*, and *aa* are examples of genotypes; round and wrinkled seeds are examples of phenotypes. Mendel's observation of dominance can be summarized by saying that the phenotypes of *AA* and *Aa* seeds were round, whereas the phenotype of *aa* seeds was wrinkled.

The hybrid that results from a mating of an *AA* plant with an *aa* plant has the genotype *Aa*. If one crosses such an *Aa* hybrid with an *aa* plant, nearly half the resulting seeds will be round and the other half wrinkled (see Figure 3.5*b*). The proportions of the two types reflect the segregation of the *A* and *a* alleles during meiosis. Half the gametes from the *Aa* plant carry *A*, and half carry *a*. All the gametes from the *aa* plant carry *a*. Therefore, the seeds that arise from the mating of *Aa* with *aa* will be *Aa* half the time (and therefore round) and *aa* the other half (and therefore wrinkled), just as a mating of Mr. Smith-Brown with Ms. Brown-Brown in Haldane's "strange savage nation" would lead to $\frac{1}{2}$ Smith-Brown and $\frac{1}{2}$ Brown-Brown offspring.

SEGREGATION OF GENES IN HUMANS

Segregation of genes occurs in humans as well as in peas. However, segregation cannot easily be observed unless the trait under study is determined by the alleles at a single locus without complications due to the effects of either alleles at other loci or the environment. Such simple traits that are determined by the alleles at a single locus are known as **simple Mendelian traits**, and they are said to show **simple Mendelian inheritance**.

One example of a simple Mendelian trait that illustrates segregation in humans is the harmless trait *woolly hair*. (The woolly hair trait is rare; examples of more common simple Mendelian traits will be discussed a little later.) Woolly hair refers to a curly, fuzzy texture of the hair, superficially similar to the texture of the hair in many Negro families, although woolly hair is not exactly the same (see the family in Figure 3.6). Woolly hair is brittle and tends to break off at the tip, so it never gets very long. This trait is due to the presence of a dominant gene, which may be called *W* (for woolly), located on one of the autosomes (which autosome is not known). People with straight, nonwoolly hair are genotypically *ww*. People with woolly hair are *Ww*, so only one copy of the *W* allele is sufficient to produce the trait and the *W* allele is therefore dominant. (It should be emphasized that the genetic basis of the differences in hair texture between Caucasians and Negroes and other racial groups is rather more complicated than the difference between straight and woolly hair.)

Figure 3.7 is a pedigree depicting the occurrence of woolly hair in a Norwegian kindred. A **kindred** is a group of people related by marriage or ancestry, and a **pedigree** is a diagram showing the ancestral history of a group of relatives. The pedigree in Figure 3.7 uses certain conventions:

1. Matings are represented by horizontal lines connecting the individuals involved.
2. The children of a mating are represented in order of birth from left to right beneath a horizontal line connected to the parental mating.
3. Males are represented by squares, females by circles.
4. Affected individuals (in this case, affected with woolly hair) are represented by solid symbols, nonaffected by open symbols.
5. Arrows call attention to certain individuals in the pedigree, usually to the first individual who came to the investigator's attention and through

Figure 3.6 A portrait of a Norwegian family showing woolly hair. The mother (seated) has the trait, whereas the father (also seated) does not. The three older children standing in the rear all have woolly hair; the three younger standing near the front have nonwoolly hair. The photograph of a woman with uncut woolly hair demonstrates that such hair breaks off naturally before getting very long.

Figure 3.7 Pedigree of extensive Norwegian kindred involving the woolly hair trait. Solid symbols indicate woolly hair; open symbols indicate nonwoolly hair. Circles indicate females; squares indicate males. The roman numerals refer to generations. Individuals designated by arrows are shown in the family portrait in Figure 3.6.

whom the pedigree was discovered; in the case of Figure 3.7, however, the arrows point to the people in the family portrait in Figure 3.6.

6. In many cases, as in Figure 3.7, in order to conserve space it is convenient to depict the successive generations (indicated by roman numerals) as a series of concentric semicircles rather than as a series of horizontal lines.

Figure 3.7 illustrates certain features that are characteristic of rare, simple Mendelian traits due to an autosomal allele. These characteristics are as follows:

1. Males and females are equally likely to be affected with the trait.
2. Affected individuals will almost always have just one affected parent. (Because the trait is assumed to be rare, matings between two affected individuals almost never occur.)
3. An affected parent may be either male or female.
4. Half the children from an affected parent will be affected.

This last characteristic of simple dominant inheritance—that half the children of affected parents will be affected—is a result of segregation in the affected parent. Because the trait is due to a rare dominant allele, matings involving one affected parent will be between the genotypes Ww (the woolly haired parent) and ww (the normal parent). The offspring expected from such a mating are illustrated in the Punnett square in Figure 3.8. (In terms of the surname analogy, the mating is like one between Mr. Smith-Smith and Ms. Brown-Smith.) As seen in Figure 3.8, half the offspring are expected to be Ww, and the other half are expected to be ww.

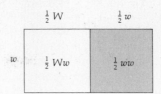

Figure 3.8 Punnett square illustrating segregation of the dominant allele *W* for woolly hair.

Does the pedigree in Figure 3.7 exhibit half woolly haired and half non-woolly haired offspring from matings in which one of the parents is affected? If so, then this would be a direct demonstration of segregation in humans. There are 20 woolly × nonwoolly matings in Figure 3.7, and these matings produced 81 offspring of whom 38 were woolly haired (*Ww*) and 43 were nonwoolly haired (*ww*). This is quite close to the 1:1 distribution of children predicted from the law of segregation (see Figure 3.8). Of course, the very closest fit to 1:1 in this case would be 40:41 or 41:40, but one cannot realistically expect the closest possible fit because of **chance variation** in the observed numbers.

Chance variation refers to the element of randomness or chance that occurs in the outcome of any experiment involving probabilities. From a statistical point of view, the distribution of woolly and nonwoolly hair among 81 offspring is the same as the distribution of heads and tails in 81 flips of a coin, because in each case the two possible outcomes are equally likely to occur. As it turns out, the exact probability of obtaining an observed distribution of 40:41 or 41:40 in 81 flips of a coin is about 18 percent, which means that such a good fit could be expected to occur with a frequency somewhat less than once in five trials (1 out of 5 corresponds to a probability of 20 percent). An observed distribution of 38:43 is well within the range that would be expected because of chance variation in 81 flips of a coin, or among 81 offspring when (as in the woolly hair example) each class of offspring is equally likely to occur.

HOMOZYGOTES AND HETEROZYGOTES

The emphasis in the examples of woolly versus straight hair in humans and round versus wrinkled seeds in peas has been on what happens when segregation occurs in only one parent. When there are two alleles at a locus, call them *A* and *a*, three genotypes will be present in the population: *AA*, *Aa*, and *aa*. Altogether six different kinds of matings can occur (the cross symbolizes the mating):

$$AA \quad \times \quad AA$$
$$AA \quad \times \quad Aa$$
$$AA \quad \times \quad aa$$
$$Aa \quad \times \quad Aa$$
$$Aa \quad \times \quad aa$$
$$aa \quad \times \quad aa$$

Segregation can be observed in all three matings in which one or both partners is *Aa*. Genotypes that carry two different alleles at a pair of homologous loci—the *Aa* genotype in this example—are thus the ones that undergo segregation, and geneticists have a special term for them: **heterozygous genotypes**, or **heterozygotes**. Conversely, the *AA* and *aa* genotypes are called **homozygous *A*** and **homozygous *a***, respectively; either genotype is sometimes referred to as a **homozygote**. Woolly hair has been chosen as an example of segregation of genes in humans because the trait is rare, and every woolly haired person will therefore almost certainly be heterozygous. The *W* and *w* alleles will therefore undergo segregation in every mating that involves a woolly haired parent. The vast majority of woolly haired people come from matings between heterozygotes and normal homozygotes. This is characteristic of traits caused by rare dominant alleles in humans; many other examples could be cited.

SIMPLE MENDELIAN INHERITANCE IN HUMANS

Woolly hair is a harmless curiosity that is useful for illustration because it is a simple Mendelian trait due to an autosomal dominant allele. Woolly hair is difficult to relate to one's own hereditary endowment, however, because it is so rare. Other simple Mendelian traits, however, are sufficiently common that you may be able to identify one or more of them in yourself, among your relatives, or among your friends.

Surprisingly, and perhaps paradoxically, *most phenotypic variation in humans is not due to simple Mendelian traits*. The phenotypic variation that is obvious to everyone—variation in height, weight, hair color, eye color, skin color, facial appearance, and so on—has a more complex genetic basis than do simple Mendelian traits. Traits such as height are determined by the collective or aggregate effects of alleles at tens or hundreds of loci and are also influenced by environmental factors (such as diet, in the case of weight, or exposure to sunlight, in the case of skin color). Traits that are determined by the aggregate effects of alleles at many loci are called **polygenic** or **multifactorial** traits. For multifactorial traits, the segregation of individual loci is virtually impossible to detect because the effect of any individual locus is so small when compared with the background effects of all the other loci and the environment. The background effects are like noise on a telephone line, noise so extreme that individual words become garbled or lost altogether. It is therefore important to keep in mind that *most readily observed traits are multifactorial*.

Perhaps the most widespread fallacy concerning human inheritance relates to eye color, specifically to the assertion that brown eyes are due to a simple Mendelian dominant and blue eyes to a simple Mendelian recessive. This error is even perpetuated in some textbooks. Were it true, it would be extremely useful because eye color is an easy trait to observe, and students could verify Mendelian segregation for themselves by observing eye color among families. Unfortunately, it is not true. There are many shades of eye color besides "brown" and "blue," and the inheritance of eye color is complex.

What's worse, the oversimplification of eye-color inheritance can cause considerable grief, as evidenced by two experiences that I became involved in some years ago in Minnesota. After reading in a textbook that blue eyes are recessive, a student, an undergraduate, became convinced that she had been adopted or was illegitimate; her parents both had blue eyes, but hers were brown. In another case a blue-eyed man from a small town in southern Minnesota became irate and suspicious when his blue-eyed wife delivered a brown-eyed baby. Resolution in both cases was easy. I simply quoted three sentences from Victor A. McKusick's *Mendelian Inheritance in Man,* a massive critical compendium of knowledge of human genetics. Under the heading "Eye color," McKusick writes: "This is almost certainly a polygenic trait. The early view that blue is a simple recessive has been repeatedly shown to be wrong by observation of brown-eyed offspring of two blue-eyed parents. Blue-eyed offspring from 2 brown-eyed parents is a more frequent finding."* The student and the husband were wise to have consulted a geneticist. How many others have not, and gone on believing something bad about themselves, their parents, or their spouses?

In spite of the complexity usually involved in the inheritance of common traits, a number of such traits that do seem to have simple Mendelian modes of inheritance have been identified (see the list in Table 3.1 and the corresponding illustrations in Figure 3.9). Certain venerable textbook traits such as ability to roll the tongue and pattern of hand clasping are not included in the list because modern research has shown that these traits are not simple Mendelian traits, if, indeed, they are inherited at all. Although the mode of inheritance of the traits in Table 3.1 is relatively simple, these traits illustrate a number of important and complicating features that are often encountered in the study of human inheritance. Brief comments about each trait are therefore in order.

Common Baldness (see Figure 3.9a)

This trait is the common form of hair loss in Caucasians, made famous because President John Adams and several of his illustrious descendants had the trait. (Figure 3.9a is another affected president, Dwight David Eisenhower.) Beginning at about age 20 in males and age 30 in females, hair begins to fall out at the sides near the front of the scalp, producing an M-shaped hairline. Recession of the hairline in front is usually followed by the development of a bald spot at the rear of the crown. Enlargement of the bald spot and continued recession of the front hairline lead to a final stage where only a peripheral fringe of scalp hair remains. The trait is a simple Mendelian trait due to an autosomal allele; however, the allele is *dominant* in males but *recessive* in females. The trait is most common in Caucasians: estimates are that at least 50 percent of Caucasian men and 25 percent of Caucasian women will be affected, although the extreme progression to a remaining fringe of hair with a bald pate occurs in only about one-third of affected men (affected women typically experience less hair loss than affected men). The trait is less common among Negroes than among Caucasians, and it is rare among Orientals.

* McKusick, V. A. 1983. *Mendelian Inheritance in Man,* 6th ed. The Johns Hopkins University Press, Baltimore, p. 704.

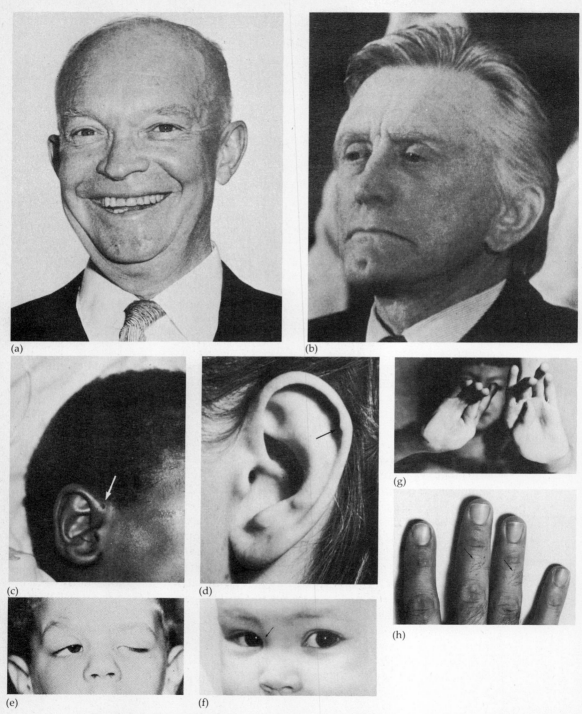

Figure 3.9 Examples of some simple Mendelian traits in humans that are relatively common: (*a*) common baldness, (*b*) chin fissure, (*c*) ear pits, (*d*) Darwin tubercle, (*e*) congenital ptosis, (*f*) epicanthus, (*g*) camptodactyly, and (*h*) mid-digital hair. See Table 3.1 for descriptions and modes of inheritance.

TABLE 3.1 CHARACTERISTICS OF SOME RELATIVELY COMMON SIMPLE MENDELIAN TRAITS IN HUMANS*

Trait	Illustration	Phenotype
Common baldness	Figure 3.9a	M-shaped hairline receding with age
Chin fissure	Figure 3.9b	Vertical cleft or dimple in chin
Ear pits	Figure 3.9c	Tiny pit in external ear
Darwin tubercle	Figure 3.9d	Extra cartilage on rim of external ear
Ear cerumen		Wet and sticky ear wax versus dry and crumbly ear wax
Congenital ptosis	Figure 3.9e	Droopy eyelid
Epicanthus	Figure 3.9f	Fold of skin near bridge of nose leading to almond-shaped eyes
Camptodactyly	Figure 3.9g	Crooked little finger due to too short tendon
Mid-digital hair	Figure 3.9h	Hair growth on middle segment of fingers
Phenylthiocarbamide tasting		Ability to taste phenylthiocarbamide; tasters report it as bitter
S-methyl thioester detection		Detection by smell of odiferous substances excreted in urine after eating asparagus
ABO blood group		Type A versus type B versus type O versus AB
Rh blood group		Type Rh$^+$ (positive) versus Rh$^-$ (negative)

* For photographs, see Figure 3.9.

The trait of common baldness illustrates several complications. First, it is a **sex-influenced** trait—that is, the trait is expressed differently in males and females (recall that the baldness allele is dominant in males but recessive in females). Extreme cases of sex-influenced traits, in which expression is limited to one sex, are known as **sex-limited** traits; examples of sex-limited traits in humans are growth of facial hair, which ordinarily occurs only in males, and development of breasts, which ordinarily occurs only in females. Common baldness also illustrates **variable age of onset**; individuals who have the baldness genotype have the genotype from the time of birth, but the trait does not occur until much later, and its time of occurrence is variable, beginning in the early 20s for some men but later for other men. The trait also exhibits **variable expressivity**, which means that it is expressed to different degrees in different people. The severity of expression of common baldness may vary from a slightly receding frontal hairline to an almost completely bald pate.

The same genotype may be expressed differently in different individuals because of differences in **genetic background** (i.e., the genotypes at other loci that also influence the trait) or because of differences in nongenetic or environmental factors. The extreme end of varible expressivity is known as **incomplete penetrance**, in which individuals who have a genotype associated with a particular trait do not express the trait at all. Incomplete penetrance is a serious complication in human genetics because it alters the phenotypic ratios from what would be expected with simple Mendelian inheritance. Finally, common baldness illustrates **variation among populations** in its incidence. As noted, the trait is common among Caucasians, less common among Negroes, and rare among Orientals. Variation among populations is a characteristic feature of many inherited traits, and additional examples will be found in later chapters of this book.

Chin Fissure   (see Figure 3.9*b*)

The most common type of chin fissure is a perpendicular furrow in the middle of the chin, and it is due to an autosomal dominant allele. The trait is variable in expression (ranging from a chin dimple to a Y-shaped furrow), but it has almost complete penetrance in males (i.e., all males who carry the allele will have some sort of chin fissure). Females are affected only about half as frequently as males, so the trait is sex influenced. Chin fissures have a variable age of onset; although sometimes present at birth, they more commonly appear in childhood or early adulthood. Among people of German extraction, approximately 20 percent of males and 10 percent of females have the trait. (The man with the handsome chin dimple in Figure 3.9*b* is actor Kirk Douglas.)

Ear Pits   (see Figure 3.9*c*)

These harmless and usually shallow pits may occur in one ear or both. Ear pits are due to an autosomal dominant allele, which is variable in expression (one ear or both affected, shallow or deep pits), has incomplete penetrance (only about half of those individuals who are heterozygous or homozygous for the dominant allele will have ear pits), and is sex influenced (twice as prevalent in females as in males). The incidence of the trait varies among populations—1 percent in American Caucasians, 5 percent in American Negroes, and said to be "very common" in Chinese.

Darwin Tubercle   (see Figure 3.9*d*)

The Darwin tubercle is a variable-sized thickening of cartilage near the upper rim of the ear, and it is probably due to an autosomal dominant allele (i.e., homozygote and heterozygote are both affected). The allele has a high penetrance and the trait is not sex influenced because it has approximately equal frequency in males and females. Prevalence varies among populations, being about 20 percent in Germany and about 50 percent in England and Finland.

Ear Cerumen

In the Japanese language, the gray, dry, and brittle type of ear wax is called "rice-bran ear wax," whereas the brown, sticky, and wet type is called "honey ear wax" or "cat ear wax." The dry type is due to an autosomal recessive allele. Type of ear cerumen has remarkable variation among poulations; the dry type (which occurs in homozygous recessives) has the following frequencies: Koreans (93 percent), Japanese (85 percent), American Indians (66 percent), American whites (25 percent), American blacks (5 percent). The trait illustrates two important phenomena. First, most Koreans and Japanese who have wet cerumen will actually be heterozygous for the dominant "wet" allele; they are thus heterozygous for the recessive "dry" allele. Such individuals who are heterozygous for a recessive allele are said to be **carriers** of the allele. Second, ear cerumen illustrates what is called **pleiotropy**, which refers to the fact that genes can have effects on several or many traits and not just one trait. In the case of ear cerumen, the alleles influence secretions of certain glands in the ear canal; these very same alleles also influence the chemical makeup of secretions from certain other glands, notably those involved in perspiration. Thus, the alleles that affect ear cerumen are **pleiotropic**: they affect several traits (type of ear wax and chemical makeup of perspiration).

Congenital Ptosis   (see Figure 3.9e)

The word **congenital** means "present at birth," and congenital ptosis refers to a drooping eyelid, usually affecting only one eye, that is in most cases due to weakness in upward movement of the eyelid. The condition is usually present at birth and may in some cases require surgical correction. The trait is caused by an autosomal dominant allele with incomplete penetrance (about 60 percent of homozygous dominants and heterozygotes actually express the trait). Although the exact prevalence of congenital ptosis is unknown, it is thought to be quite common.

Epicanthus   (see Figure 3.9f)

Epicanthus refers to a condition in which the inner corner of the eye is bridged by an arching fold of skin extending from the bridge of the nose and giving the eye an almond-shaped or "Oriental" appearance. The trait is due to an autosomal dominant allele that is extremely variable in its expression. The trait often tends to disappear with age, being present in about 20 percent of 1-year-old Caucasians but in only about 3 percent of those 12 years and older. Approximately 70 percent of adult Orientals have epicanthus, but it is not clear whether the Oriental form of the trait has the same genetic basis as in other racial groups. (As a point of interest, epicanthus is a normal characteristic of the *fetus* in all racial groups.)

Camptodactyly   (see Figure 3.9g)

Camptodactyly is a minor hand abnormality usually expressed as an inability to completely straighten the little finger. It is due to an autosomal dominant allele with incomplete penetrance and variable expressivity, but its prevalence is not known.

Mid-Digital Hair   (see Figure 3.9h)

Presence of hair on the middle segment of the fingers is thought to be due to an autosomal dominant allele. Homozygotes for the recessive allele lack hair on these finger segments, even if they otherwise have abundant body hair. The prevalence of the trait is unknown.

Phenylthiocarbamide Tasting

The ability to taste this chemical is due to an autosomal dominant allele usually symbolized as T; the recessive allele is symbolized as t. Thus, genotype TT and Tt are tasters and report that the chemical is bitter, whereas tt genotypes are nontasters and cannot detect the chemical. (There are different thresholds of taste sensitivity, which seem to be multifactorial.) Approximately 30 percent of Western Europeans are nontasters. Tasting ability is of some interest because it illustrates that genes can influence almost any trait, including, in this case, our sensory perceptions.

S-methyl Thioester Detection

Certain odiferous substances are excreted into the urine by at least some people after eating asparagus. At one time, excretion was thought to be due to a single dominant allele, nonexcretors being homozygous recessives, but now it appears that the primary genetic variation involves the ability to smell the substances. Approximately 10 percent of Caucasians can smell the diluted asparagus-related substances.

ABO Blood Groups

The ABO blood groups refer to certain constituents found on the surface of red blood cells. These constituents can be identified by means of the appropriate chemical reagents (see Chapter 12). An individual may have one of four blood types. Those individuals whose red cells have only the A substance are called type A individuals; those with only the B substance are called type B; those with neither A nor B are said to be of type O; and those with both the A and B substances are said to be of type AB.

The ABO blood types are due to three alleles at a single locus near the tip of the long arm of chromosome 9. These alleles are designated I^A, I^B, and I^O. Individuals of genotype I^AI^A or I^AI^O have blood type A; those of genotype I^BI^B or I^BI^O have blood type B; those of genotype I^OI^O have blood O; and those of genotype I^AI^B have blood type AB. Thus, I^A is dominant to I^O, and I^B is also dominant to I^O. However, because I^AI^B individuals express phenotypic characteristics associated with *both* the I^A and I^B alleles, I^A and I^B are said to be **codominant**. The ABO blood group locus illustrates the phenomenon of **multiple alleles**, which refers to the occurrence of more than two alleles at a locus. (Of course, multiple alleles refer to an entire population; any one individual can carry at most two alleles at a locus.)

Considerable variation in the frequency of the ABO blood types occurs among populations. Among Iowans of primarily Northern European extraction, for example, the frequencies of A, B, O, and AB blood types are 42 percent, 9 percent, 46 percent, and 3 percent, respectively; among Chinese, the comparable frequencies are 31 percent, 28 percent, 34 percent, and 7 percent, respectively. The chimpanzee (our closest living relative) also has the ABO blood group substances, with about 87 percent of chimps having blood type A and the rest having blood type O; the I^B allele is evidently rare in chimps or does not occur at all.

Rh Blood Groups

Individuals who are Rh^+ (Rh positive) have a certain substance present on the surface of their red blood cells, whereas Rh^- (Rh negative) individuals lack this substance. (The substance is unrelated to those involved in the ABO blood groups.) Presence of the Rh^+ substance is due to a dominant allele at a locus on the short arm of chromosome 1. We will denote the dominant allele as *D* and its recessive counterpart as *d*. Thus, *DD* and *Dd* genotypes are phenotypically Rh^+, whereas *dd* genotypes are Rh^-. Details of the genetics are unclear, however. The system may involve two or more alleles at each of three loci that are extremely close together on the chromosome, certain combinations of which include the *D* allele at one locus. Alternatively, the system may involve multiple alleles (at least eight) at a single locus, some alleles (collectively designated *D*) associated with the presence of the Rh^+ substance and other alleles (collectively designated *d*) with its absence.

As with many other traits, there is variation in the incidence of Rh^+ and Rh^- among populations. Among American Caucasians, the frequencies of Rh^+ and Rh^- are about 86 percent and 14 percent, respectively. Among the Basques, a group of people who live in the Pyrenees Mountains between France and Spain, the frequencies of Rh^+ and Rh^- are about 58 percent and 42 percent, respectively; the Basques have the highest known frequency of Rh^-. (For further discussion of the Rh blood groups, see Chapter 12.)

HUMAN HEREDITY: SOME CAUTIONARY REMARKS

It should be clear from the preceding section that determining the genetic basis of human traits can be fraught with difficulties, such as incomplete penetrance, variable expressivity, variable age of onset, and sex influences. Because of these complications, some traits that may seem to be inherited may not be genetic; and some traits that may seem to be caused by the environment may nevertheless be determined by genes. Determination of the actual genetic basis of human traits therefore requires careful study and an awareness of certain pitfalls and complications. Here is a brief compilation of some of these pitfalls.

1. *Traits that are* **familial** *(i.e., tending to "run in families") are not necessarily genetic*. Children learn to speak from their parents, for example, and they may pick up certain peculiarities in pronunciation or accent. Such peculiarities are certainly familial, but this is due to learning, not genetics. Familial traits that are transmitted from generation to generation by means of the learning process are said to be due to **cultural inheritance**.

2. *Traits caused by environmental factors may nevertheless have an underlying genetic basis.* This means that genes can determine the degree of **liability** (risk) toward the expression of environmentally triggered traits. For example, genes can cause an extreme sensitivity and violent side reactions to certain drugs, such as antibiotics. Yet, unless the individual is exposed to the drug, sensitivity will not be expressed. In this case, the drug is the environmental trigger for a genetically determined trait.

3. *Genetic traits may have no obvious pattern of simple Mendelian inheritance.* This is true because genetic traits need not be determined by the alleles at a single locus. Many traits are multifactorial (influenced by alleles at many loci). Many traits are more or less strongly influenced by the environment. Pedigrees for such traits will therefore be incompatible with simple Mendelian inheritance, but the traits are nonetheless genetic.

4. *A genetic trait may consist of several seemingly unrelated symptoms.* The cause of this is pleiotropy—the fact that genes can influence many traits at the same time. For example, the blood disorder sickle cell anemia has many seemingly unrelated symptoms, including general weakness, susceptibility to infections, fever, and recurrent pain in the joints and other body parts. However, all these symptoms are pleiotropic effects of the allele that causes the abnormal hemoglobin protein in red blood cells. (See Chapter 4 for further discussion.)

5. *Genetic traits can have different modes of inheritance in different kindreds.* Cleft palate (failure in the fusion of the halves forming the roof of the mouth) is a good example. In most kindreds in which cleft palate occurs, the trait is multifactorial. In some rare kindreds, however, the trait is due to an autosomal dominant allele.

6. *Genetic traits with well-established modes of inheritance can be mimicked by environmentally caused disorders.* For example, the kinds of malformations in the embryo that can result from certain drugs or from infection with German measles virus (rubella) may strongly resemble those caused by certain genetic disorders. Phenotypes that resemble known genetic conditions but are caused by environmental factors are known as **phenocopies**.

7. *Genetic traits may not be expressed in all individuals who have the relevant genotype.* This is true of genes with incomplete penetrance, and such examples as congenital ptosis were discussed in the previous section. Nonexpression of certain genotypes is also characteristic of sex-limited traits because, by their very nature, these traits can be expressed in only one sex.

8. *Individuals who are affected with the same genetic trait may not be affected to the same degree.* This is an important principle, and its causes include variable expressivity, variable age of onset, and sex influence on expression, all of which were discussed earlier in connection with common baldness.

9. *Genetic traits with well-established modes of inheritance may nevertheless be expressed differently in different kindreds.* This can occur because kindreds may differ in **modifier genes** (alleles at other loci that affect expression of the trait in question). Also, different kindreds may

have different alleles at the **major locus** (the locus that has the major effect on expression of the trait).

10. *Seemingly identical traits with the same mode of inheritance may nevertheless be caused by alleles at different loci.* For example, production of the pigment **melanin** is a multistep process, with each step controlled by the product of a different locus. Absence of melanin leads to **albinism**, in which affected individuals have stark white hair and pinkish skin (see Chapter 4 for a more detailed discussion). Two almost identical forms of albinism are known, both due to autosomal recessive alleles. The recessive alleles are at different loci, however, and represent defects at different steps in pigment production. Hence, some kindreds have one form of autosomal-recessive albinism, and other kindreds have the alternative form of autosomal-recessive albinism.

SIMPLE MENDELIAN DOMINANCE

At latest count, some 934 human traits were known to be simple Mendelian traits due to autosomal dominant alleles, and another 893 traits were strongly suspected to have this sort of simple Mendelian dominance as their mode of inheritance. Most of these traits are rare, usually much rarer than those listed in Table 3.1. Nevertheless, the traits are important because many of them involve inherited abnormalities of various kinds, and understanding their genetic basis permits proper **genetic counseling** (genetic advising) of parents or other relatives of afflicted individuals about what chance they may have of transmitting the genes that are responsible for the trait (or what chance they have of developing the trait themselves). Moreover, understanding the genetic basis of a trait often reveals clues about the trait that are important in developing new clinical treatments of affected individuals. A few examples of simple Mendelian dominance are discussed below; many others will be found in later chapters, and further discussion of genetic counseling will be taken up in Chapter 14.

One simple Mendelian dominant trait is of major importance in human genetics because it affects so many people. The trait is **familial hypercholesterolemia**, and it is the *single most frequent simple Mendelian disorder*. In most populations, about 1 person in 500 is heterozygous for the dominant allele. Heterozygotes for this allele have elevated levels of cholesterol in their blood serum that are untreatable by dietary restriction, early onset of atherosclerosis ("hardening of the arteries") due to deposits of fatty substances in the arteries, and a high risk of heart attack (myocardial infarction)—at least 25 times the risk in normal individuals. The allele has a pleiotropic effect leading to the development of yellowish nodules in the tendons, especially near the knuckles of the hands (see Figure 3.10 for an extreme example). These characteristic nodules have an age of onset between 30 and 40 in males (somewhat later in females), and they appear at approximately the same time as symptoms of coronary artery disease. Homozygotes for the high-cholesterol allele are much rarer than heterozygotes (the frequency of homozygotes is about one per million in most populations), and they are much more severely affected than the heterozygotes. The yellowish nodules appear earlier in life and may even be present at birth, serum

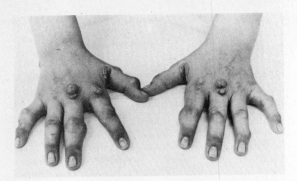

Figure 3.10 Hands of an individual with familial hypercholesterolemia showing nodules in tendons near the knuckles.

cholesterol levels are greatly elevated, and death from heart disease usually occurs by age 20. We can underscore the importance of familial hypercholesterolemia by noting that there are approximately 500,000 heterozygotes in the United States alone!

Another example of simple Mendelian dominance in humans is a rare dominant gene that causes a kind of dwarfism known as **achondroplasia** (Figure 3.11). Achondroplasia is a rare condition, affecting only about 1 in 10,000 people. One kindred that was segregating for the dominant gene causing achondroplasia—call the gene *A* and its normal allele *a*—was a Mormon family in Utah dating back to 1833, when the region was pioneered. This form of dwarfism is not incompatible with a relatively normal life. Indeed, in Utah during the late 1800s, achondro-

Figure 3.11 An achondroplastic dwarf.

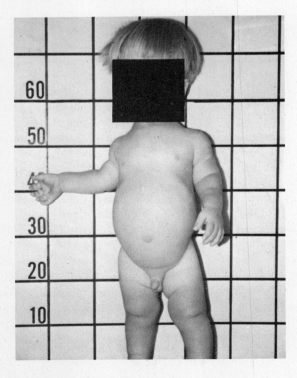

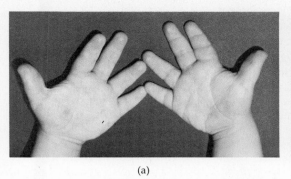 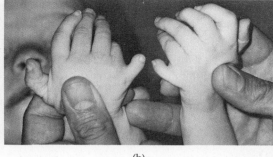

(a)   (b)

Figure 3.12 (a) Hands of a child with brachydactyly (short fingers); note that the fingers are only slightly longer than the thumb. (b) Hands of a child with polydactyly (extra fingers); the extra finger on each hand of this child is a small but well-formed digit next to the little finger.

plastic dwarfs of this type were able to function well in an agrarian society, and some became community leaders and men and women of substance. Also, although the pelvis of achondroplastic women is flattened somewhat (front to back), childbearing is not the problem it is in many other kinds of dwarfism. As in the case of woolly hair, most marriages of interest are heterozygous $Aa \times$ homozygous aa, and of 76 children born to such marriages in the Utah kindred, 34 were achondroplastic and 42 were normal, a ratio in satisfactory agreement with the expected 38:38.

Still another example of simple Mendelian dominance is **brachydactyly**, or shortfingeredness (Figure 3.12a); the fingers of affected individuals are about two-thirds the length of normal fingers, owing primarily to the shortness of the outermost long bones. (There are other skeletal abnormalities in the hands as well.) This condition is of historical interest; studied in 1905, it was the first demonstration of simple Mendelian dominance in humans.

One final example of simple dominance should be mentioned because the trait is relatively common; it is **polydactyly**, or extra fingers (see Figure 3.12b). In some kindreds, the trait that occurs is a fully formed extra finger; it does no harm, so it is ordinarily not even removed. This form of polydactyly is a simple Mendelian dominant. In other kindreds, a second form of polydactyly is found, which is multifactorial. The extra finger is a tiny one, just a nub, not fully formed. The tiny extra digit, which may be next to either the thumb or the little finger, is eliminated by cutting of its blood supply with a loop of silk thread. The nub eventually degenerates, much like what happens to the remnants of the umbilical cord of a newborn baby. The two forms of polydactyly taken together are fairly common. About 1 percent of American Negroes have the trait, making it about seven times more common among them than among whites.

SUMMARY

1. Alleles are alternative forms of a gene that can occupy a particular locus. The word **gene** is a general term used to refer to those elements that are transmitted in inheritance. A **locus** is the physical position of a gene on

a chromosome. An **allele** is a particular form of a gene; that is, an allele is a particular sequence of nucleotides in the DNA corresponding to the gene.

2. Alternative alleles at a particular locus are often designated by upper and lower case letters, such as A and a. For an autosomal locus that has two alleles, three genetic constitutions (called **genotypes**) are possible: AA, Aa, and aa. The Aa genotype is said to be **heterozygous**; the AA and aa genotypes are said to be **homozygous** A and **homozygous** a, respectively.

3. The physical appearance of an individual is known as his or her **phenotype.** When Mendel studied the phenotypes round seeds versus wrinkled seeds in garden peas, he discovered that seeds of genotype AA and Aa were phenotypically round, whereas seeds of genotype aa were phenotypically wrinkled. This situation is described by saying that the A allele is **dominant** to the a allele; conversely, we could say that the a allele is **recessive** to the A allele.

4. Mendel's **law of segregation** states that, during the formation of gametes, the alleles in an individual separate (segregate) from each other in such a way that each gamete will carry one or the other; moreover, half the gametes will carry one of the alleles and half will carry the other. Thus, for example, a heterozygous Aa individual will produce 50 percent A-bearing gametes and 50 percent a-bearing gametes. Said another way, the **probability** that a gamete from an Aa individual carries A is $\frac{1}{2}$, and the probability that it carries a is $\frac{1}{2}$. Segregation of alleles results from the separation of homologous chromosomes during meiosis.

5. Because of segregation in Aa individuals, the mating $Aa \times aa$ is expected to produce 50 percent Aa and 50 percent aa offspring. If the A allele is dominant, then such matings are expected to produce 50 percent of offspring having the phenotype associated with the dominant allele and 50 percent having the phenotype associated with the homozygous recessive. In any particular family, however, there need not be equal numbers of the two offspring phenotypes because of **chance variation**.

6. **Simple Mendelian traits** are traits with inheritance that is due to the alleles at a single locus. In spite of considerable phenotypic variation for many traits in humans, most of these traits are not simple Mendelian. Many of the most easily observed traits (such as height, weight, hair color, eye color, and skin color) are **multifactorial**, which means that the traits are due to the combined effects of alleles at many loci. Many traits are also influenced by environmental factors (such as diet, in the case of weight).

7. Studies of simple Mendelian traits in humans (such as common baldness, chin fissure, ear cerumen, and others listed in Table 3.1) reveal complications in the traits or in their genetic basis. For example, genes may have **pleiotropic effects**, which refers to the fact that a gene can simultaneously influence several, even seemingly unrelated, traits. Complications in trait expression include **incomplete penetrance** (nonexpression of the trait in some individuals who have the relevant genotype), **variable expressivity** (variation in severity of expression of the trait in different individuals), and **variable age of onset** (variation in the age at which the trait appears during life). Traits may also be **sex influenced** (different

expression in males and females) or even **sex limited** (expressed in only one sex). Genetic complications include **multiple alleles** (more than two alleles at the major locus affecting the trait) and **variation among populations** (incidence of the trait differing in different populations).

8. A number of additional cautions should be kept in mind regarding human heredity. The most important are: (*a*) traits that are familial (especially behavoral traits) need not be genetic, (*b*) traits caused by environmental factors may nevertheless have an underlying genetic basis, and (*c*) genetic traits with well-established modes of inheritance can be mimicked (**phenocopied**) by environmentally caused disorders.

9. **Familial hypercholesterolemia**, a trait due to a simple Mendelian dominant allele, is the most frequent simple Mendelian disorder in humans. In most populations, about 1 individual in 500 is heterozygous for the allele. These individuals, in addition to other symptoms, have a greatly elevated risk of heart attack.

KEY WORDS

Achondroplasia	Heterozygous	Pleiotropy
Allele	Homozygous	Probability
Chance variation	Kindred	Punnett square
Codominant	Locus	Recessive
Common baldness	Modifier gene	Segregation
Dominant	Multifactorial trait	Sex-influenced trait
Evolution	Multiple alleles	Sex-limited trait
Familial	Pedigree	Simple Mendelian
Familial	Penetrance	inheritance
hypercholesterolemia	Phenocopy	Variable age of onset
Genotype	Phenotype	Variable expressivity

PROBLEMS

3.1. How can a familial trait not be genetic?

3.2. Explain how "traits caused by environmental factors may nevertheless have an underlying genetic basis."

3.3. Why can't common, easily observed traits such as skin color, eye color, and hair color be used to illustrate simple Mendelian inheritance?

3.4. What is a phenocopy?

3.5. What is the difference between a kindred and a pedigree?

3.6. In the Haldane analogy for the mating of Mr. Smith-Brown with Ms. Smith-Brown, what fraction of the offspring will have at least one Smith in their names? What fraction will be Smith-Smith?

3.7. Which of the following are genotypes and which are phenotypes: *AA, Ww,* woolly hair, homozygous, bald, O blood type, phenocopy, dry ear cerumen, heterozygous.

3.8. Which of the following are homozygotes and which heterozygotes? *Ww, Aa, $I^O I^O$, aa, Bb, $I^A I^O$.*

3.9. In the mating *Aa* × *Aa,* what fraction of offspring are expected to be *AA? Aa? aa?*

3.10. If *A* is a dominant allele for which *AA* homozygotes survive and are affected, what is the expected proportion of affected offspring from the mating *Aa* × *Aa?* What is the expected proportion if *AA* homozygotes undergo early spontaneous abortion?

3.11. A certain dominant allele has a penetrance of 50 percent (i.e., $\frac{1}{2}$ of the individuals of genotype *Aa* express the trait). In this situation, what is the probability that an offspring of an *Aa* × *aa* mating will be affected?

3.12. A newborn boy has a bald mother and a nonbald father. What is the boy's genotype? Will he develop baldness? Is it possible to state the age at which baldness will begin and the eventual degree of baldness?

3.13. Could the trait shown in the pedigree be due to a dominant allele with complete penetrance? With incomplete penetrance? Why or why not?

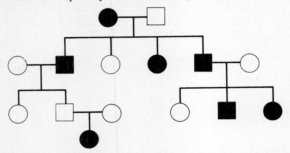

3.14. There is a relatively simple formula for calculating the probability that a sibship of size *n* offspring will have exactly *i Aa* individuals when the mating is *Aa* × *aa*. The formula is

$$\text{Prob (exactly } i \text{ } Aa \text{ offspring)} = \frac{n!}{i!(n-i)!}(\tfrac{1}{2})^n$$

where the symbol *n!* (read *factorial*) means the product of all whole numbers up to and including the number in question. Thus, $n! = 1 \times 2 \times 3 \times \ldots \times n,$ and so on. (0! is defined to equal 1.) Using this formula, calculate the probability that a family of 6 children consists of exactly 4 *Aa* and 2 *aa*.

3.15. A heterozygous taster mates with a nontaster and they have 4 children. Use the formula in problem *14* to calculate the probability that all 4 are tasters. Use the formula to calculate the probability that exactly 2 are tasters.

3.16. To illustrate chance variation, use the formula in problem *14* to calculate the probability that 8 offspring of the mating *Aa* × *aa* will *not* have an exactly 4:4 distribution.

FURTHER READING

Barnes, P., and T. R. Mertens. 1976. A survey and evaluation of human genetic traits used in classroom laboratory studies. J. Heredity 67: 347–52. Many of the "classical" human traits used to illustrate simple Mendelian inheritance actually have complex inheritance.

Bergsma, D. 1979. Birth Defects Compendium. 2d, ed. Alan R. Liss, New York. This lengthy catalogue of birth defects, although primarily designed for physicians and counselors, contains a wealth of useful information. A good medical dictionary at your elbow may be necessary.

Crow, J. F. 1979 (February). Genes that violate Mendel's rules. Scientific American 240: 134–46. A readable account of certain genes that do not segregate as expected, and why.

Goldstein, J. L., and M. S. Brown. 1979. The LDL receptor locus and the genetics of familial hypercholesterolemia. Ann. Rev. Genet. 13: 259–89. More details on this important and relatively common inherited disorder.

Hsia, V. E., K. Hirschhorn, R. L. Silverburg, and L. Godmilow (eds.). 1979. Counseling in Genetics. Alan R. Liss, New York. Meant for specialists, but illustrates that counseling involves much more than instruction.

Iltis, H. 1932. Life of Mendel. (Trans. by E. and C. Paul.) W. W. Norton, New York. A fine old biography, and still the best.

Lison, M., S. H. Blondheim, and R. N. Melmed. 1980. A polymorphism of the ability to smell urinary metabolites of asparagus. Brit. Med. J. 281: 1676–78. At one time this trait was thought to be excretion of the substances, not the ability to smell them.

Martin, J. B. 1982. Huntington's disease: genetically programmed cell death in the human central nervous system. Nature 299: 205–06. The title speaks for itself, and the article addresses why the cell death might occur.

Myers, R. H., J. J. Madden, J. L. Teague, and A. Falek. 1982. Factors related to onset age in Huntington's disease. Am. J. Hum. Genet. 34: 481–88. Age of onset varies from 4 to 65.

chapter 4

Mendel's Laws and Recessive Inheritance

- Mechanics of Recessive Inheritance
- Characteristics of Autosomal-Recessive Inheritance
- Familial Emphysema
- Cystic Fibrosis
- Sickle Cell Anemia
- Tay-Sachs Disease
- Albinism
- Independent Assortment, Recombination, and Linkage
- Linkage in Humans
- Somatic Cell Genetics
- Summary

The examples of dominant inheritance of rare human traits discussed in the preceding chapter illustrate segregation directly, because most individuals who carry the dominant allele result from matings between heterozygotes and homozygotes and because the segregation of alleles in the heterozygotes during the formation of gametes leads to a 1:1 distribution of phenotypes among the offspring. In another class of traits, attention should be focused on matings between two unaffected heterozygotes. These traits are caused by recessive alleles, which are not expressed unless a person is homozygous.

Dominance and recessiveness are opposite sides of the same coin. Dominant genes are dominant with respect to some other allele at the homologous locus, and this other allele is therefore recessive. Conversely, recessive genes are recessive with respect to some other allele at the homologous locus, and this other allele is therefore dominant. Ordinarily, the terms **dominant** and **recessive** are used to refer to abnormal alleles, and the dominance or recessiveness is intended to mean with respect to the normal allele at the locus. The use of "abnormal" in this context should not be taken to imply "harmful" or "undesirable," especially in regard to the sorts of traits listed in Table 3.1; the word is meant to imply only "atypical," "unusual," or "not common." Mid-digital hair and chin fissure are examples of traits that are abnormal in the sense of unusual but certainly not in the sense of harmful.

Most of the examples of recessive inheritance to be discussed in this chapter, however, are very serious and harmful indeed. In some cases, affected homozygous individuals can never lead a normal life and can never produce children. Consequently, seriously affected children result almost exclusively from matings between two heterozygous people, people who are themselves normal (because they carry one dominant allele) and usually completely unaware that they carry the deleterious recessive allele. Unlike the traits inherited as dominants, mutation cannot be suggested as the cause of recessively inherited traits. This is because two simultaneous mutations at corresponding loci on homologous chromosomes would be required to produce the genotype of an affected individual, and the odds against the simultaneous occurrence of events that are each so very rare are astronomical. Like the conditions inherited as dominants, however, traits inherited as recessives can be mimicked (phenocopied) almost perfectly by environmentally induced abnormalities or by other hereditary conditions. This situation emphasizes again the need for caution in the study of inherited traits.

Many human traits are known to be caused by recessive genes. At the present time, 588 conditions are known to be inherited as autosomal recessives, and another 710 traits are strongly suspected to be. In the more serious of these conditions, affected individuals almost always result from matings—frequently consanguineous matings—between heterozygotes.

MECHANICS OF RECESSIVE INHERITANCE

The most important matings in the case of rare, harmful recessives are between heterozygotes, symbolically written as $Aa \times Aa$. Segregation therefore occurs in both parents. Half the eggs carry the A allele and half the a allele; likewise for the sperm. The eggs and sperm join at random to give rise to the genotypes of the children. Children of three genotypes can be produced: $AA, Aa,$ and aa; these are expected to occur in the proportions $\frac{1}{4} AA, \frac{1}{2} Aa,$ and $\frac{1}{4} aa$. The reason for these proportions can be seen by considering the surname analogy of the last chapter. The analogous mating is Mr. Smith-Brown with Ms. Smith-Brown. Four surnames of offspring can result: Smith-Smith, Smith-Brown, Brown-Smith, and Brown-Brown. Since the selection of offspring surnames is random, the four possibilities must be equally alike.

This equal likelihood can be expressed numerically in three ways. One is to say that the ratio of the four possibilities (or the "odds," as the bookmakers

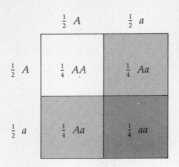

Figure 4.1 Punnett square showing segregation in a mating of two *Aa* heterozygotes. In both parents, half the gametes carry *A* and half carry *a*. The overall distribution of offspring is $\frac{1}{4}$ *AA*, $\frac{1}{2}$ *Aa*, and $\frac{1}{4}$ *aa*.

would put it) is 1:1:1:1. (This is read as "one to one to one to one.") Another way is to give the proportions, which is 25 percent for each of the four possible offspring surnames. Note that the percentages associated with all the possibilities add up to 100, which is the numerical way of saying that these four offspring surnames are the *only* possibilities. Still another way to represent the likelihood of the four outcomes is to express the percentages as probabilities, in this case $\frac{1}{4}$ for each of the four outcomes. Note here (as in the last chapter) that all the probabilities add up to 1.

If you designate the name Smith as *A* and Brown as *a*, then the four possible offspring surnames can be seen to correspond to the four ways that eggs and sperm can unite in the mating *Aa* × *Aa*. The four possibilities are *AA*, *Aa*, *aA*, and *aa*. These possibilities are equally likely, so their ratio is 1:1:1:1 (see the Punnett square in Figure 4.1). However, the *Aa* and *aA* genotypes are identical in that each has one dominant and one recessive allele; they only seem different because the alleles are written in opposite order. The *Aa* and *aA* genotypes can therefore be lumped together. This reduces the number of outcomes to three: *AA*, *Aa*, and *aa*. The ratio of the three is 1:2:1. Expressed as percentages, 25 percent of the children will be genotypically *AA*; 50 percent will be *Aa*; and the remaining 25 percent will be *aa*. Or, what is equivalent, the chance that a particular unborn child will be *AA* is $\frac{1}{4}$; the chance that it will be *Aa* is $\frac{1}{2}$; and the chance that it will be *aa* is $\frac{1}{4}$.

If the *a* allele is recessive, then only the *aa* individuals will be affected; the *AA* and *Aa* individuals will all be normal and indistinguishable. The ratio of normal to affected in the children of matings of *Aa* × *Aa* therefore boils down to 3:1 or, as fractions, $\frac{3}{4}$ normal and $\frac{1}{4}$ affected. This is the 3:1 ratio that many people associate with Mendelism. Note that it is a ratio of *phenotypes*. The underlying ratio of *genotypes* is still 1:2:1.

One of Mendel's own experiments provides a good illustration. Mendel carried out matings of *Aa* × *Aa,* where *A* again designates the dominant gene for round seeds in the garden pea. He observed a total of 7324 seeds from the mating; 5474 of these were round and 1850 were wrinkled—a ratio in very close agreement with the expected 5493:1831 (that is, 3:1). By subsequent breeding tests Mendel was able to show that, *among those seeds that were phenotypically round,* the ratio of *AA* to *Aa* genotypes was 1:2. Stated another way, among

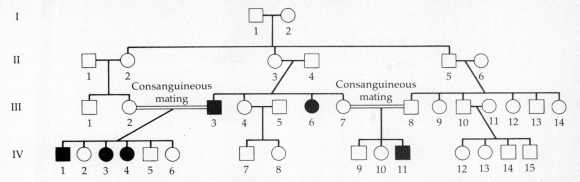

Figure 4.2   Pedigree of one form of autosomal-recessive albinism showing consanguineous matings between III-2 and III-3 and between III-7 and III-8.

those seeds that received at least one *A* allele and were therefore phenotypically round, $\frac{1}{3}$ were genotypically *AA* and $\frac{2}{3}$ were genotypically *Aa*. These experiments verified the underlying 1:2:1 ratio of *AA:Aa:aa*.

CHARACTERISTICS OF AUTOSOMAL-RECESSIVE INHERITANCE

Figure 4.2 is a pedigree of one form of autosomal-recessive albinism (lack of pigmentation) that illustrates certain characteristic features of this mode of inheritance. The pedigree symbols are as described in Chapter 3, but here each person can be uniquely designated by specifying his or her generation number (I to IV in Figure 4.2) and the position (number from the left) in the generation. For example, the male in generation I is referred to as individual I-1, and his mate is designated I-2; similarly, the affected male in generation III is individual III-3, and his affected sister is individual III-6. Some characteristic features in pedigrees of traits due to autosomal recessives are as follows:

1. There will often be a **negative family history** for the trait; that is, the trait may not occur among the ancestors of affected individuals. In Figure 4.2, for example, there is a negative family history for albinism in generations I and II.
2. Males and females are equally likely to be affected with the trait.
3. Affected individuals *may* have unaffected parents. In Figure 4.2, for example, the parents of III-3 and III-6 are both unaffected, as are the parents of IV-11. If the trait in question is rare, then affected individuals will *usually* have unaffected parents (i.e., the parents will be heterozygous for the recessive allele and therefore phenotypically normal).
4. Affected individuals will often arise from matings between relatives. (Matings between relatives are called **consanguineous matings**.) In Figure 4.2, for example, individuals III-2 and III-3 are first cousins, as are III-7 and III-8. As shown in the pedigree, **consanguinity** (i.e., genetic relationship) between mating individuals is represented by a double horizontal line connecting those involved.

Consanguinity among the parents of affected individuals is a typical feature in pedigrees involving autosomal-recessive traits, especially rare traits. An individual who is heterozygous for a rare recessive allele can transmit the allele to many of his or her descendants. If two of the descendants mate, the mating may be of the type $Aa \times Aa$, and 25 percent of the offspring will be expected to be homozygous aa and therefore affected. On the other hand, if the heterozygous descendant mates with an unrelated individual, the unrelated individual will usually be genotypically AA (because the a allele is rare); the nonconsanguineous mating will usually be of type $Aa \times AA$, therefore, and, although half the offspring will be heterozygous, none will be affected.

In the following sections, some of the most widely known and important harmful autosomal-recessive conditions will be discussed.

FAMILIAL EMPHYSEMA

Emphysema refers to an overinflation and distension of the air sacs in the lungs marked by continuous shortness of breath (even when at rest), a wheezy cough, and increased blood pressure in the arteries of the lungs, leading to an enlargement of the right side of the heart. Without proper treatment, the disease progresses to obstructive lung disease and eventual death due to respiratory failure or congestive heart failure. Onset of the disease usually occurs in middle age, and the life span of affected indivduals is shortened by 10 to 30 years, depending on the success of treatment.

Emphysema can be caused by environmental factors, such as heavy smoking, but one form of the disease, called **familial emphysema**, is inherited as an autosomal recessive. In most populations, the incidence of familial emphysema is about 1 in 1700 individuals. As with rare autosomal-recessive traits in general, heterozygotes are much more common than homozygotes (because it takes two copies of the rare allele to produce a homozygote but only one to produce a heterozygote). With familial emphysema, the frequency of heterozygotes is about 1 in 20 individuals. The high frequency of heterozygotes is particularly important in this case because there is evidence that heterozygotes may also have an increased risk of lung disease. In one study of 103 patients with obstructive lung disease, for example, there were 5 homozygotes and 25 heterozygotes.

Familial emphysema is also known as α_1-**antitrypsin deficiency**. α_1-antitrypsin is a protein found in blood serum. Its function is to inhibit an enzyme (**trypsin**) that breaks down other proteins. The locus for α_1-antitrypsin, called the Pi locus, is apparently on chromosome 2, and at least 23 different Pi alleles have been identified. One of these alleles, designated Pi^Z, leads to an inactive form of α_1-antitrypsin. This allele is the one responsible for familial emphysema because $Pi^Z Pi^Z$ homozygotes have an extremely high risk of developing the disease. As noted, however, Pi^Z heterozygotes may also have an increased risk.

CYSTIC FIBROSIS

Among Caucasians, one of the most frequent simple Mendelian recessive diseases of childhood is **cystic fibrosis**, which is due to a recessive allele on the long arm of chromosome 5. The incidence of the condition itself (homozygous recessives) is about 1 in 2500 individuals, but about 1 in 25 individuals is heterozygous for the allele. Although relatively frequent in Caucasians, cystic fibrosis is extremely rare in other racial groups.

Cystic fibrosis is characterized by the malfunction of the pancreas and other glands, resulting in the production of abnormal secretions. Its chief symptoms are recurrent respiratory infections, malnutrition resulting from incomplete digestion and absorption of fats and proteins, and cirrhosis of the liver. Patients with cystic fibrosis have an accumulation of thick, sticky, honeylike mucus in their respiratory tract, which often leads to respiratory complications. The victim becomes a prime target for secondary infections such as pneumonia or bronchitis. Cystic fibrosis is a disease of childhood, and the symptoms may appear early. The disease usually leads to death in childhood or adolescence, but the lives of affected children can be prolonged somewhat by intensive respiratory and dietary treatment. Left untreated, 95 percent of affected children will die before age 5. With treatment, the average life expectancy of affected girls is more than 12 years, and that of affected boys is more than 16 years. If treatment is begun before any appreciable lung damage has occurred, the child's chance of living to age 21 or older is now greater than 50 percent. The treatment includes a special diet, daily administration of antibiotics, administration of extracts of the pancreas of animals, and special daily lung care (sometimes including the flushing out of the lungs with an aerosol mist). The treatment is continual and expensive.

SICKLE CELL ANEMIA

Among blacks, the most common disorder inherited as a simple Mendelian recessive is **sickle cell anemia**, due to an allele on chromosome 11. The hereditary defect shows up in the red blood cells—the cells that carry oxygen from the lungs to the tissues. More specifically, the defect is in the **hemoglobin**, a major component of the red blood cells. (Hemoglobin is the protein that physically binds with oxygen molecules and transports them.) People homozygous for the sickle cell allele have a form of hemoglobin that tends to crystallize or stack together when exposed to lower than normal levels of oxygen. The crystallization of hemoglobin causes the entire red cell to collapse from its normal ellipsoidal shape into the shape of a half-moon or "sickle." (See Figure 4.3.) In this form the red cell cannot carry the normal amount of oxygen. More important, the sickled cells tend to clog the tiny capillary vessels, interrupting the nutrient blood supply to vital tissues and organs.

Children with the disease tire easily, may be retarded in their physical development, and tend to be susceptible to infections of all kinds—owing to their general weakened condition caused by the severe chronic anemia brought on by the reduced amount of normal hemoglobin. At intervals the affected people experience sickle cell crises, marked by fever and severe, incapacitating pains in the joints, particularly in their extremities, and in their chest, back, and abdomen, caused by the clogging of the blood supply to these vital areas. These painful episodes may last from hours to days to weeks. They may be provoked by

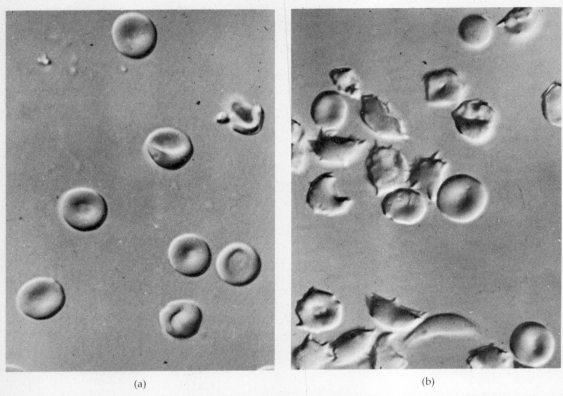

(a)   (b)

Figure 4.3 (a) Micrograph of red blood cells from a patient with sickle cell anemia when the cells are saturated with oxygen; under these conditions, the appearance of the cells is the same as that of normal red blood cells. (b) Micrograph of red blood cells from a patient with sickle cell anemia when the cells are subjected to conditions of low oxygen; note the extensive collapsing or "sickling" of the cells. Normal red blood cells do not collapse with low amounts of oxygen.

anything that lowers the oxygen supply: overexertion, high altitude, respiratory ailments, and so on.

Pregnancy in a woman with sickle cell anemia is especially dangerous, not only for the mother because of the increased stress but also for the baby. The sickling of cells in the placenta can decrease or block the oxygen supply to the baby and cause a spontaneous abortion, miscarriage, or stillbirth. (The expulsion of a fetus from the time of fertilization to three months of pregnancy is an **abortion**; expulsion between three and seven months of pregnancy is a **miscarriage**; expulsion of a nonliving fetus thereafter is a **stillbirth**.)

The lifespan of people affected with sickle cell anemia is often short. Frequently they die in their teens or twenties, almost always before age 45. At present there is no cure for the disease, although the symptoms can be treated, transfusions can be given, and the pain can be decreased or relieved.

A confusing terminology about sickle cell anemia has come into widespread use. The carriers of the gene, the heterozygotes, are often said to have the sickle cell "trait." These people are actually quite healthy. Rare cases are known, however, of heterozygotes suffering symptoms of the disease brought on by extreme and prolonged reduction of oxygen in the blood. The affected people who are homozygous for the sickle cell gene are said to have the sickle cell "disease." Sickle cell anemia is the condition that afflicts *homozygous* people. These distinctions should be kept in mind to avoid confusing the condition of the heterozygotes, who are normally healthy, with that of the homozygotes, who are not.

Sickle cell anemia is extraordinarily common as compared with other serious conditions that have an equally simple inheritance. In most of West Africa, about 1 per 100 children has the disease, and about 1 person in 6 is a carrier. Among American blacks, the frequencies are lower because only part of their ancestry traces back to West Africa; about 1 per 400 children has the disease and approximately 1 person in 10 is heterozygous. This disease causes the young to die in enormous numbers; the yearly toll worldwide has been estimated at 100,000.

Sickle cell anemia is so common because of an infectious disease, **malaria**, and because of the law of segregation. As shown in Figure 4.4, the geographic distribution of sickle cell anemia in Africa almost coincides with the geographic distribution of falciparum malaria, so named because of the protozoan parasite that causes it (*Plasmodium falciparum*). This overlap occurs because the carriers of the sickle cell allele have an enhanced resistance to the disease. The parasites

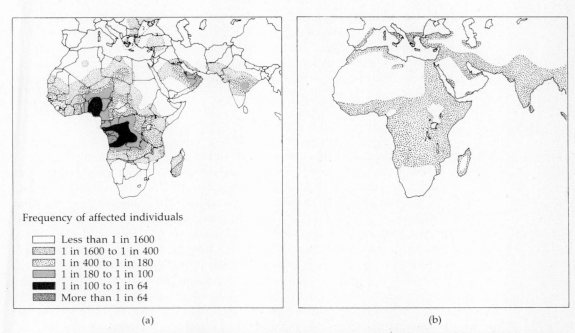

Frequency of affected individuals

- Less than 1 in 1600
- 1 in 1600 to 1 in 400
- 1 in 400 to 1 in 180
- 1 in 180 to 1 in 100
- 1 in 100 to 1 in 64
- More than 1 in 64

(a)   (b)

Figure 4.4   (a) Map of distribution and frequency of individuals affected with sickle cell hemoglobin disease in Africa, India, the Middle East, and southern Europe. (b) Map of distribution of malaria caused by *Plasmodium falciparum* in about the 1920s (before extensive control programs were instituted). Note the extensive overlap of the shaded areas on the two maps.

causing malaria are mosquito-borne; they are transferred from bloodstream to bloodstream by mosquitoes and get transported throughout the body by inserting themselves into red blood cells. The red blood cells with abnormal hemoglobin tend to sickle when infested with parasites; they clump together and are thereby effectively removed from circulation and destroyed. It may also be that the parasite cannot infect these red cells as readily as the normal ones.

Still other factors may be at work, but, in any case, the heterozygous carriers of the sickle cell gene tend to be more resistant to malaria than normal homozygotes. In one study, the risk of severe malarial infections in children who were carriers was found to be only half as great as the risk to homozygous normal children. Partly because of this, carriers of the sickle cell allele have an enhanced ability to survive and reproduce in an environment in which malaria is widespread. Inevitably, since the carriers have an advantage over even the normal homozygotes, many matings occur between two carriers, and the law of segregation foreordains that $\frac{1}{4}$ of the children will have the disease, $\frac{1}{4}$ will be homozygous normal (and therefore more susceptible to malaria than their parents), but $\frac{1}{2}$ of the children will themselves be carriers and will perpetuate the sickle cell allele because of their advantage.

Almost as if to emphasize the connection between malaria and abnormal hemoglobin, another inherited blood disease is found in other parts of the world where malaria is common, particularly in the Mediterranean basin. In this case the abnormality is in the amount of hemoglobin. The Mediterranean condition is also inherited as a simple recessive, and homozygous recessive individuals suffer a severe and debilitating anemia that is frequently fatal in early childhood. The disease itself is called **thalassemia major**. Carriers of the gene have at worst a mild anemia known as thalassemia minor. Thalassemia major is found at frequencies of up to 1 percent in certain Italian populations and among other peoples in the Mediterranean region—for example, among Greeks, Sardinians, Armenians, and Syrians.

TAY-SACHS DISEASE

Like cystic fibrosis among Caucasians and sickle cell anemia among Negroes, Tay-Sachs disease is a simple Mendelian recessive that has an increased incidence in a particular population—in this case, Jews. (Incidentally, many diseases are named after the person or persons who first recognized the symptoms as recurring clinical entities. In this case, Tay was a British ophthalmologist and Sachs an American neurologist; both described the disease in the 1880s, although they were not aware of its genetic basis.) **Tay-Sachs disease**, formerly called "familial infantile amaurotic idiocy," is due to a recessive allele on chromosome 15. The normal form of the gene is responsible for producing an enzyme, called *hexosaminidase A*, which functions in the breakdown of a sugary-fatty substance (*ganglioside GM$_2$*) in the central nervous system. The role of the enzyme in breaking down ganglioside GM$_2$ is outlined in Figure 4.5. Homozygotes for the Tay-Sachs allele lack a functional form of hexosaminidase A. The absence of this enzyme causes an abnormal accumulation of ganglioside GM$_2$ in the central

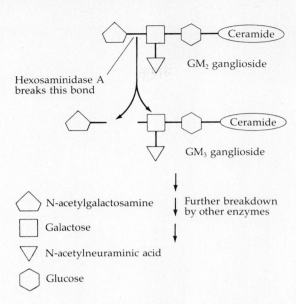

Figure 4.5 Hexosaminidase A is involved in the breakdown of ganglioside GM_2. A defective enzyme leads to accumulation of GM_2 in the central nervous system and causes Tay-Sachs disease.

Hexosaminidase A breaks this bond

GM_2 ganglioside

Ceramide

GM_3 ganglioside

Ceramide

⬠ N-acetylgalactosamine

▢ Galactose

▽ N-acetylneuraminic acid

⬡ Glucose

Further breakdown by other enzymes

nervous system, which leads to blindness, seizures, and a complete degeneration of mental and motor function. The disease symptoms are usually evident by six months of age; the disease becomes progressively more severe and is invariably fatal within the first five years of life.

Among Jews from Central Europe (Ashkenazi), the incidence of the condition is about 1 per 4000 births. This incidence is about 100 times higher than that among non-Jews or among Jews from the Mediterranean basin (Sephardic); the incidence of Tay-Sachs disease among non-Jews is about 1 per 400,000. Approximately 1 in 30 Ashkenazi Jews is a carrier of the recessive allele.

ALBINISM

One final example of variation among populations in the incidence of inherited disorders is albinism among the Hopi Indians of Arizona and the Jemes and Zuni Indians of New Mexico. **Albinism** results from an inherited defect in the ability of certain specialized cells to produce normal amounts of the brown-black pigment melanin. Albinos therefore lack pigmentation (Figure 4.6). Their hair, skin, and the iris of the eyes are very light or white, although their eyes look pink due to reflected light passing through the blood vessels. The condition is inherited as a simple Mendelian recessive. It is not as serious a disorder as many others, but albinos tend to have vision problems, sometimes including blindness, and are extremely susceptible to sunburn and prone to skin cancer because they have no melanin to protect them from the sun's ultraviolet rays. The incidence of the condition among the Indians mentioned is about 1 in 200, and about 1 person in 8 is a carrier of the albino allele. Among people of European ancestry, by contrast, the incidence is about 1 in 40,000.

The social position of albinos in Hopi society is interesting because tradi-

Figure 4.6 Photograph of three Hopi girls, taken about 1900. The girl in the middle is an albino.

tional Hopis do not recognize the condition as inherited; they believed that albinism and other abnormalities at birth are the result of some previous action or incident in the life of a parent or relative. (This explanation of birth defects is encountered not only in tribal societies; I was once solemnly informed by a friend that the cerebral palsy of a mutual acquaintance was caused by the victim's father having been drunk at the time of conception.)

The traditional explanation of albinism in Hopi children varies from case to case. One man was said to have loved a white donkey so dearly that two of his granddaughters were born white. The Indians have come to accept the high incidence of albino children as a fact of life and even to admire them. It is thought to be good luck to have one or more albinos in a village; for this reason, some Hopi women desire to have an albino baby. Albinos are thought to be clean, smart, and very pretty, although many of them never marry.

As discussed in Chapter 3, there is a second form of albinism due to a recessive allele at a different autosomal locus. This form also has an incidence of about 1 in 40,000, but it seems to be no more frequent in Southwest American

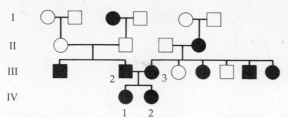

Figure 4.7 Partial pedigree of albinism in a Hopi kindred. Note that the mating between homozygotes (III-2 and III-3) produces exclusively homozygous offspring.

Indians than in other populations. Albino Hopis almost always have the first form of albinism; they are all homozygous for a recessive allele at the same locus. That the same locus is involved is indicated by the fact that when two albinos mate, they produce only albino children. The pedigree in Figure 4.7 is a Hopi pedigree illustrating that all of the children of albino parents are albinos.

Since there are two autosomal-recessive forms of albinism, it is possible that two unrelated albino parents will have different forms of the condition. In this case, in contrast to Figure 4.7, all the offspring will be normal because each parent will contribute a normal allele that is dominant to the albino allele contributed by the other parent; that is to say, the offspring will be heterozygous for a recessive allele at each of two distinct loci and therefore will be phenotypically normal. Figure 4.8 is a hypothetical pedigree of this situation, and it is in clear contrast to Figure 4.7. Matings between two homozygous-recessive parents are often carried out in experimental organisms to determine whether the parents are homozygous for recessive alleles at the same locus.

At this point, it might be useful to review the types of mating that can occur when there are two alleles at an autosomal locus and the genotypes of offspring that result:

$$AA \times AA \rightarrow \text{all } AA$$
$$AA \times Aa \rightarrow \tfrac{1}{2} AA \text{ and } \tfrac{1}{2} Aa$$
$$AA \times aa \rightarrow \text{all } Aa$$
$$Aa \times Aa \rightarrow \tfrac{1}{4} AA, \tfrac{1}{2} Aa, \text{ and } \tfrac{1}{4} aa$$
$$Aa \times aa \rightarrow \tfrac{1}{2} Aa \text{ and } \tfrac{1}{2} aa$$
$$aa \times aa \rightarrow \text{all } aa$$

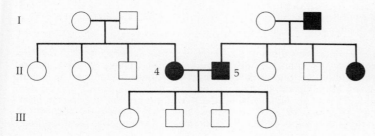

Figure 4.8 Hypothetical pedigree showing the offspring of parents who are affected with phenotypically similar but genetically distinct disorders. Both II-4 and II-5 are homozygous, but homozygous for autosomal-recessive alleles at different loci. All of their offspring are phenotypically normal.

INDEPENDENT ASSORTMENT, RECOMBINATION, AND LINKAGE

Now we must consider what happens when the segregation of *two* loci is followed simultaneously in the same individual. This is the problem addressed by Mendel's second law, the **law of independent assortment**, which asserts that the segregation of the first locus during meiosis has no influence on the segregation of the second one. Like the discovery of dominance, this generalization of Mendel's is of secondary importance compared with the law of segregation. Indeed, *independent assortment* occurs only between certain loci and not others.

A principal factor in determining whether independent assortment occurs between two loci is the relative position of the loci on the chromosomes. Loci on nonhomologous chromosomes must assort independently because of the mechanics of meiosis. Individuals of genotype *AaBb* are segregating for two loci at once. If the *A* locus and the *B* locus are on different chromosomes, then the four chromatids of one bivalent will carry *AAaa* and those of the other bivalent will carry *BBbb* (Figure 4.9). The movement of the bivalents during metaphase I to the imaginary plane cutting across the cell and the orientation of the bivalents on this plane occur with total indifference to the alignment of nonhomologous centromeres. The result is that in anaphase II, an *A*-bearing chromosome is as likely to go to the same pole with a *B*-bearing chromosome as with a *b*-bearing one; likewise, the *a*-bearing chromosome may end up at a pole with either a *B*- or a *b*-bearing chromosome (see Figure 4.9). Therefore, the *AaBb* genotype will produce four kinds of gametes—*AB, Ab, aB, ab*—and these will be equally likely;

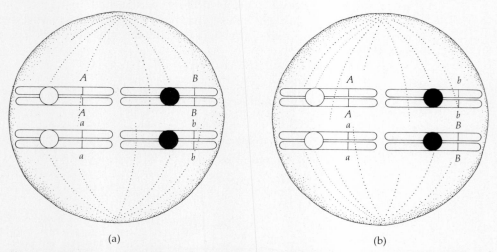

(a)   (b)

Figure 4.9 Independent alignment of nonhomologous chromosomes at metaphase I of meiosis produces independent assortment of loci on nonhomologous chromosomes. Shown here are two possible alignments of nonhomologous chromosomes at metaphase I. Alignment as in part (a) produces two *AB* and two *ab* gametes; alignment as in part (b) produces two *Ab* and two *aB* gametes. Since both alignments are equally likely, the overall proportion of gametes from an *AB/ab* individual will be ¼ *AB*, ¼ *Ab*, ¼ *aB*, and ¼ *ab*. In these diagrams, the possible occurrence of crossing-over between the *A* and *B* loci and their respective centromeres has been ignored.

each of the four types will have a $\frac{1}{4}$ chance of being present in a particular gamete. It is this 1:1:1:1 distribution of the four gametic types that characterizes independent assortment.

The situation is quite different when the A and B loci are on the same chromosome. It is important in the first place to know exactly which alleles in the $AaBb$ genotype are linked on the same chromosome of the pair of homologues. There are two possibilities: The A and B alleles may have been inherited from one parent, a and b from the other, in which case the genotype should be written as AB/ab; the slash is intended to separate and identify which alleles are on which chromosome. Alternatively, the A and b alleles may have come from one parent, a and B from the other, and in this case the genotype should be denoted as Ab/aB. Both genotypes will produce four kinds of gametes; the AB/ab parent will produce AB, ab, Ab, and aB; the Ab/aB parent will produce Ab, aB, AB, and ab. These gametic types are genetically the same. But in each case the first two gametes listed are genetically like the parental chromosomes; these gametes are known as **parental** or **nonrecombinant** gametes. The last two gametes mentioned in each case are known as **recombinant** gametes.

The specific ratio of parental to recombinant gametic types depends on the distance between the loci on the chromosome. A bivalent in which one crossover occurs between the A and B loci will, after separation at the two anaphases, be resolved into four gametes—one of each of the parental and recombinant types [Figure 4.10(a)]. A bivalent in which no crossover occurs in the region will be resolved into four gametes—two of each of the parental types [Figure 4.10(b)]. If the A and B loci are very close together on the chromosome, few bivalents will actually experience a crossover in the region between them, so relatively few recombinant gametes will be formed. Thus, the proportion of recombinant gametes produced between two loci will increase as the physical distance between the loci on the chromosome increases.

The distance between loci on a chromosome can be measured and expressed in terms of the amount of recombination between the loci. The proportion of recombinant gametes produced by an individual is known as the **recombination fraction** between the loci. (The recombination fraction between two loci is calculated as the number of recombinant gametes divided by the total number of gametes.) The genetic distance between loci is expressed in terms of **map units** (so called because gene arrangements on chromosomes are typically illustrated by diagrams called **genetic maps**). *For loci that are not too far apart on the chromosome, 1 map unit corresponds to 1 percent recombination between the loci.* For example, if two loci are separated by a distance of 3 map units, there will be a 3 percent recombination between them.

Although the frequency of recombination increases as the physical distance between the loci increases, the upper limit is 50 percent, the same as observed for loci on nonhomologous chromosomes. The upper limit is reached when the loci are so far apart on the chromosome that at least one crossover almost always occurs between them. This is quite easy to understand if exactly one crossover always occurs, because the crossover involves only two of the four strands of a

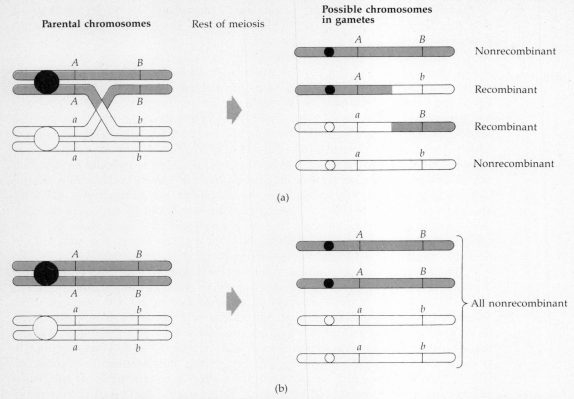

Figure 4.10 A measure of the distance between loci on the same chromosome can be obtained from the frequency of recombination. (*a*) When a crossover occurs between the loci, two gametes will be nonrecombinant and two will be recombinant. (*b*) When a crossover does not occur between the loci, all the gametes will be nonrecombinant. Thus, the frequency of recombinant gametes is proportional to the probability of a crossover between the loci, and this, in turn, is proportional to the distance between the loci on the chromosome.

bivalent, and therefore two of the strands will be recombinant but the other two will be parental [see Figure 4.10(*a*)]. This corresponds to a recombination frequency of two out of four, or 50 percent. The same is true when two, three, or more crossovers occur, provided that the nonsister chromatids involved in any crossover are random.

Loci that are on the same chromosome are said to be **syntenic**, and syntenic loci with less than 50 percent recombination are said to be **linked**. (However, the terminology is sometimes misused because syntenic loci that show 50 percent recombination are sometimes also referred to as *linked*.) One further point about linkage: The **map distance** between two loci (i.e., the number of map units separating them) can be greater than 50, which seems to imply that more than 50 percent recombination can occur. However, the map distance between two unlinked syntenic loci is obtained by adding together the map distances between intervening loci that are linked. Summation of many such numbers can produce map distances greater than 50, even though the actual recombination fraction between the loci would be 50 percent. This is why map distances correspond to recombination fractions only for loci that are sufficiently close together.

LINKAGE IN HUMANS

The mapping of human chromosomes by means of recombination is generally a difficult matter. Nevertheless, in recent years rapid progress has been made in filling in the chromosome map of human genes (especially by means of such methods as somatic cell genetics, which will be discussed shortly). The study of linkage between loci using pedigrees is difficult because families suitable for study tend to be rare, which makes the total number of offspring available small. What is required for the study of human linkage using pedigrees is a group of parents who are double heterozygotes (like *AaBb*). If one of the alleles is rare, then finding such people will be difficult; if two of the alleles are rare, then it will be even more so.

The individual's family history must also be known well enough to be able to infer whether the genotype is actually *AB/ab* or *Ab/aB*; otherwise, the recombinant chromosomes might be confused with the nonrecombinant ones. In addition, the person's mate must have a genotype that will allow the investigator to determine in each of the children which alleles were inherited on the chromosome from the doubly heterozygous parent. Finally, a large number of such families must be pooled together and the recombination frequency determined.

Although family studies of segregation and recombination can reveal linkage, they cannot by themselves identify which of the 22 pairs of autosomes a locus is on. (The sex chromosomes are a special case.) The identification of the particular chromosome can often be accomplished by other methods. Sometimes a particular allele will be found to be inherited simultaneously with a particular chromosomal abnormality such as a giant satellite; that is, the allele will tend to be present whenever the abnormal chromosome is. (A perfect association between the allele and the chromosomal abnormality cannot be expected because of crossing-over, of course.) When such a parallelism in inheritance is found, then the locus of the allele can be inferred to be physically located on the abnormal chromosome.

In these studies the relatively common and innocuous chromosome "abnormalities" such as giant satellites and unusually long short arms are extremely useful. In some families, for example, an "uncoiler" region is found on chromosome 1. Chromosomes with this region look unusually long, and certain segments stain less intensely than is normal. Because many family studies have been carried out to detect particular alleles that are inherited simultaneously with this chromosome, chromosome 1 is the best mapped human autosome. From the combined results of several methods of studying linkage, more than 30 loci have now been assigned to chromosome 1. Figure 4.11 is a genetic map of 11 loci on chromosome 1. The numbers between loci refer to the map distances (recombination fraction) between them. Note that the map distance between *Amy* and *FUCA* (or any other pair of loci) would be calculated as the sum of the intervening

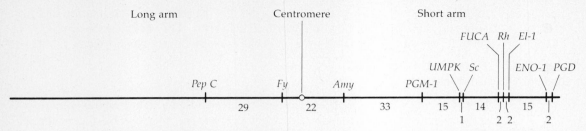

Figure 4.11 Partial genetic map of chromosome 1. The numbers between the indicated positions of the loci refer to the distance between the loci in map units. For map distances of about 15 or less, the percentage of recombination equals the map distance. For longer distances, the percentage of recombination is smaller than the map distance. Most of the map distances here have been inferred from family studies. (Abbreviations: *PepC*, peptidase C enzyme; *Fy*, Duffy blood group; *Amy*, amylase enzyme; *PGM-1*, phosphoglucomutase-1 enzyme; *UMPK*, uridine monophosphate kinase enzyme; *Sc*, Scianna blood group; *FUCA*, α-L-fucosidase enzyme; *Rh*, rhesus blood group; *El-1*, elliptocytosis-1 disease; *ENO-1*, enolase-1 enzyme; and *PGD*, 6-phosphogluconate dehydrogenase enzyme.)

distances—in this case, $33 + 15 + 1 + 14 = 63$, which is larger than 50, even though the recombination fraction between *Amy* and *FUCA* would be 50 percent.

SOMATIC CELL GENETICS

An effective method of detecting linkage in humans involves tissue culture and **cell fusion**—the physical fusing of two somatic cells. The fused cells are called **hybrid cells**, and, because the cells divide by mitosis, their genetic study is called **somatic cell genetics**. The methods of somatic cell genetics can be used only to study genes whose presence and action can be detected in cells grown in laboratory cultures. This includes many genes that direct the production of specific enzymes (protein molecules that accelerate particular chemical reactions), but many traits cannot be detected in tissue cultures. For example, the gene that causes common baldness cannot be detected in tissue culture (at least not at this time). The procedure for identifying the chromosome on which a gene is located involves the fusion of human cells with cells of another species, often mouse cells. Formation of human-mouse hybrid cells can be stimulated with special techniques, and the hybrid cells can be selected from mixtures of cells by growing the cells in environments that discriminate in favor of the hybrids.

The principles involved in selecting out hybrid cells are illustrated in Figure 4.12. Here the nucleus of the human cell is indicated by horizontal lines and that of the mouse cell by vertical lines. The symbols *HGPRT* and *TK* represent loci coding for enzymes that are involved in selecting out the hybrid cells, the superscript $+$ denoting the normal form of the gene and the superscript $-$ indicating a mutation in the gene. In a mixture of cells, some of the cells will fuse. (A number of methods can be used to increase the rate of fusion.) Not only do the cells fuse, but their nuclei fuse as well. Thus, in terms of genetic information, the human-human fusions will result in $HGPRT^-$; TK^+ cells (because both progenitors are $HGPRT^-$; T^+). Similarly, the mouse-mouse fusions will result in $HGPRT^+$; TK^- cells. But the human-mouse hybrids will be $HGPRT^+$; TK^+

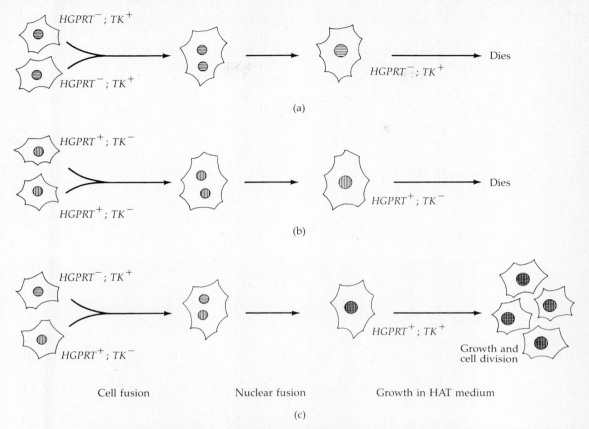

Figure 4.12 HAT selection of hybrid cells. One cell line is *HGPRT⁻*; TK⁺ and the other is *HGPRT⁺*: *TK⁻*. In HAT medium, the only cells that can survive are the ones that have undergone fusion. *HGPRT* codes for the enzyme hypoxanthine guanine phosphoribosyl transferase, and *TK* codes for the enzyme thymidine kinase. HAT medium contains hypoxanthine (H), aminopterin (A), and thymine (T).

because the human cell will contribute *TK⁺* and the mouse cell will contribute *HGPRT⁺*.

As illustrated in Figure 4.12, when such mixtures are grown in a special medium called HAT medium, only the human-mouse hybrids (nucleus indicated by crossed lines) will survive. The reason is that HAT medium contains an ingredient (symbolized by the H in HAT), which will allow *HGPRT⁺* but not *HGPRT⁻* cells to grow, and this ingredient will eliminate the fused human-human *HGPRT⁻* hybrids as well as any unfused human cells. HAT medium also contains another ingredient (symbolized by the T in HAT) which will allow *TK⁺* but not *TK⁻* cells to grow, thus eliminating the fused mouse-mouse *TK⁻* hybrids as well as any unfused mouse cells. The upshot is that only the human-mouse hybrids will survive in HAT medium. This scheme may remind you of the happy accommodation between Jack Sprat, who could eat no fat, and his wife, who could eat no lean; and, indeed, the principles involved are quite the same: Only the

$HGPRT^+$ and TK^+ genes acting in concert in the same nucleus can lick the HAT platter clean.

The cells obtained by the HAT procedure are true human-mouse hybrids. They have 86 chromosomes—the full somatic complement of the mouse (40) plus the full somatic complement of humans (46). Figure 4.13 shows a quinacrine-stained metaphase spread of one such cell, and some of its human chromosomes are indicated. Although the hybrid cells perpetuate their chromosomes as they undergo successive mitotic divisions, sometimes chromosomes are lost—especially human chromosomes. After a while, a culture of hybrid cells will become genetically heterogeneous because different cells will have lost varying numbers of human chromosomes. When individual cells from such a culture are isolated and placed in nutrient medium, each cell will give rise to a **clone**—a group of genetically identical cells. By screening the karyotypes of the clones, using such procedures as Giemsa staining (see Chapter 2), an investigator can identify which human chromosomes have been retained in the various clones. Then, upon

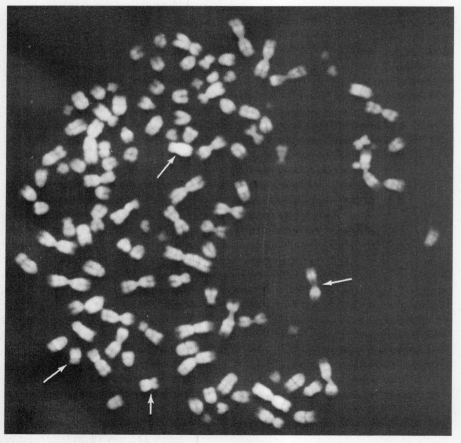

Figure 4.13 Arrows indicate several human chromosomes in a human-mouse hybrid cell stained with quinacrine.

Table 4.1 PRESENCE OR ABSENCE OF HUMAN CHROMOSOMES AMONG EIGHT HUMAN-MOUSE HYBRID CELL LINES, AND PRESENCE OR ABSENCE OF HUMAN UMPK ENZYME*

Clone†	Chromosome																							Enzyme UMPK	Expected pattern
	1	2	3	4	5	6	7	8	9	10	11	12	13	14	15	16	17	18	19	20	21	22	X		
a	A	P	P	P	A	A	A	A	A	A	P	P	A	P	A	P	A	A	A	A	A	A	P	−	A
b	P	P	A	A	P	A	A	P	A	A	A	P	A	A	P	P	P	A	A	A	A	P	P	+	P
c	P	P	A	P	A	A	P	P	A	P	A	P	P	A	P	P	P	P	P	P	P	P	P	+	P
d	P	A	P	A	A	P	P	A	A	P	P	P	A	P	P	P	P	A	A	P	A	P	P	+	P
e	A	P	A	P	P	A	P	P	A	P	P	P	P	P	A	A	P	P	A	P	A	A	P	−	A
f	A	A	A	A	A	P	P	A	P	A	P	P	P	P	P	P	P	P	P	P	P	P	P	−	A
g	A	P	P	A	A	A	P	A	A	P	A	P	A	P	A	A	A	A	A	A	A	A	P	−	A
h	P	P	P	P	P	A	P	P	A	P	A	P	A	P	P	P	A	P	A	A	P	P	P	+	P

Source: Data from A. Satlin, R. Kucherlapati, and F. H. Ruddle, 1975, Cytogenet. Cell Genet. 15: 146–52.

* A = absence of chromosome; P = presence of chromosome; − = absence of human enzyme; + = presence of human enzyme.

† Each of these clones has a unique combination of human chromosomes.

assaying each clone for the product of a particular human gene, it can easily be determined which human chromosome carries the locus corresponding to the gene. For example, if the human gene product is present in every clone that has retained chromosome 3 but absent in every clone that has lost chromosome 3, then the locus of the gene can confidently be assigned to chromosome 3. Examples of the successful use of this technique include the assignment of the locus of hexosaminidase A (the locus involved in Tay-Sachs disease) to chromosome 15. Indeed, more than 200 genes have now been assigned to the human autosomes, and at least two genes are known for every autosome.

An actual example of the use of hybrid cell clones to determine the chromosome that carries the gene for UMPK (**uridine monophosphate kinase**) is shown in Table 4.1. The original hybridization involved an *HGPRT*⁺; *TK*⁻ human cell and an *HGPRT*⁻; *TK*⁺ mouse cell. Individual clones were cultured in HAT medium, so the one human chromosome that must be retained is the one that carries the locus of *HGPRT*. (If this chromosome were lost, the resulting cell would be *HGPRT*⁻; *TK*⁺ and would die in the HAT medium.) Note in Table 4.1 that the one human chromosome retained in all clones is the X chromosome; the locus for *HGPRT* is thus on the X chromosome. For localizing UMPK, the strategy is to examine the chromosomal constitutions of the clones to identify which single chromosome has the same pattern of presence and absence (P's and A's) among the clones as the observed pattern of UMPK presence and absence (+'s and −'s). That is to say, one wishes to find a chromosome that has the same pattern of P's and A's as in the last column of Table 4.1. The appropriate pattern is exhibited by chromosome 1 and only by this chromosome; the UMPK locus must therefore be on chromosome 1.

SUMMARY

1. Matings between two heterozygotes—*Aa* × *Aa*—produce a ratio of genotypes among the progeny of $\frac{1}{4}$ *AA*, $\frac{1}{2}$ *Aa*, and $\frac{1}{4}$ *aa*. If the *a* allele is recessive, then *AA* and *Aa* will be phenotypically indistinguishable, and

the ratio of phenotypes among the progeny will be $\frac{3}{4} \cdot \frac{1}{4}$. This 3:1 ratio of phenotypes is characteristic of autosomal recessive inheritance.

2. Pedigrees of rare traits with autosomal recessive inheritance have characteristic features, including: (*a*) there is a negative family history for the trait, (*b*) both sexes are equally likely to be affected, (*c*) affected individuals usually have nonaffected parents, and (*d*) there is a relatively frequent occurrence of **consanguineous mating** (mating between relatives) among the parents of affected individuals. For rare, autosomal-recessive traits, heterozygous genotypes will greatly outnumber homozygous recessive genotypes.

3. Among the most common autosomal recessive disorders is **familial emphysema** (α_1-**antitrypsin deficiency**), for which about 1 individual in 20 is heterozygous. The high frequency of heterozygotes is especially important in this case because heterozygotes seem to have a higher than average risk of developing pulmonary emphysema. Like other traits, the incidence of autosomal recessive disorders can vary among populations. Among Caucasians one of the most common is **cystic fibrosis**. Among blacks one of the most frequent is the hemoglobin disorder **sickle cell anemia**, for which heterozygotes are more resistant to malaria than are both types of homozygotes. Ashkenazi Jews have a relatively high incidence of **Tay-Sachs disease** (hexosaminidase A deficiency), and Southwest American Indians have a high frequency of one form of **albinism**.

4. **Independent assortment** refers to the independent segregation of alleles at two loci. If an *A/a*; *B/b* double heterozygote produces equal frequencies of the four possible gametic types (*AB, Ab, aB,* and *ab*), the loci are said to undergo independent assortment.

5. **Linkage** refers to a situation in which two loci fail to undergo independent assortment; linked loci are necessarily on the same chromosome. An individual of genotype *AB/ab* (i.e., *A* and *B* alleles on one chromosome and *a* and *b* alleles on the homologue) will produce four gametic types— two **nonrecombinant** types (*AB* and *ab*) and two **recombinant** types (*Ab* and *aB*). An individual of genotype *Ab/aB* will produce the same four gametic types, but the *Ab* and *aB* gametes are the nonrecombinants, whereas the *AB* and *ab* gametes are the recombinants. The amount of linkage between two loci is measured by the **recombination fraction**, which equals the proportion of recombinant gametes produced by a double heterozygote. If the recombination fraction is 6 percent, for example, a double heterozygote will produce $\frac{6}{2} = 3$ percent of each type of recombinant gamete and $(100 - 6)/2 = 47$ percent of each type of nonrecombinant gamete. The maximum possible recombination fraction is $\frac{1}{2}$ (i.e., 50 percent), which corresponds to independent assortment.

6. Loci on the same chromosome are called **syntenic**. Syntenic loci that are sufficiently far apart can undergo independent assortment. A **chromosome map** is a diagram showing the arrangement of syntenic loci, and the **map distance** between loci is measured in terms of **map units**. For loci that are sufficiently close together, the number of map units between them equals the recombination fraction in percent (e.g., 6 map units corresponds to 6 percent recombination). The map distance between more distant loci is calculated by summing the map distances between intervening loci.

7. Cell cultures are especially useful in the study of human genetics. **Somatic cell genetics** is the study of the inherited characteristics of somatic cells in laboratory cultures. A principal method uses **cell fusion** between human cells and cells of another species (often mouse) to produce **hybrid cells**. Hybrid cells sometimes lose various human chromosomes. Studies of **clones** (cultures of genetically identical cells) that have lost particular chromosomes can be used to identify the chromosome on which a gene is located. The product of a gene on a particular chromosome will be expressed in clones that retain the chromosome but will be absent in clones that have lost the chromosome.

KEY WORDS

Abortion	Emphysema	Map distance
Albinism	Enzyme	Recombinant chromosome
α_1-antitrypsin deficiency	Genetic map	Recombination fraction
Cell fusion	HGPRT	Sickle cell anemia
Clone	Hybrid cells	Somatic cell genetics
Consanguineous mating	Independent assortment	Syntenic loci
Consanguinity	Linkage	Tay-Sachs disease
Cystic fibrosis	Malaria	

PROBLEMS

4.1. For an autosomal-recessive allele, what is the genotype of a carrier?

4.2. The reasons for genetic differences among populations are seldom known for certain, but there are exceptions. Why is the sickle-cell hemoglobin allele so frequent among populations in West Africa and their descendants?

4.3. If a locus has 3 alleles, how many genotypes are possible? How many are possible with 4 alleles? With 5 alleles?

4.4. What feature of this pedigree suggests recessive inheritance?

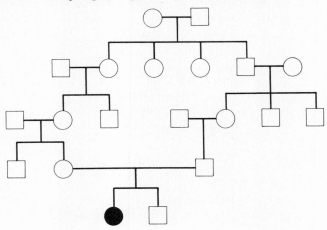

4.5. Relative to an autosomal-recessive allele, what type of mating gives rise to $\frac{3}{4}$ normal: $\frac{1}{4}$ affected? What are the genotypes among the offspring?

4.6. Relative to an autosomal-recessive allele, what type of mating gives $\frac{1}{2}$ normal: $\frac{1}{2}$ affected offspring?

4.7. For rare autosomal recessives, why are heterozygotes more frequent than recessive homozygotes?

4.8. For the clones in Table 4.1, what would be the pattern of expression (+) or nonexpression (−) of an enzyme locus on chromosome 9? Of a locus on the X?

4.9. For three enzymes, A, B, and C, the clones in Table 4.1 show the following pattern on presence or absence:

A: − − − + − + − −

B: − + + − + + − −

C: − − + + − + − +

Identify the chromosome carrying each enzyme locus.

4.10. If two linked loci are 4 map units apart, what is the percent of recombination between them?

4.11. What is the recombination fraction between unlinked syntenic loci?

4.12. For three loci, A, B, and C, suppose that A and B are 5 map units apart and B and C are 15 units apart. Which map orders of the genes are consistent with these data? Suppose A and C are 20 map units apart. Does this tell you the order?

4.13. An individual heterozygous for two loci (A and B) produces the following proportions of gametes:

(a) A B   4 percent

(b) A b 46 percent

(c) a B 46 percent

(d) a b   4 percent

Which are nonrecombinant and which recombinant? What was the genotype of the individual?

4.14. For the mating Aa × Aa, where a is a recessive allele, the probability that a sibship of size n consists of exactly i affected and n-i nonaffected is given by

$$\frac{n!}{i!\,(n-i)!}\left(\tfrac{1}{4}\right)^{i}\left(\tfrac{3}{4}\right)^{n-i}$$

where the exclamation point (factorial) means the product of all whole numbers up to and including the number in question (0! = 1 by definition). In a family of 4 children, what is the probability that exactly 1 will be affected?

4.15. Using the formula in Problem 14, what is the probability that, among 4 offspring of an Aa × Aa mating, none will be aa?

FURTHER READING

Anderson, W. F., and E. G. Diacumakos. 1981 (July). Genetic engineering in mammalian cells. Scientific American 245:106–121. Prospects of somatic cell genetics and recombinant DNA for the treatment of disease.

Dickerson, R. E., and I. Geis. 1983. Hemoglobin: Structure, Function, Evolution and Pathology. Benjamin/Cummings, Menlo Park, California. An introduction to a remarkable molecule. Beautifully illustrated.

Goodman, R. 1979. Genetic Disorders Among Jewish People. Johns Hopkins University Press, Baltimore. Genetic studies of Jewish populations have been extensive.

Gordon, R., and A. G. Jacobson. 1978 (June). The shaping of tissues in embryos. Scientific American 238:106–113. Computer-assisted study of the forces that sculpture the developing embryo.

Harper, M. E., A. Ullrich, and F. G. Saunders. 1981. Localization of the human insulin gene to the distal end of the short arm of chromosome 11. Proc. Natl. Acad. Sci. U.S.A. 78:4458–60. Technical, but illustrates application of sophisticated mapping procedures to an important gene.

Mourant, A. E., A. C. Kopec, and K. Domaniewska-Sobczak. 1978. The Genetics of the Jews. Clarendon Press, Oxford. Collection of genetic data relevant to Jewish populations.

Ruddle, F. H. 1981. A new era in mammalian gene mapping: somatic cell genetics and recombinant DNA methodologies. Nature 294:115–20. Excellent discussion of new avenues opened up by the combination of two powerful techniques.

Shay, J. (ed.). 1982. Techniques in Somatic Cell Genetics. Plenum, New York. Technical, but a good reference.

Spyropoulos, B., P. B. Moens, J. Davidson, and J. A. Lowden. 1981. Heterozygote advantage in Tay-Sachs carriers? Am. J. Hum. Genet. 33:375–80. Why is the Tay-Sachs allele so common in certain Jewish populations?

Woolf, C. M., and F. C. Dukepo. 1969. Hopi Indians, inbreeding and albinism. Science 164:30–37. On the role of albinos in Hopi society.

chapter 5

The Genetic Basis of Sex

Although Mendel single-handedly worked out the statistical rules of inheritance of genes in peas, and although these rules are as valid for genes on the autosomes in humans as they are for genes in peas, the genes on the sex chromosomes in humans—the X and the Y—follow different rules. Mendel can hardly be faulted for not discovering these rules, because peas and most other plants do not have sex chromosomes. Flowering plants such as peas are sexual, but both the male and female structures are present in all the flowers on the plant. Nevertheless, it is ironic that Mendel did not interpret sex as an inherited trait; if he had, he might have realized that sex itself provides one of the most convincing demonstrations of segregation.

THE SEX RATIO

Human females have two X chromosomes and 22 pairs of autosomes; during meiosis these become parceled so that the egg contains one X and one member of each of the autosome pairs. Males have an X and a Y chromosome plus 22 pairs of autosomes; during meiosis the X and the Y segregate and the autosomes are distributed so that half the sperm carry an X and one member of each of the autosome pairs, whereas the other half carry a Y and the other member of each of the autosome pairs. As illustrated in Figure 5.1, if an X-bearing sperm fertilizes an egg, the resulting zygote will have two X's and 22 pairs of autosomes and will develop into a female. If a Y-bearing sperm fertilizes an egg, the zygote will have an X, a Y, and 22 pairs of autosomes and will develop into a male. The segregation of X and Y during meiosis means that a mating between a normal male and a normal female (XY × XX) is somewhat analogous to a mating between a heterozygote and a homozygote (e.g., $Aa \times aa$).

In discussing the **sex ratio** (the ratio of males to females), one must distinguish the **primary sex ratio** (the sex ratio at fertilization) from the **secondary sex ratio** (the sex ratio at birth), because spontaneous abortion is relatively frequent and may well affect one sex more than the other and thus alter the sex ratio among survivors. Little is known about the human primary sex ratio because the earliest stages of fertilization and development are inaccessible to large-scale study. Very early spontaneous abortuses are amenable to study, but such studies present problems. First, the very earliest spontaneous abortions (including failure of the zygote to implant in the uterine wall) usually are unrecognized as such. Second, the number of such abortuses available for study is very limited. Third, spontaneous abortuses, by their very nature, are not a random sample of the corresponding developmental stage in normal embryos. And fourth, very early embryos are exceedingly difficult to sex accurately.

Theoretically, of course, the primary sex ratio should be 1:1 because of segregation of the X and Y chromosomes in the father. On the other hand, segregation in the father is not the only determinant of the primary sex ratio. It could be, for example, that the Y-bearing sperm (or the X-bearing sperm) has a

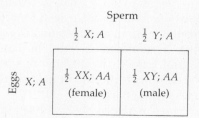

Figure 5.1 Punnett square showing segregation of X and Y chromosomes in the male. The symbol *A* represents a haploid autosomal complement (i.e., one each of chromosomes 1 through 22).

slightly greater capacity to reach the site of fertilization or to participate in fertilization. It could even be the case that the X-bearing and Y-bearing sperm have different efficiencies depending on the precise physiological conditions in the mother or the father or both. In short, the primary sex ratio in humans is unknown. Nevertheless, what little evidence is available suggests that it is close to (but perhaps not exactly) 1:1.

In contrast to the limited data available about the primary sex ratio, the secondary sex ratio in humans is well documented in birth records. (Birth records with the sex of the newborn recorded were first kept in France in the mid-1700s.) When the sex of each offspring is recorded, it is possible to compare the observed sex distribution among sibships with its theoretical expectation. (A *sibship* is a group of brothers and sisters.)

Table 5.1 shows various theoretical calculations of the expected sex distributions in sibships of size 1 to 4. In the second column, M indicates male and F indicates female, and the third column provides the general formula for the proportions applicable to any value of m, where the symbol m refers to the proportion of males at the time of birth. Given any value of m, the theoretically expected sex distributions can be obtained merely by plugging this value of m into the formulas. The fourth column presents the expected sex distributions (as percentages) for the case $m = 0.5$, and the final three columns pertain to actual cases. Among U.S. whites, for example, the observed secondary sex ratio has a proportion of males equal to $m = 0.5135$. (This proportion corresponds to 1055 male births for every 1000 female births.) The expected percentages of various sex distributions in sibships of size 1 to 4 for $m = 0.5135$ are indicated in the column corresponding to U.S. whites. One consequence of the slight excess of males in the secondary sex ratio is that all-male sibships will be somewhat more frequent than all-female sibships; among sibships of size 3 in U.S. whites, for example, 13 percent are expected to have all males, whereas 12 percent are expected to have all females.

As indicated in Table 5.1, the secondary sex ratio varies among populations, usually with a slight excess of males. The differences in the secondary sex ratio are quite small, ranging from 1026 males per 1000 females in U.S. nonwhites to 1097 males per 1000 females in the Philippines, but the differences are not due to chance variation because they are based on millions of births. (The designation ''nonwhite,'' incidentally, is a U.S. Census Bureau term referring to a heterogeneous population consisting largely of blacks but also including many other ethnic groups.)

The calculations in Table 5.1 are based on the assumption that individual births are random in regard to sex. That is to say, among U.S. whites, the probability that a newborn will be male is 0.5135 irrespective of whether the newborn is the first or the tenth in a sibship and irrespective of the sexes of prior children. Whether this assumption is valid or not depends on the observed distribution of males and females in sibships of various sizes. If the observed distributions are in accord with the theoretical ones in Table 5.1, then the assumption of randomness is supported.

In fact, the agreement between observed sex distributions and the theoretical ones in Table 5.1 is remarkably good, not only for sibships of four or fewer

Table 5.1   EXPECTED SEX DISTRIBUTIONS IN SIBSHIPS OF SIZE 1 TO 4 FOR VARIOUS VALUES OF m*

Sibship size	Sex distribution	General formula	Sex distribution, percentage†			
			$m = 0.5000$	$m = 0.5063$ (U.S. nonwhites)	$m = 0.5135$ (U.S. whites)	$m = 0.5233$ (Philippines)
1	1M:0F	m	50.0	50.6	51.4	52.3
	0M:1F	$1 - m$	50.0	49.4	48.6	47.7
2	2M:0F	m^2	25.0	25.6	26.4	27.4
	1M:1F	$2m(1 - m)$	50.0	50.0	50.0	49.9
	0M:2F	$(1 - m)^2$	25.0	24.4	23.7	22.7
3	3M:0F	m^3	12.5	13.0	13.5	14.3
	2M:1F	$3m^2(1 - m)$	37.5	38.0	38.5	39.2
	1M:2F	$3m(1 - m)^2$	37.5	37.0	36.5	35.7
	0M:3F	$(1 - m)^3$	12.5	12.0	11.5	10.8
4	4M:0F	m^4	6.2	6.6	7.0	7.5
	3M:1F	$4m^3(1 - m)$	25.0	25.6	26.4	27.3
	2M:2F	$6m^2(1 - m)^2$	37.5	37.5	37.4	37.3
	1M:3F	$4m(1 - m)^3$	25.0	24.4	23.6	22.7
	0M:4F	$(1 -)^4$	6.2	5.9	5.6	5.2

* m is the proportion of males in the secondary sex ratios.
† Some of the columns do not add to exactly 1 because of round-off error.

children but for larger sibships as well. Thus, *there do not seem to be tendencies for certain kinships to have boys or for certain other ones to have girls; the distribution of boys and girls in sibships seems to be random.* Many people are mildly surprised at this conclusion, for they know of one or more large kinships that consist of mostly boys or mostly girls. But a few kinships of predominantly one sex would be expected simply by the laws of chance. When one occurs, its curious sex distribution commands attention, so these kinships receive publicity in disproportion to their numbers. In short, sibships with extremely odd sex distributions are not more frequent than would be expected by chance.

The secondary sex ratio is not a constant like the speed of light, however. The secondary sex ratio varies not only among populations but in other ways as well. Although the effects are real, they are small and their causes are unknown. Four correlates of variation in the secondary sex ratio have been well documented:

1. *Birth order.* The secondary sex ratio decreases with order of birth in sibships, from a high of 1066 males per 1000 females among firstborn offspring to a low of 1045 males per 1000 females among seventh and later offspring. (These and the data below pertain to U.S. whites.)
2. *Father's age.* Young fathers tend to have slightly more male offspring than older fathers; the secondary sex ratio for fathers aged 15 to 19 is about 1070 males per 1000 females, but for fathers aged 45 to 49 it is about 1049 males per 1000 females.
3. *Seasonal effects.* The secondary sex ratio changes cyclically throughout the year, from a low of 1048 males per 1000 females in February to a high of 1062 males per 1000 females in July.
4. *Temporal variation.* The secondary sex ratio changes slightly from year

to year. It rose from 1056 males per 1000 females in 1935 to a high of 1063 in 1945, then dropped to 1052 in 1962 and rose again to 1060 in 1968. Peaks in the secondary sex ratio in 1945 and 1968 coincided with the Second World War and the Vietnam War, respectively, which has led some investigators to suggest that there may be a causal connection. On the other hand, male births declined steadily during the Korean War in the early 1950s.

This variation in the secondary sex ratio seems to conflict with the conclusion reached earlier that the sex of each birth is independent of previous ones and that the distribution of sexes within sibships is random. However, variation in the secondary sex ratio due to the sources discussed above is so small in magnitude that it does not invalidate the earlier calculations.

Y-LINKED GENES

Since males transmit a replica of their Y chromosome to all of their sons but to none of their daughters, pedigrees of traits due to genes on the Y chromosome (**Y-linked genes**) should be extremely simple.

1. Only males are affected with the trait.
2. Females never transmit the trait, irrespective of how many affected male relatives they may have.
3. All sons of affected males are also affected.

In spite of the relative ease with which these pedigree characteristics could be detected, few human traits have these characteristics. The principal reason for the rarity of such traits is that *the human Y chromosome carries very few genes*. (This relative paucity of Y-linked genes is also found in mice, fruit flies, and many other organisms.) Of course, one trait that is inherited along with the Y chromosome is maleness. Maleness, like the Y chromosome, is passed from father to son to grandson and so on. This mode of transmission is of great genetic importance, but it is so commonplace that its importance is easily overlooked. Moreover, as we shall see in the next chapter, individuals with abnormal sex-chromosome constitutions are phenotypically male or malelike if they have one or more Y chromosomes, but female or femalelike if they lack a Y. Thus, *the Y chromosome in humans carries the genes that trigger the embryonic development of maleness*. One Y-linked gene, apparently near the centromere, is involved in the production of a substance called the **H-Y antigen**, which appears very early in embryonic development. It is currently thought that the H-Y antigen is important in sex determination.

Are there any traits other than maleness itself that are determined by Y-linked genes? The inheritance of such traits could easily be detected because of their striking pedigree characteristics. Nevertheless, of the several thousand hereditary traits known, only one seems to follow the pattern of Y-linked inheritance. This is a gene for the trait **hairy ears**, which refers to the growth of stiff hair an inch or longer on the outer rim of the ears of males beyond the age of about 20 (Figure 5.2). The trait is most commonly found in men of India (incidence

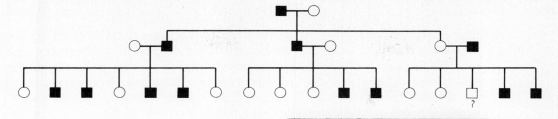

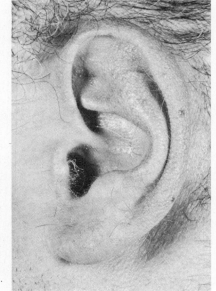

Figure 5.2 Partial pedigree and photograph of hairy pinna, a trait thought to be due to a Y-linked gene because of its male-to-male transmission, with all sons of an affected male being affected. (The male indicated with the question mark died at the age of 1 yr, so whether he would have developed the trait is unknown.) Although the pattern of inheritance of hairy pinna suggests Y linkage, many geneticists regard the assignment as tentative.

about 20 percent) and Israel, but it also occurs in Caucasians, Japanese, and other ethnic groups. Y-linked inheritance of hairy ears is not entirely certain because the trait is difficult to study. Not only is it extremely variable in age of onset (in some men it is not expressed until 60 or 70), but the trait is also extremely variable in its expression. Nevertheless, pedigrees of hairy ears (see Figure 5.2) do suggest Y linkage because the trait occurs only in males and is transmitted only through males. Whether the hairy ear gene is truly Y linked or not, it is clear that the Y carries few genes.

X-LINKED GENES

The X chromosome, in contrast to the Y, carries as many genes as would be found on an autosome of comparable size. Such genes are called **X-linked** (less frequently **sex-linked**) genes. At present, 115 loci are known to be X linked and 128 others are strongly suspected to be. Among these loci are the *HGPRT* locus used routinely in somatic cell genetics, loci involved in color perception and blood clotting, and a locus that, when mutated, causes a degenerative disease of muscle known as Duchenne-type muscular dystrophy.

Pattern of Inheritance

The pattern of inheritance of X-linked genes is unique because males receive their X chromosome only from their mothers and transmit it only to their daughters. This pattern is often called **crisscross** inheritance because an X chromosome can crisscross between the sexes in successive generations. A surname analogy for X-linked inheritance can be devised if we imagine a "strange savage nation" in which males have one surname, but females have two. In this nation, a male's surname is one chosen at random from his mother, but a female always receives her father's surname in addition to one randomly chosen from her mother. Thus, for example, a mating between Mr. Smith and Ms. Brown-Robinson would produce sons named Brown or Robinson with equal likelihood and daughters named Smith-Brown or Smith-Robinson with equal likelihood (see Figure 5.3).

In this analogy, of course, the names correspond to X-linked alleles. With just two alleles at an X-linked locus (call them A and a), there will be three possible genotypes in females (AA, Aa, and aa) but only two possible genotypes in males (A and a). Since males have only one X chromosome, the terms *homozygous* and *heterozygous* do not apply to their X-linked loci. Instead, males are said to be **hemizygous** for X-linked loci, and males who carry A or a are referred to as hemizygous A or hemizygous a, respectively.

If the a allele is a recessive allele associated with some disorder, then homozygous aa females and hemizygous a males will both be affected. The hemizygous a males will be affected because, having only one X chromosome, they lack a dominant allele to compensate for the recessive.

Punnett squares for the matings $AY \times Aa$ and $aY \times AA$ are shown in Figure 5.4. Here A and a represent A-bearing and a-bearing X chromosomes,

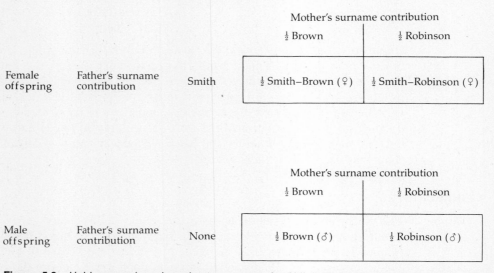

Figure 5.3 Haldane analogy based on surnames for X-linked genes. A male has just one surname and contributes it to all his daughters but none of his sons. A female has two surnames, and both daughters and sons receive one of these at random.

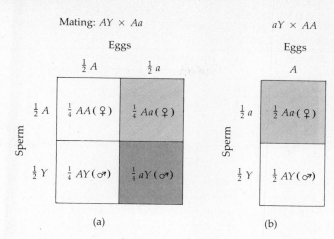

Mating: $AY \times Aa$

Figure 5.4 Punnett squares showing inheritance of an X-linked pair of alleles, A and a. (a) Mating between a hemizygous A male and a heterozygous female produces $\frac{1}{4}$ homozygous A daughters, $\frac{1}{4}$ heterozygous daughters, $\frac{1}{4}$ hemizygous A sons and $\frac{1}{4}$ hemizygous a sons. (b) Mating between a hemizygous a male and a homozygous A female produces $\frac{1}{2}$ heterozygous daughters and $\frac{1}{2}$ hemizygous A sons. The pattern of inheritance results from the fact that a male transmits his X chromosome only to his daughters and his Y chromosome only to his sons.

respectively, and Y represents the Y chromosome. If a trait due to an X-linked recessive is rare, the matings in Figure 5.4 will be the most frequent ones in which the recessive allele is involved. Figure 5.4(a) shows that heterozygous females have 50 percent affected sons and 50 percent carrier daughters; Figure 5.4(b) shows that affected males have all normal sons and all carrier daughters.

Pedigree Characteristics

Corresponding to the unique nature of X-linked inheritance are certain unique features in pedigrees of traits due to X-linked recessive alleles. These features are illustrated in Figure 5.5, which is a pedigree of hemophilia, a trait characterized by excessive bleeding following an injury due to the absence of an essential blood-clotting factor. To be noted in the pedigree are the following points:

1. Predominantly males are affected with the trait. Indeed, in Figure 5.5, only males are affected, which happens frequently when the trait in question is rare.
2. Affected males have phenotypically normal offspring. As noted in connection with Figure 5.4(b), however, all their daughters are carriers.
3. Affected males usually have phenotypically normal parents. This characteristic of rare, X-linked recessives occurs because most affected males arise from heterozygous (and therefore phenotypically normal) mothers.

Incidence in Males and Females

For traits due to X-linked recessive alleles, affected males occur much more frequently than affected females. Indeed, the relationship between the frequency of affected males and females is given by a simple formula, namely,

$$\text{Freq of affected females} = (\text{Freq of affected males})^2$$

Table 5.2   RELATIVE FREQUENCY OF MALES AND FEMALES AFFECTED WITH AN X-LINKED RECESSIVE CONDITION

Freqency of affected males	Frequency of affected females [= (Frequency of affected males)^2]
0.10	0.01
0.05	0.0025
0.01	0.0001
0.005	0.000025
0.001	0.000001

Since the square of a number less than 1 is always smaller than itself, the frequency of affected females will be smaller than the frequency of affected males. The disparity in the frequencies is illustrated for several cases in Table 5.2. For a trait that affects 0.01 (i.e., 1 percent) of males, for example, the frequency of affected females will be 0.0001 (i.e., 0.01 percent), so there will be 100 times fewer affected females than affected males. Although the mathematical basis for this difference in incidences between the sexes will be discussed in Chapter 13, the reason for the differences can be seen intuitively: Whereas two copies of a recessive X-linked allele are required to produce an affected female, only one copy is required to produce an affected male; and an individual is less likely to inherit two copies of a rare allele than only one copy.

Some of the most well-known and important inherited traits are due to X-linked alleles, and these warrant brief individual discussions.

GLUCOSE 6-PHOSPHATE DEHYDROGENASE (G6PD) DEFICIENCY

One well-known gene on the X chromosome controls the production of an enzyme, **glucose 6-phosphate dehydrogenase** (G6PD), which is involved in carbohydrate metabolism and is important in maintaining the stability of red blood cells. Individuals who have abnormally low amounts of this enzymatic activity are prone to a severe anemia which occurs when many of their red blood cells cannot function normally and therefore break down and are destroyed. The anemia can be provoked by a number of environmental triggers such as inhaling pollen of the broad bean *Vicia faba* or eating the bean raw, in which case the illness is known as **favism**. Such individuals are also sensitive to certain chemicals (such as naphthalene, used in mothballs), certain sulfa antibiotics (such as sulfanilamide), or the antimalarial drug primaquine. In the absence of the offending substances these individuals are completely normal, and they recover from the anemia when the agents are eliminated. G6PD deficiency is found in high frequency in people of Mediterranean extraction (10 to 20 percent or more of males are affected) and among Asians (about 5 percent of Chinese males are affected), and it occurs in about 10 percent of black American males. As noted, the locus of G6PD is X linked, and well over 50 alleles coding variant forms of the enzyme are known. However, only a few of these alleles lead to a sufficiently defective form of G6PD to cause the drug sensitivity and anemia associated with G6PD deficiency.

COLOR BLINDNESS

Almost everyone is familiar with the common form of color blindness, called **red-green color blindness**, which is inherited as an X-linked recessive condition. Several other kinds of defects in color vision are known; they differ according to which of the three pigments in the retina of the eye—red, green, or blue—is defective or present in an abnormally low amount. The most common types of color blindness involve the red or green pigments; these are collectively known as red-green color blindness, but they are not a single entity. Both conditions are X linked, however, and in different Caucasian populations the frequency of red-green color-blind males is between 5 and 9 percent. Color blindness in females is much rarer, of course, but it does occur.

Two loci on the long arm of the X chromosome seem to be involved in red-green color blindness. Defects in green vision are due to mutations at one of the loci; defects in red vision are due to mutations at the other locus. Among Western European males, about 5 percent have defects in green perception and another 1 percent have defects in red perception.

Of course, females having a red-defect allele on one X chromosome and a green-defect allele on the other will have normal color vision, but almost half their sons will be red color-blind and almost half will be green color-blind. Because of rare recombination between the loci, a small proportion of sons will be both red and green color-blind, and an equally small proportion will have normal color vision.

HEMOPHILIA

Recall that **hemophilia** is a bleeding disorder due to an X-linked recessive that results from the excessively long time required for the blood to clot following an injury, owing to the absence of one of more than a dozen **clotting factors** in the blood (Figure 5.5). In **hemophilia A**, which is perhaps the most widely recognized hereditary disease in humans and is often called royal hemophilia because of its presence in certain European royal families, the missing clotting factor is factor

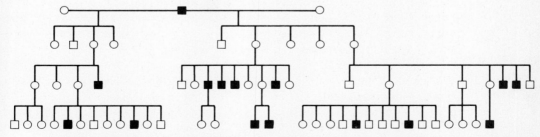

Figure 5.5 Pedigree of hemophilia in a kinship from Scotland showing several typical features of the inheritance of rare X-linked mutations. Note that only males are affected. Because males transmit their X chromosome only to their daughters, and because the rare mutation is recessive, affected fathers have normal sons and carrier daughters [see Figure 5.4(*b*)]. Moreover, half the sisters of all affected males are expected to be carriers [see Figure 5.4(*a*)]. Carrier females are expected to have 50 percent affected sons [see Figure 5.4(*a*)].

VIII. Affecting about 1 in 7000 males, hemophilia A is much rarer than red-green color blindness, yet it is certainly as well known. This is partly because hemophilia affects the blood; and even though we now know that blood has no magical or hereditary properties, we still carry a vestige of the old beliefs in such expressions as "pure blood," "blood lines," and "blood relation." A second reason hemophilia is so well known is that it occurred in many members of the European royalty who descended from Queen Victoria of England (1819–1901). She was a carrier of the gene and, by the marriages of her carrier granddaughters, the gene was introduced into the royal houses of Russia and Spain. Ironically, the present royal family of Great Britain is free of the gene because it descends from King Edward VII, one of Victoria's four sons, who was not himself affected (Figure 5.6).

Where Queen Victoria's hemophilia-A mutation came from is not known. She herself was a carrier, but her father was completely normal and nothing in her mother's family suggests that the gene was present. The best guess is that

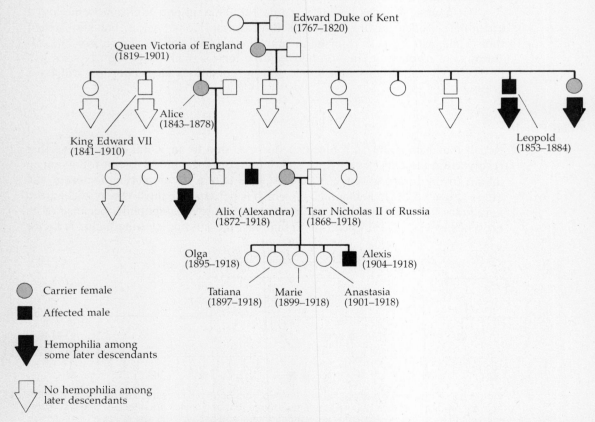

Figure 5.6 Partial pedigree of hemophilia A among the descendants of Queen Victoria, including Alexandra, Empress of Russia, and her five children.

either the egg or the sperm that gave rise to Queen Victoria carried a new mutation. Perhaps it was in the sperm of her father, Edward Duke of Kent, because it is thought that mutations are somewhat more likely to occur in the sperm of older men, and he was 52 when Victoria was born. However it happened, Victoria was a carrier. She had nine children (see Figure 5.6): five daughters (two were certainly carriers, two were almost certainly not, and the remaining daughter may or may not have been as she left no children) and four sons (one a hemophiliac—Leopold Duke of Albany—and three normal). In the five generations since Queen Victoria, 10 of her male descendants have had the disease. Virtually all died very young. Those who survived childhood often died in their 20s or early 30s, usually from excessive bleeding following injuries.

The most famous of Victoria's affected descendants is undoubtedly her great-grandson Alexis, the son of her granddaughter Alix (Empress Alexandra of Russia) and Tsar Nicholas II (see the pedigree in Figure 5.6 and the photograph in Figure 5.7).

Alix—more properly, Princess Alix of Hesse-Darmstadt—was a bright, educated, lovely woman who on November 26, 1894, at the age of twenty-two, married Nicholas, the Tsarevich of Russia. She Russianized her name to Alexandra and embraced the Russian Orthodox faith. A year and a half later, on May 25, 1896, Nicholas was crowned Tsar Nicholas II of Russia, and Alexandra became Empress. The first child of the marriage was the Grand Duchess Olga, born in 1895. Then, at two-year intervals, the Grand Duchesses Tatiana, Marie, and Anastasia were born. Finally, in 1904, Alexandra bore a son, an heir to the throne, Tsarevich Alexis. Alexandra, unbeknown to her, was a carrier of the

Figure 5.7 Tsar Nicholas II of Russia, the Empress Alexandra (granddaughter of Queen Victoria), and their son, Tsarevich Alexis. Alexandra was a carrier of the hemophilia-A mutation, and her son was afflicted with the disease. For a pedigree of the family, see Figure 5.6.

hemophilia-A allele, and her baby boy had inherited it. Ann Viroubova, one of Alexandra's favorite ladies-in-waiting, relates:

> . . . [T]he heir was born amid the wildest rejoicings all over the Empire. . . . [A]fter many prayers, there was an heir to the throne of the Romanoffs. The Emperor . . . was quite mad with joy. His happiness and the mother's, however, was of short duration, for almost at once they learned that the poor child was afflicted with a dread disease. The whole short life of the Tsarevich, the loveliest and most amiable child imaginable, was a succession of agonizing illnesses due to this congenital affliction. The sufferings of the child were more than equaled by those of his parents, especially of his mother. . . .*

How the dread disease affected the character and upbringing of the boy himself, and how his affliction scarred the parents psychologically and changed their lives, is a case history of the anguish of families who have a hopelessly sick and suffering child. The mother doted on him almost from the day of his birth. In the words of Viroubova, she became "more of a nurse than an Empress." The parents tried to conceal the illness: "No parents . . . are ready all at once to reveal a misfortune such as that. . . ." They became reclusive, "closing their eyes and ears to the growing unpopularity of the Empress. . . . [She] appeared . . . cold, haughty, and indifferent." The Tsar, too, was preoccupied with his son's illness. At one point he observes in his diary how difficult it was to live through the worry resulting from Alexis's excessive and life-threatening bleeding from minor injuries and bruises that all children unavoidably experience.

Worst of all, of course, were the feelings of guilt that the parents themselves were the cause of the boy's disease. One day in the autumn of 1913 Alexis slipped and bruised his right knee against the corner of a piece of furniture. Bleeding began and lasted for days, the whole leg swelling with blood and causing the boy excruciating pain. Pierre Gilliard, Alexis's tutor, witnessed the episode:

> One morning I found the mother at her son's bedside. He had had a very bad night. Dr. Derevenko was anxious, as the hemorrhage had not been stopped and his temperature was rising. The inflammation had spread further and the pain was even worse than the day before. The Tsarevich lay in bed groaning piteously. His head rested on his mother's arm, and his small, deathly white face was unrecognizable. At times the groans ceased and he murmured the one word "Mummy!" in which he expressed all his sufferings and distress. His mother kissed him on the hair, forehead, and eyes, as if the touch of her lips could have relieved his pain and restored some of the life which was leaving him. Think of the tortures of that mother, an impotent witness of her son's martyrdom in those hours of mortal anguish—a mother who knew that *she herself* was the cause of his sufferings, that *she* had transmitted to him the

---

* Viroubova (Vyrubova), A. 1923. Memories of the Russian Court. Macmillan, New York, p. 10.

terrible disease against which human science was powerless! *Now* I understand
the secret tragedy of her life!*

The Tsarevich recovered from this episode, as he did from many others, but
even in his periods of health his life was hardly normal.

As Alexis grew older his parents carefully explained to him the nature of his
illness and impressed on him the necessity of avoiding falls and blows. But
Alexis was a child of active mind, loving sports and outdoor play, and it was
almost impossible for him to avoid the very things that brought him suffering.
"Can't I have a bicycle?" he would beg his mother. "Alexis, you know you
can't." "Mayn't I play tennis?" "Dear, you know you mustn't." Often these hard
denials of the natural play impulse were followed by a gush of tears as the
child cried out: "Why can other boys have everything and I nothing?" Suffering
and self-denial had their effect on the character of Alexis. Knowing what pain
and sacrifice meant, he was extraordinarily sympathetic towards other sick
people.†

In their grief and worry the parents turned to whoever could give the child relief.
Enter Rasputin, "that strange and ill-starred being about whom almost nothing
is known to the multitude but against whom such horrible accusations have been
made that he is universally classed with such monsters of iniquity as Cain, Nero,
and Judas Iscariot."‡ From Viroubova:

In behalf of the suffering little Tsarevich the Emperor and Empress constantly
asked the prayers of Rasputin, and the incident which I shall now relate will
appeal to any mother or father of a suffering child and will render less childlike
the faith of the afflicted parents of the heir to the throne. One day during the
War [December 16, 1914], the Emperor left Tsarskoe Selo for general head-
quarters, taking with him as usual the Tsarevich. The child seemed to be in
good condition, but a few hours after leaving the palace he was taken with a
nosebleed. This is ordinarily a harmless enough manifestation, but in one
suffering from Alexis's incurable malady it was a very serious thing. The doctors
tried every known remedy, but the hemorrhage became steadily worse until
death by exhaustion and loss of blood was threatened. I was with the Empress
when the telegram came announcing the return of the Emperor and the boy to
Tsarskoe Selo, and I can never forget the anguish of mind with which the poor
mother awaited the arrival of her sick, perhaps her dying child. Nor can I ever
forget the waxen, gravelike pallor of the little pointed face as the boy with
infinite care was borne into the palace and laid on his little white bed. Above
the blood-soaked bandages his large blue eyes gazed at us with pathos un-
speakable, and it seemed to all around the bed that the last hour of the unhappy
child was at hand. The physicians kept up their ministrations, exhausting every

* Gillard, P. 1970. Thirteen Years at the Russian Court. (Trans. by F. A. Holt.) Arno Press,
New York, p. 43.
   † Viroubova, *op. cit.*, pp. 81–82.
   ‡ Viroubova, *op. cit.*, p. 149.

means known to science to stop the incessant bleeding. In despair the Empress sent for Rasputin. He came into the room, made the sign of the cross over the bed and, looking intently at the almost moribund child, said quietly to the kneeling parents: "Don't be alarmed. Nothing will happen." Then he walked out of the room and out of the palace. That was all. The child fell asleep, and the next day was so well that the Emperor left for his interrupted visit to the Stavka. Dr. Derevenko and Professor Fedoroff told me afterwards that they did not even attempt to explain the cure.*

(Gilliard, who detests Rasputin, tells a different version. He says, "The doctors ultimately succeeded in cauterizing the scar which had formed at the spot where a little blood vessel had burst. Once more the Tsarina attributed the improvement in her son's condition that morning to the prayers of Rasputin, and she remained convinced that the boy had been saved thanks to his intervention."†)

Whatever happened at the palace, the boy lived. How long he might have lived will never be known, for along with his mother and father, his four sisters, and several servants he was brutally murdered by the Bolsheviks in Ekaterinburg fortress on July 17, 1918. This tragic end, and the whole story of Nicholas and Alexandra, are told compassionately and beautifully in the book by R. K. Massie. He concludes:

Infinitely . . . remarkable and . . . fatefully enigmatic . . . is the awesome, overwhelming drama of the Russian Revolution. . . . This is the legacy of Lenin. And also the legacy of Rasputin and hemophilia. Kerensky once said, "If there had been no Rasputin, there would have been no Lenin." If this is true, it is also true that if there had been no hemophilia, there would have been no Rasputin. This is not to say that everything that happened in Russia and the world has stemmed entirely from the personal tragedy of a single boy. It is not to overlook the backwardness and restlessness of Russian society, the clamor for reform, the strain and battering of a world war, the gentle, retiring nature of the last Tsar. All of these had a powerful, bruising impact on events. Even before the birth of the Tsarevich, autocracy was in retreat. Here, precisely, is the point. Had it not been for the agony of Alexis's hemophilia, had it not been for the desperation which made his mother turn to Rasputin, first to save her son, then to save the pure autocracy, might not Nicholas II have continued retreating into the role of constitutional monarch so happily filled by his cousin King George V? It might have happened, and, in fact, it was in this direction that Russian history was headed.‡

LESCH-NYHAN SYNDROME (HGPRT DEFICIENCY)

As a final example of X-linked recessive inheritance we consider the rare condition known as **Lesch-Nyhan syndrome**, which is due to a deficiency of the enzyme HGPRT. (Recall from Chapter 4 that HGPRT is one of the enzymes frequently used to select hybrid cells in somatic cell genetics.) Persons with this

* Viroubova, *op. cit.*, pp. 169–70.
† Gilliard, *op. cit.*, p. 156.
‡ Massie, R. K. 1967. Nicholas and Alexandra. Dell, New York, pp. 529–30.

condition, virtually all male, have two groups of symptoms. One group results from excessive accumulation of **uric acid** in the blood. Accumulation of excessive uric acid leads to hyperuricemia (elevated levels of uric acid in the blood), bloody urine, crystals in the urine, urinary tract stones, severe inflammation of the joints (**arthritis**), and the extremely painful swelling and inflammation of joints in the hands and feet (particularly the big toe) known as **gout**. (Although gout is always associated with elevated levels of uric acid, the inheritance of most forms of gout is multifactorial; only a minority of patients with excessive uric acid and gout have HGPRT deficiency.)

Symptoms of excessive uric acid can be treated with appropriate drugs, but Lesch-Nyhan syndrome has another group of symptoms that are of unknown origin and are not alleviated by treatment for excessive uric acid. These symptoms involve the nervous system and are associated with jerky, unwilled, and uncoordinated muscular movement. The most bizarre symptom of Lesch-Nyhan syndrome is extreme aggressive behavior, most often expressed as self-mutilation of the hands and arms, usually by biting. At present, such self-mutilation can be prevented only by appropriate binding of the hands and arms. Lesch-Nyhan syndrome is an extremely serious condition. Most patients die before the age of five, and almost all die before adulthood.

DOSAGE COMPENSATION

In many instances, the amount of a gene product in cells is directly related to the number of copies of the corresponding gene: a cell with one gene copy will produce half as much gene product as a cell with two copies, and a cell with two copies will produce two-thirds as much gene product as a cell with three copies of the gene. On the basis of this simple reasoning, one would expect the amount of gene product of X-linked genes in males to be half as large as the amount in females, because males have only one X chromosome whereas females have two. This expectation is not realized, however. For the X-linked locus *G6PD,* for example, a cell from a female contains the same amount of G6PD enzyme as does a cell from a male. Similarly, to take another example, a female cell contains the same amount of clotting factor VIII as does a male cell.

In all known organisms in which the sexes differ in the **dose** (number of copies) of sex-chromosomal genes, the activity of genes on the sex chromosome in one of the sexes is adjusted by a special type of regulation known as **dosage compensation**; the net effect of dosage compensation is to offset ("compensate") the sex differences in gene dosage. In mammals, dosage compensation occurs in the female, and, *in humans, dosage compensation involves the turning off of one X chromosome in each somatic cell.* This X-chromosome inactivation is known as the **single active X principle**. In any somatic cell of an adult female, only one X chromosome is genetically active, and only the genes on that X chromosome are working; the other X chromosome is genetically inactive ("turned off"), and the genes on that X chromosome are not expressed. However, the X chromosome that is the active one in some cells may be the inactive one in other cells. Every female is therefore a kind of **mosaic** with respect to X-linked genes. (Recall from

Chapter 2 that a mosaic is an individual composed of two or more genetically different types of cells.) The decision as to which X chromosome is to be inactivated in each cell is made early in embryonic development. Once the decision is made, the same X remains inactive in all of a cell's descendants.

The genetic consequences of X-chromosome inactivation are easily observed in cells from females who are heterozygous for an X-linked mutation. One experimental procedure used to study the situation is outlined in Figure 5.8,

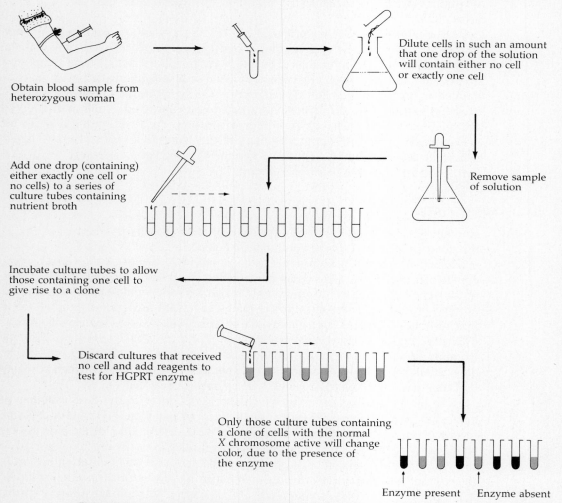

Obtain blood sample from heterozygous woman

Dilute cells in such an amount that one drop of the solution will contain either no cell or exactly one cell

Add one drop (containing) either exactly one cell or no cells) to a series of culture tubes containing nutrient broth

Remove sample of solution

Incubate culture tubes to allow those containing one cell to give rise to a clone

Discard cultures that received no cell and add reagents to test for HGPRT enzyme

Only those culture tubes containing a clone of cells with the normal X chromosome active will change color, due to the presence of the enzyme

Enzyme present   Enzyme absent

Figure 5.8 One procedure for showing that only one X chromosome is active in each somatic cell of a normal female. The woman here is heterozygous for a recessive mutation that causes a nonfunctional HGPRT enzyme to be produced. Cells from the woman are cultured one by one, each giving rise to a clone. Clones in which the normal X chromosome is active will produce the functional enzyme; clones in which the normal X is inactive will produce only the nonfunctional enzyme. Approximately half of each type of clone is found, including that half the cells have the normal X chromosome active and half have only the mutation-bearing X chromosome active.

where the woman whose cells are to be studied is assumed to be heterozygous for an allele that leads to a nonfunctional form of HGPRT. The genotype of such a female is *HGPRT*⁺/*HGPRT*⁻—the + and − designating the normal and defective alleles of *HGPRT*, respectively. Half the cells in a heterozygous female are expected to have their *HGPRT*⁺-bearing X chromosome active, and these cells will produce the normal HGPRT enzyme. In the remaining cells, the *HGPRT*⁻-bearing X chromosome will be active, and these cells will produce a nonfunctional HGPRT enzyme. (In the latter case, the *HGPRT*⁺ allele is still physically present in the cells, of course, but this allele and apparently most other genes on the same X chromosome are inactive.) When white blood cells from an *HGPRT*⁺/*HGPRT*⁻ female are separated from each other and laboratory cultures are started from single cells placed in nutrient medium, each cell gives rise to a *clone* (a group of genetically identical cells). Because of the single active X principle, half the clones are expected to produce a functional HGPRT, and the remaining clones are expected to produce a nonfunctional HGPRT. As indicated in Figure 5.8, this expectation is borne out in practice.

In certain cells of the body, the inactive X chromosome can be seen directly in the microscope. Nuclei of nondividing cells scraped off the surface layer of skin on the inside of the cheek often have a densely staining blob near the nuclear envelope [Figure 5.9(*b*)]. This blob is the coiled and inactive X chromosome, and it is called a **sex-chromatin body** or a **Barr body**. Actually, the staining procedure does not work perfectly, and for this reason the Barr body cannot be seen in all cells scraped from the inside of the cheek. Even though all the cells of normal females have one inactive X chromosome, not all the cells have a visible Barr body. The important finding is that cells from normal females frequently have a Barr body [see Figure 5.9(*b*)], whereas cells from normal males never do [see Figure 5.9(*a*)].

In the next chapter we shall discuss certain relatively rare individuals who have an abnormal sex-chromosome constitution. Here we should note one important finding relative to the somatic cells of such individuals: *irrespective of*

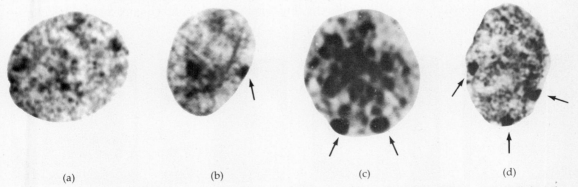

(a)   (b)   (c)   (d)

Figure 5.9 Micrographs of nuclei showing zero, one, two, or three Barr bodies (arrows). Inactive X chromosomes are tightly condensed in the nucleus and strongly absorb certain dyes, giving rise to Barr bodies. Cells that have zero, one, two, or three Barr bodies have respectively one, two, three, or four X chromosomes.

the number of X chromosomes in an individual, all X chromosomes except for one are inactivated in each somatic cell. Thus, for example, XO females (who have only one X chromosome instead of two) have no X-chromosome inactivation, XX females have one X inactivated, XXX females have two X's inactivated, XXXX females have three X's inactivated, and so on. Consequently, cells of XO females have no Barr bodies, those of XX females have up to one [see Figure 5.9(*b*)], those of XXX have up to two [see Figure 5.9(*c*)], and those of XXXX have up to three [see Figure 5.9(*d*)]. The single active X principle also applies to male or malelike individuals who have more than one X chromosome. That is to say, males with more than one X chromosome have all except one X chromosome inactivated, and these inactivated X's are frequently visible as Barr bodies. Consequently, XXY or XXYY or XXYYY males have cells with up to one Barr body, XXXY and XXXYY males have cells with up to two, XXXXY males have cells with up to three Barr bodies, and so on. Note that the number of Y chromosomes makes no difference in the number of Barr bodies because extra Y chromosomes are not inactivated.

EMBRYONIC AND FETAL DEVELOPMENT

In the early stages of embryonic development, all embryos are identical regardless of whether they will ultimately develop as male or female. Indeed, early embryos might be thought of as "bisexual" because they contain internal structures characteristic of both sexes.

Before discussing this fascinating aspect of sexual development, we should perhaps begin with the beginning—**fertilization**, which normally occurs in one of the two **fallopian tubes**. About 4 inches long and less than $\frac{1}{2}$ inch in diameter, the fallopian tubes stretch from the upper corners of the uterus to the ovaries on either side; there the tubes flare out with tiny fingerlike projections to catch the egg released at the time of ovulation. The fertilized egg (**zygote**) travels down the fallopian tube toward the uterus, dividing by mitosis as it moves along. The one cell becomes two, two become four, four become eight, eight become sixteen, and so on, with all the cells sticking together in a clump (Figure 5.10). This clump

  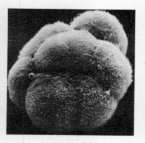 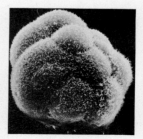

Two cells   Four cells   Eight cells   Morula

Figure 5.10 Scanning electron micrographs of the earliest stages in mouse development. The tiny cell visible most noticeably in the photograph of the four-cell stage is a polar body.

of cells, called the **morula**, reaches the uterus about three or four days after fertilization.

Cell division continues in the morula until the cells number a few thousand. The cells then begin to move in relation to one another; they push against and expand the envelope that encloses them and rearrange themselves to form a hollow sphere called the **blastocyst** (Figure 5.11).

In one region of the inner wall of the blastocyst there is a thickening composed of a mass of cells known as the **inner cell mass**. The body of the embryo itself is destined to develop from this group of cells, as will several **embryonic membranes**, including the **amnion**, which enshroud the developing embryo. Once the blastocyst has been formed, it implants itself in the uterine wall; this occurs on approximately the eighth day after fertilization.

After implantation, the relationship between the blastocyst and the mother

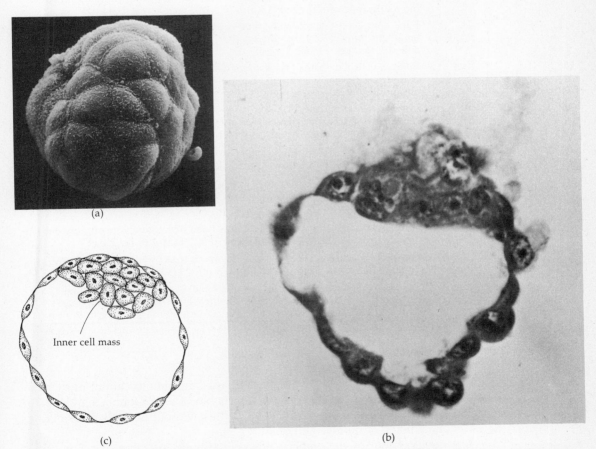

(a)

(c)

Inner cell mass

(b)

Figure 5.11 Three views of the blastocyst. (*a*) Scanning electron micrograph of mouse blastocyst. (*b*) Cross section of human blastocyst showing inner cell mass that gives rise to the body of the embryo. (*c*) Schematic of the cross section in (*b*). In humans, the blastocyst forms about one week after fertilization.

Table 5.3   TIMETABLE OF HUMAN EMBRYONIC AND FETAL DEVELOPMENT

Week after fertilization	Major features of development*
0	Fertilization
1	Egg travels down fallopian tube into uterus; begins to divide
2	Implantation of blastocyst in wall of uterus; embryonic membranes begin to form
3†	Early stage of skeleton, nervous system, and blood vessels
4†	Head, heart, and tail visible; heart begins to beat; gill-pouch rudiments; paddle-shaped rudiments of arms and legs; length 4 mm (0.2 in); see Figure 5.12(*a*)
5†	Chest and abdomen formed; fingers and toes appear; head increases in size; eyes developing; length 8 mm (0.3 in)
6†	Face, features, and external ears developing; upper lip formed; gill rudiments disappearing; length 13 mm (0.5 in), weight 1 g (0.04 oz)
7†	Face completely developed—now resembles a human child; palate forms; length 18 mm (0.7 in), weight 2 g (0.07 oz); from now on embryo usually called *fetus*; see Figure 5.12(*b*)
8†	Fingers distinct; testes and ovaries distinguishable; beginning of differentiation of external sex organs; beginnings of all essential internal and external structures present; length 30 mm (1.2 in), weight 4 g (0.2 oz)
12†	Limbs—including fingers, toes, nails—fully formed; external sex organs developed; sex can be determined by a trained person without microscopic examination; 77 mm (3.0 in), 30 g (1.1 oz); see Figure 5.12(*c*)
16	Movements ("quickening") begin; heart can be heard with stethoscope; hair all over body; eyebrows and eyelashes present; 190 mm (7.5 in), 180 g (6.4 oz)
21	Head hair appears; 300 mm (11.8 in), 450 g (1 lb)
25	Eyes open; 350 mm (14 in), 875 g (2 lb)
30	Fetus pink and scrawny—little fat has been deposited; if born now can survive, given special care; 400 mm (16 in), 1425 g (3 lb, 2 oz)
32	Considerable fat deposited—fetus begins to look "plump"; excellent chance of survival if born; 450 mm (18 in), 2375 g (5 lb, 4 oz)
38	Full term; skin covered with cheeselike material; head hair typically 25 mm (1 in) long; may be other hair on shoulders, but soon disappears; head still very large relative to body; 500 mm (20 in), 3250 g (7 lb, 3 oz)

* All measuresments are averages; there are many wide departures from them, especially in the later months.
† Critical stages for teratogenic agents.

is crucial. The **placenta**, which develops in the region of implantation, must help transfer nutrients and oxygen from the mother's blood to the embryo and waste from the embryo back into the mother, all of this without actually intermixing the circulatory fluids of mother and child.

The development of the embryo proceeds rapidly after the blastocyst implants in the uterine wall. (Table 5.3 is a timetable of human embryonic development highlighting certain landmark events.) In the third week after fertilization, the early stages of the skeleton and nervous system are formed. By the fourth

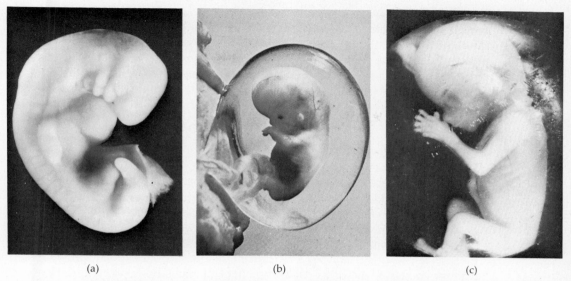

(a)   (b)   (c)

Figure 5.12 The human embryo at (a) 4 weeks, (b) 7 weeks, and (c) 12 weeks after fertilization. Facial features are formed by the seventh week of development; the limbs, including fingers, toes, and nails, are fully developed by the twelfth week. The ages of the embryos shown here are approximate.

week the head of the embryo is visible and the embryo has a beating heart [Figure 5.12(a)]. The arms and legs of the embryo begin to develop; these start as tiny paddle-shaped bumps of cells that grow outward and elongate. Also in the fourth week after fertilization, the embryo acquires a tail, which later disappears. Early stages of gill development can be recognized; these are much the same as in fishes, but in humans the gills do not develop completely and they eventually recede. (The tail and primitive gills are reminders of our evolutionary ancestry.)

By the fifth week of development, the chest and abdomen of the embryo are formed, fingers and toes are beginning to develop, and the eyes appear. The embryo is less than $\frac{1}{2}$ inch long. By six weeks after fertilization, the embryo is still less than 1 inch long. The gills, which never developed completely, are disappearing. The face and ears of the embryo are developing, and it begins to look like a human embryo. By the seventh week after fertilization, the embryo is unmistakably a human embryo [see Figure 5.12(b)]. Many of the internal organs have begun to take shape.

From this point on the embryo is properly called a **fetus**. By the end of 12 weeks, the major outlines of embryonic development are almost complete [see Figure 5.12(c)]. The fetus is only 3 inches long and weighs little more than 1 ounce, yet the last months of development will be spent, in a sense, in putting on the finishing touches and, most especially, growth.

The first three months of embryonic development are the most critical because so much happens so fast. The cells in the embryo divide extremely rapidly. Different types of cells with special functions and abilities arise. (It has been estimated that humans may have up to 200 distinct cellular types, such as muscle cells, red blood cells, and so on.) Several important events occur at nearly the same time in different parts of the embryo. Everything seems to be in constant motion. Outfoldings of groups of cells occur here, infoldings there, and bulges somewhere else. Fusions of tissues happen in one place, splittings in another. Some cells migrate from up to down, some from down to up. The embryo seems intent on its business, as if in a hurry to be born. The last six months of development seem almost leisurely by contrast.

TERATOGENS

Because so much cell division and differentiation occur in the first 12 weeks, these are the weeks in which the embryo is most sensitive to external influences. The cells are extremely sensitive to radiation, for example, and to many drugs. Any agent that can cause malformations during development is called a **teratogen**. One dramatic example of a teratogen is the sedative **thalidomide**, which was in widespread use in certain parts of Europe in the late 1950s. The drug seemed harmless enough at first, but then the effect of thalidomide on the embryos of pregnant women was recognized. When present during the first critical months of embryonic development, thalidomide interferes with the normal growth of the long bones in the arms and legs (Figure 5.13). The children born to women who took the drug early in pregnancy were normal in every respect, except that they had abnormally short or absent arms or legs. Before the tragic mistake was discovered, approximately 5000 children had been maimed. Not everyone suffered, however. Those pregnant women who started to take the drug *after* the first critical months of embryonic development gave birth to normal babies.

Developing embryos are also extremely sensitive to certain kinds of infections. Of these, **rubella**, or **German measles**, is the best-known example. Rubella is caused by a virus, a very tiny internal parasite of cells, which multiplies inside infected cells to produce hundreds of virus particles. These particles may, in turn, infect other cells. Consequently, infection by a single virus particle can eventually lead to the presence of millions of them in the body.

Adults are not greatly threatened by German measles. Even small children are not gravely endangered. A few days of fever, red spots, and discomfort, and the infection usually runs its course. If a woman contracts German measles in the early stages of pregnancy, however—in the first three months—the virus can penetrate the placental barrier and infect cells in the embryo. Such infection destroys embryonic cells and can lead to spontaneous abortion of the embryo or fetus. Among fetuses that survive the infection, a great danger is blindness—or deafness, or heart defects, or severe mental retardation, or all of these.

After an adult or child has been infected with German measles virus, the body develops an immunity that prevents subsequent infection by destroying the invading virus before its cycle of reproduction can get started. In times past, young girls were encouraged to visit friends who had the disease in the hope they would contract it and thereby become naturally immune to later infection

Figure 5.13 Malformation induced by the teratogen thalidomide.

during their childbearing years. Nowadays, an antirubella vaccine is used to immunize women against the virus.

DIFFERENTIATION OF THE SEXES

Development in the first seven weeks following fertilization is virtually identical in both male and female embryos. The sex of embryos in these early weeks cannot be outwardly distinguished. Their sex chromosomes reveal which is which, of course, but the bodies of the embryos themselves give no hint of whether they will ultimately develop as male or female. But in the latter half of the critical first 12 weeks of development, the body of the embryo takes on a sexual identity as its internal and external sex organs are formed. In the seventh week after fertilization, the embryo physically commits itself to one direction or the other. Sexual development then occurs very rapidly and by 12 weeks it is essentially completed. In the twelfth week a careful observer can visually determine the sex of the fetus, then only 3 inches long.

Important events in sexual development do occur before the seventh week, but these events are the same in male and female embryos. The origin of the sexual structures is intimately connected with the urinary system, as might well be expected from the anatomical relationship of the two systems in adults. The gonads of the embryo, one on either side—which in females will develop into ovaries and in males into testicles—are formed as ridgelike thickenings in a pair of ducts or tubes that form a sort of primitive kidney, a structure that is fully

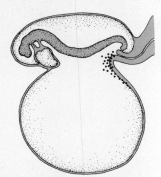

Figure 5.14 Drawing of human embryo at about $3\frac{1}{2}$ weeks after fertilization showing that the germ cells (black dots) arise outside of the body of the embryo in the margin of the yolk sac. From there, they migrate into the body of the embryo and reside in the embryonic gonad.

formed in the embryo by the sixth week. (The permanent kidney originates separately in development.) The embryonic gonads rapidly enlarge as the cells divide repeatedly and increase in size.

The embryonic gonad contains the **primordial germ cells**, the cells whose descendants will eventually give rise to eggs or sperm. Although the germ cells are present in the gonads in the sixth week of development and probably earlier, their actual place of origin is far removed from the gonad. Indeed, the primordial germ cells do not arise in the body of the embryo itself. Rather, they arise at an early stage in the margin of the yolk sac outside the embryonic body (Figure 5.14)! From there they migrate into the body of the embryo and proceed to the embryonic gonad.

The germ cells from the yolk sac are genetically identical to all the other cells in the embryo and its membranes, of course, because all these cells are descendants of the fertilized egg. Yet it is remarkable how early in development the germ cells are set apart. From a genetic point of view this is not so surprising, however. Inheritance in animals proceeds fundamentally from germ cell to germ cell to germ cell down through the generations. A person's body is a sort of carrier or receptacle of the germ cells, although it is important to keep in mind that the genes expressed in the body are the same genes carried by the germ cells. But, in a not wholly facetious sense, a hen is just an egg's way to make another egg.

By the time the embryonic gonads are formed, a second pair of ducts has developed in the embryo [Figure 5.15(a)]. The embryo thus has two pairs of ducts that are associated with sex. Internally, in the sixth and seventh week of development, the embryo is neither male nor female. In a certain sense it is both, because the pair of ducts that develops last—the **Müllerian ducts**—is associated with the internal sex organs of females, while the primitive kidney—the **Wolffian ducts**—is associated with the internal sex organs of males. The external parts of the body are also identical in both male and female embryos of this age. In roughly the position where the external sexual structures will be formed is a raised bump or elevation of cells that in males will develop into the penis and in females into the clitoris. Flanking the elevation of cells in the genital region of the embryo is a pair of vaguely outlined ridges that in males will grow out and

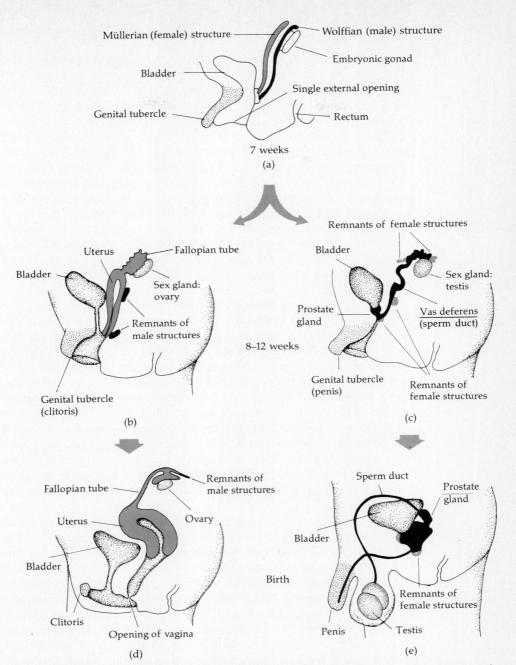

Figure 5.15 Outline of differentiation of the sex organs in humans. Until about two months after fertilization, the development of the sex organs in male and female embryos is identical; the embryo has two sets of internal sexual structures: the Müllerian ducts (female) and the Wolffian ducts (male). Presence of the male hormone testosterone in later embryonic development causes development of the male structures and degeneration of the female structures; absence of testosterone causes development of the female structures and degeneration of the male ones. Note that the penis and the clitoris develop from the same embryonic tissue, a small bump of cells called the genital tubercle.

fold toward each other and fuse, forming the scrotum; in females these ridges of tissue become the labia majora.

Development of the two pairs of sex ducts in the embryo and formation of the subtle bumps and swellings in the embryo's genital region set the stage for subsequent sexual development. Internally the embryo is both male and female; externally it is neither. The embryo is capable of developing in either direction, male or female, and only now will it commit itself. Hereafter the sexual structures of male embryos develop in one direction and those of female embryos in another.

What tips the balance of sexual development is whether the male hormone **testosterone** is secreted by the embryonic gonad. This hormone is produced in the gonads of normal males, not in normal females, and the sex-chromosome constitution of the embryo usually determines whether it will be produced. The gonads of XY embryos secrete testosterone; the gonads of XX embryos do not. The genes involved in the synthesis of testosterone seem to be located on the autosomes. They somehow respond to the presence of the Y chromosome by becoming active.

In a sense, every human embryo is actually "preprogrammed" to develop as a female. It is the presence of testosterone in male embryos that causes the course of sexual development to deviate from femaleness to maleness [see Figure 5.15(c) and (e)]. The effect of the hormone is twofold: first, it stimulates development of the male parts of the embryo; second, it prevents differentiation of the female parts. In the presence of testosterone, the inner part of the embryonic gonad expands and the outer part recedes; these changes accompany the development of the embryonic gonads into testes. The parts of the Wolffian ducts that are in intimate contact with the developing testes differentiate into tubules that will serve as a conduit and a place of storage for sperm. Simultaneously with these events, the Müllerian ducts, under the influence of testosterone, gradually degenerate [see Figure 5.15(c)]. While all this is happening inside the embryo, the external genital structures are developing into the typical external sex organs of males.

In embryos that do not possess the male hormone testosterone, sexual development proceeds the way it had been preprogrammed [see Figure 5.15(b) and (d)]. In the development of a human female, the outer part of the embryonic gonad expands; the inner part recedes. The gonads differentiate into the ovaries. The upper parts of the Müllerian ducts develop into the fallopian tubes, whereas the lower parts of the Müllerian ducts fuse together. The upper portion of this fused segment becomes the uterus; the lower portion becomes part of the vagina. As this is going on, the Wolffian ducts slowly degenerate [see Figure 5.15(b)]. External female sexual structures begin to develop at about the same time. (Here's a little memory aid that may help to keep the Müllerian/Wolffian distinction clear: men have lost their Müllerian ducts; women have lost their Wolffian ducts.)

Sexual development is essentially completed by the twelfth week after fertilization. Indeed, most of the major events in human development are completed by the twelfth week. The fetus is still utterly dependent on the placenta for nutrition and removal of wastes, however. If aborted, naturally or intentionally, the fetus has no hope of survival. By the sixteenth week the "quickening" occurs; the fetus begins to move. But it is still totally dependent on its mother. Not until about 30 weeks after fertilization does the fetus have a reasonable chance of surviving on its own, although in rare cases a somewhat younger fetus has survived.

AMBIGUOUS SEX

Although most human beings are unambiguously male or female, about 1 percent have abnormalities in sexual development. These include several of the abnormal sex-chromosome constitutions to be described in Chapter 6, such as XO and XXY, but they also include a wide variety of developmental abnormalities in children who have normal sex chromosomes. Some of these conditions are known to be caused by mutations in particular genes; other conditions have extremely complex or unknown causes. Among the latter group is a particularly common class of abnormalities in sexual development known as hermaphroditism.

Hermaphrodites are neither entirely male nor entirely female but have a mixture of male and female sex organs; that is, their sex is ambiguous. (The name *hermaphrodite* is a compound of the names of the god Hermes and the goddess Aphrodite.) In most of biology, the term **hermaphrodite** refers to individuals that possess both *functional* ovaries and *functional* testes, so that, theoretically, they could fertilize themselves. The term is used differently in human genetics, where hermaphrodites never have functional sex organs of both sexes, and, indeed, in most cases, can produce neither functional sperm nor functional eggs.

The overall incidence of hermaphroditism in humans is about 0.2 percent, or 1 per 500 newborns. This makes it relatively common among abnormalities of sexual development. Hermaphroditism is not a single condition, however, and its causes are diverse. Homozygosity for any one of several different autosomal recessive mutations can lead to hermaphroditism, but beyond this rudimentary knowledge the genetics of the disorder are not well worked out. The trait is extremely variable from patient to patient. In some cases the internal reproductive organs are of one sex, the external organs of the other. In other cases the reproductive organs are female on one side of the body and male on the other. Often **ovotestes** are found; these are mixtures of ovarian and testicular tissue that result from the failure of either the inner or outer part of the embryonic gonad to degenerate.

Although the underlying causes of hermaphroditism are diverse and generally not understood in detail, there are exceptions. Among these are the several abnormalities of the sex chromosomes to be discussed in the next chapter. Two types of inherited hermaphroditism not due to sex-chromosome abnormalities are also relatively well understood, and the nature of the disorder in these cases illustrates the importance of the male hormone testosterone in sexual development. One of these types is a male hermaphroditism known as the testicular-feminization syndrome; the other is a female condition known as the adrenogenital syndrome.

Testicular-Feminization Syndrome

The **testicular-feminization syndrome** is an inherited condition caused by an X-linked mutation expressed only in chromosomally XY individuals. Estimates of incidence vary from 1 in 20,000 to 1 in 64,000 male births. Affected individuals are chromosomally XY, yet they have few traces of masculinity (Figure 5.16). Although production of male hormone is nearly normal, the cells have defective male-hormone receptors and so cannot respond. The external genitalia are female, a vagina is present but incompletely developed, and rarely there is a rudimentary uterus and fallopian tubes. Tissue recognizable as testicular can be found, usually in an abnormal position in the body cavity, and the Wolffian ducts, associated with male sexuality, are somewhat developed. Affected individuals are therefore chromosomally male but for the most part phenotypically female, and they are, of course, unable to bear children. Nevertheless, many do marry as women and have normal sexual relations with their husbands.

Adrenogenital Syndrome

The **adrenogenital syndrome** is a form of hermaphroditism in humans caused in some instances by a hereditary defect in the adrenal glands. These glands, one perched atop each kidney, produce a variety of important hormones, among them **cortisol**. This hormone is produced by converting **cholesterol**, chemical step by chemical step, into the cortisol molecule (Figure 5.17). Along this kind of chemical assembly line, each step adds some atoms or takes some off or chemically rearranges the preceding molecule. Each step in this assembly pathway requires

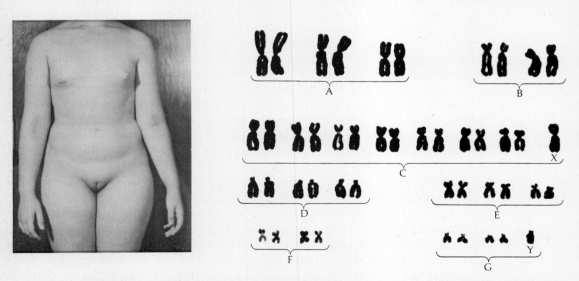

Figure 5.16 This individual is phenotypically female but chromosomally male (XY), as indicated in the karyotype. The condition is known as the testicular-feminization syndrome, due to an X-linked mutation.

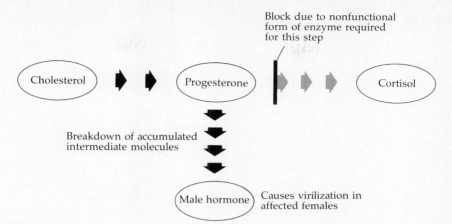

Figure 5.17 In the adrenal glands, cortisol is produced from cholesterol by means of a complex biochemical pathway; the key steps are illustrated here. Individuals with the adrenogenital syndrome have an inherited defect in one of the enzymes in the pathway, leading to a buildup in the amount of one of the intermediates (progesterone). This excess progesterone is broken down by another pathway into male hormone, which causes virilization in affected females.

the presence of a specific enzyme molecule to accomplish the chemical reaction, and the genetic information enabling the cell to manufacture each of the enzymes is present in a different gene. In the adrenogenital syndrome, one enzyme required in an intermediate step along the pathway is not functional. Thus, the chemical pathway is blocked and cortisol cannot be produced. The defective enzyme is supposed to convert one of the intermediate molecules (**progesterone**) in the pathway into the next molecule in the pathway, but the enzyme is nonfunctional due to homozygosity for a recessive mutation and the conversion cannot be accomplished. As a consequence, the intermediate molecule accumulates in cells of the adrenal gland, much as half-finished automobiles would accumulate on an assembly line if one of the workers quit abruptly.

Progesterone and related molecules cannot accumulate forever, of course. They are chemically altered by other pathways into other molecules including male hormone. In affected females, the male hormone causes a degree of **virilization** (masculinization).

Females afflicted with the adrenogenital syndrome have a normal uterus and fallopian tubes; the vagina is usually small, and the clitoris may be enlarged, giving it the appearance of a penis. Externally, these people are malelike, although the reproductive organs are small and not as well formed as in normal males. The most serious effects of the syndrome are not the physical abnormalities, however; of greatest concern is the lack of cortisol, which can lead to death. Treatment of the condition involves continual administration of cortisol to the patient. The presence of cortisol prevents the adrenals from starting more cholesterol molecules along the chemical pathway to cortisol. This, in turn, prevents the accumulation of the intermediate molecules and their conversion into male hormone, which prevents further masculinization.

SUMMARY

1. The **primary sex ratio** is the ratio of males to females at fertilization. Although the primary sex ratio in humans is not well known, it is thought to be close to 1:1, which corresponds to the ratio theoretically expected due to segregation of the X and Y chromosomes in males.

2. The **secondary sex ratio** is the ratio of males to females at birth. A great deal is known about the secondary sex ratio because sex is one of the items typically noted in birth records. Although there is variation among populations in the secondary sex ratio, most populations have a slight excess of males. Among U.S. whites, for example, the secondary sex ratio is about 1055 males for every 1000 females. There do not seem to be inherited tendencies for certain kinships to have an excess of males and for others to have an excess of females.

3. Small but real changes in the secondary sex ratio do occur, however. The secondary sex ratio decreases slightly with birth order, decreases slightly with father's age, varies cyclically during the year, and changes somewhat from year to year. The reasons for these small changes in the secondary sex ratio are not known.

4. The Y chromosome in humans and many other organisms carries few genes beyond those responsible for the determination of maleness. One Y-linked human gene codes for the **H-Y antigen**, which is thought to be of great importance in sex determination. The pattern of inheritance of traits due to Y-linked genes is exceedingly simple: the trait occurs only in males and is transmitted only through males—from father to son to grandson, and so on.

5. The X chromosome seems to carry as many genes as an autosome of comparable size. Recessive X-linked genes will be expressed in homozygous females and in the corresponding **hemizygous** males. The pattern of inheritance of X-linked genes follows the inheritance of the X chromosome itself; namely, males receive their X chromosome from their mother and transmit it only to their daughters (**crisscross inheritance**). Pedigrees due to rare, X-linked recessive alleles therefore have certain characteristic features, including: (*a*) primarily or only males are affected, (*b*) affected males have nonaffected offspring, and (*c*) affected males have phenotypically normal parents. Traits due to rare X-linked recessives are much more frequent in males than in females.

6. Examples of traits due to X-linked recessives include **G6PD deficiency** (characterized by chemical- or drug-induced anemia), **color blindness** (two X-linked forms associated with defective red or defective green perception), **hemophilia** (the form of hemophilia occurring in certain European royalty is due to a deficiency of blood clotting factor VIII), and **Lesch-Nyhan syndrome** (deficiency of HGPRT enzyme; symptoms include gout and bizarre self-destructive behavior).

7. **Dosage compensation** refers to regulation of gene activity of the sex chromosomes to adjust and equalize gene dosage in the sexes. In humans, dosage compensation is accomplished by one randomly chosen X chromosome in each somatic cell in normal females being rendered genetically inactive. X-chromosome inactivation occurs early in development, and the particular X that is rendered inactive may differ from

cell to cell. However, all descendants of a cell that has a particular X chromosome inactivated have that same X inactivated; so an adult female is a sort of **mosaic** for X-linked genes, with some cells having the genes on one X chromosome expressed and the rest of the cells having the genes on the other X chromosome expressed. In individuals with multiple X chromosomes, all X chromosomes but one are inactivated (**single active X**). The inactive X's can be recognized cytologically as **sex-chromatin (Barr) bodies**. The Y chromosome, in contrast to the X, does not seem to be subject to dosage compensation and does not form Barr bodies.

8. Human development passes through stages known as **zygote**, **morula**, **blastocyst**, **embryo**, and **fetus**. Although birth usually occurs about 38 weeks after fertilization, the beginnings of all essential internal and external structures are present after 8 weeks. The first 12 weeks of development are marked by great sensitivity to **teratogens** (agents that cause developmental abnormalities). Examples of teratogenic agents are **thalidomide** (a sedative) and **rubella** (German measles virus).

9. Sexual differentiation occurs in the seventh to twelfth week of development. Prior to the seventh week, all embryos possess two sets of sex-related ducts: the female-related **Müllerian ducts** and the male-related **Wolffian ducts**. Under the influence of the male hormone **testosterone**, the Müllerian ducts degenerate and the Wolffian ducts differentiate into the internal male sexual structures. In the absence of testosterone, development is preprogrammed in such a way that the Wolffian ducts degenerate and the Müllerian ducts differentiate into the internal female sexual structures.

10. Abnormalities in sexual development include **hermaphroditism** (presence of both male and female sexual structures in the same individual but neither functional eggs nor functional sperm produced), **testicular-feminization syndrome** (an X-linked gene expressed only in chromosomally XY individuals, causing nonresponsiveness to testosterone and consequent phenotypic femaleness), and the **adrenogenital syndrome** (block in conversion of cholesterol to cortisol, leading to accumulation of intermediate metabolites that are broken down into male hormone and related compounds causing masculinization of affected females).

KEY WORDS

Adrenogenital syndrome	Hemizygous	Secondary sex ratio
Amnion	Hemophilia	Sibship
Barr body	Hermaphrodite	Single active X principle
Blastocyst	Lesch-Nyhan syndrome	Teratogen
Color blindness	drome	Testicular-feminization syndrome
Dosage compensation	Müllerian ducts	Testosterone
Embryo	Placenta	Thalidomide
Favism	Primary sex ratio	Wolffian ducts
Fetus	Primordial germ cells	X-linked gene
G6PD	Rubella	Y-linked gene

PROBLEMS

5.1. What are the primordial germ cells?

5.2. In what way is the sedative thalidomide a teratogen?

5.3. Distinguish between the primary sex ratio and the secondary sex ratio. Why are these not necessarily identical?

5.4. Calculate the proportion of males corresponding to the following secondary sex ratios:

    (a) 1048 males per 1000 females
    (b) 1055 males per 1000 females
    (c) 1063 males per 1000 females
    (d) 1070 males per 1000 females

5.5. If a population has a secondary sex ratio of 1055 males per 1000 females, what is the number of females per 1000 males?

5.6. Considering the Haldane surname analogy in the context of autosomal-dominant inheritance, autosomal-recessive inheritance, X-linkage, and Y-linkage, which mode of inheritance best approximates the rules of transmission of surnames in the United States, and why?

5.7. An individual has the sex-chromosome karyotype XXXXYY. Is the individual femalelike or malelike? How many Barr bodies would be found?

5.8. In an XXX female, one of the X chromosomes carries the HGPRT⁻ allele. If clones are cultured from this woman, what fraction are expected to be HGPRT⁺ and what fraction HGPRT⁻? Why?

5.9. Why are the phenotypic effects of the adrenogenital syndrome similar to the effects of repeated administration of testosterone?

5.10. What features of the illustrated pedigree suggest X-linked inheritance?

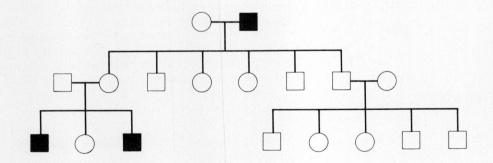

5.11. Why is X-linked inheritance sometimes called "crisscross" inheritance?

5.12. A woman whose parents were normal has a brother affected with hemophilia. How can this happen? What is the chance that the woman herself is a carrier?

5.13. A woman's father is color-blind and she has a son. What is the probability that the boy is color-blind? Does the genotype of the woman's mate make a difference? Why or why not?

5.14. In the pedigree of color blindness shown here, one individual in addition to the two indicated is affected. Which one is it? Using the symbols C and c for the normal and color-blind alleles, identify the genotype of each individual in the pedigree.

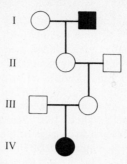

5.15. A woman whose father had the red form of color blindness marries a man with the green form of color blindness. They have a daughter, who has normal color vision. The daughter marries a normal man and they have four sons. Two have the red form of color blindness and two have the green form. How can this happen?

5.16. A certain X-linked trait has an incidence in males of 1/500 (i.e., 0.002). What is the expected incidence of the trait in females?

5.17. In some respects a normal XY × XX mating resembles an $Aa \times aa$ mating for an autosomal locus. Thus, if the secondary sex ratio consists of $\frac{1}{2}$ males and $\frac{1}{2}$ females, the probability that a sibship of size n will consist of exactly i males and $n\text{-}i$ females (see Problem 14 in Chapter 3) is

$$\frac{n!}{i!\,(n\text{-}i)!}\left(\tfrac{1}{2}\right)^n$$

What is the probability that a sibship of size 3 will consist of all girls? Of all boys? Of all girls *or* all boys?

5.18. Using the formula in Problem 17, calculate the probability that a sibship of size 4 will contain exactly, 0, 1, 2, 3, or 4 boys.

5.19. The formula in Problem 17 can readily be modified for the case when the secondary sex ratio consists of a proportion m boys and $1\text{-}m$ girls. In this case the probability of i boys and $n\text{-}i$ girls is

$$\frac{n!}{i!\,(n\text{-}i)!}\,m^i(1\text{-}m)^{n\text{-}i}$$

For a secondary sex ratio of $m = 0.51$ (i.e., 1049 males per 1000 females), calculate the probability that a sibship of size 3 contains 1 male and 2 females. Calculate the probability of 2 males and 1 female.

5.20. Using the formula in Problem 19 with $m = 0.52$ (i.e., 1083 males per 1000 females), calculate the probability that a sibship of size 6 contains only males. What is the probability that all 6 are females?

FURTHER READING

Bardin, C. W., and J. F. Catterall. 1981. Testosterone: a major determinant of extragenital sexual dimorphism. Science 211: 1285–94. The influence of testosterone on nonsexual structures.

Browder, L. W. 1980. Developmental Biology. Saunders, Philadelphia. A comprehensive text dealing with a complex subject.

Edwards, A. W. F., and M. Fraccaro. 1958. The sex distribution in the offspring of 5,477 Swedish ministers of religion, 1585–1920. Hereditas 44: 447–50. Finds no evidence that sex ratios "run in families."

Fuchs, F. 1980 (June). Genetic amniocentesis. Scientific American 242: 47–53. Details the procedure of amniocentesis and its uses.

Gordon, J. W., and F. H. Ruddle. 1981. Mammalian gonadal determination and gametogenesis. Science 211: 1265–71. On the role of the X and the Y chromosomes in mammalian sexual development.

Haseltine, F. P., and S. Ohno. 1981. Mechanisms of gonadal differentiation. Science 211: 1272–78. Excellent discussion of testosterone and testicular feminization.

Kurnit, D. M., and H. Hoehn. 1979. Prenatal diagnosis of human genome variation. Ann. Rev. Genet. 13: 235–58. An assessment of the use of studies at the DNA level in prenatal diagnosis.

MacLusky, N. J., and F. Naftolin. 1981. Sexual differentiation of the central nervous system. Science 211: 1294–1303. Gonadal hormones also influence sex-specific developmental processes in the brain.

Ohno, S. 1979. Major Sex-Determining Genes. Springer-Verlag, New York. A short but advanced discussion.

Oppenheimer, S. B. 1980. Introduction to Embryonic Development. Allyn and Bacon, Boston. A good introduction to general aspects of differentiation.

Wilson, J. D., F. W. George, and J. E. Griffin. 1981. The hormonal control of sexual development. Science 211: 1278–84. Which hormones are necessary for normal sexual development, and what they do.

chapter 6

Abnormalities in Chromosome Number

- Types of Chromosome Abnormality
- Nondisjunction
- Sex-chromosomal Abnormalities
- Trisomy X
- XYY and Criminality
- XXY: Klinefelter Syndrome
- XO: Turner Syndrome
- Autosomal Abnormalities
- Trisomy 21: Down Syndrome
- Trisomy 18: Edwards Syndrome
- Trisomy 13: Patau Syndrome
- Amniocentesis
- Death Before Birth
- Summary

In this chapter and the next we discuss genetic conditions due to an excess or deficiency of whole chromosomes or parts of chromosomes instead of mutations in particular genes. The importance of gene dosage in cells was emphasized in Chapter 5. Human cells have become adapted to having a proper and harmonious balance in the action of genes when each gene on the autosomes is present twice, which is related to the fact that autosomes occur in homologous pairs. The X and Y chromosomes are special cases, of course, but the dosage of X-linked genes is equalized by dosage compensation, and apparently the Y chromosome carries so few genes that dosage compensation is unnecessary.

Because of the importance of proper gene dosage, too many or too few copies of genes can upset the normal processes of development, even though the added or absent genes themselves are not necessarily mutated. These abnormalities result from gross imbalances in the number and action of genes, and a general rule is that the greater the imbalance, the more severe is the abnormality. Some imbalances are small enough to have almost no effects on development. Other, larger ones are **lethal**; they lead to death of the embryo or child at an early age.

TYPES OF CHROMOSOME ABNORMALITY

The balance of genes can be disrupted in many ways, but the most obvious is by variation in the number of chromosomes. In this chapter we shall consider two distinct types of abnormality in chromosome number. The first involves the presence of extra entire *sets* of chromosomes. Recall from Chapter 1 that human gametes are **haploid**; they carry one complete set of chromosomes consisting of 22 autosomes and 1 sex chromosome. Somatic cells, by contrast, are **diploid** because they carry two haploid sets of chromosomes—46 altogether. Cells that have three complete sets of chromosomes (totaling 69) are known as **triploids**, and those with four complete sets (totaling 92) are known as **tetraploids**.

The second type of abnormality in chromosome number involves individual chromosomes instead of entire sets of chromosomes. A normal somatic cell is diploid and has 46 chromosomes. A cell with a missing chromosome would have 45, and such cells are said to be **monosomic** for the chromosome represented only once instead of twice. For example, a female having only one X chromosome instead of two (i.e., an XO female) is monosomic for the X. Similarly, cells with three copies of a chromosome instead of two are said to be **trisomic** for the chromosome in question. For example, an individual having three copies of chromosome 21 is trisomic for chromosome 21; alternatively, we could say that the individual has **trisomy** 21.

Most people are astonished to learn that about 8 percent (1 out of 12) of diagnosed human pregnancies involve an embryo that has some major chromosomal abnormality. Most of these abnormalities lead to grossly deformed embryos that mercifully undergo **spontaneous abortion**, usually quite early in development. Indeed, about 50 percent of spontaneously aborted fetuses have major chromosomal abnormalities. One might say of these as the Apocrypha says of the souls of the righteous—that in the sight of the unwise they seem to die and their departure is taken for misery, but they are at peace and no torment shall touch them.

Yet many such embryos, the ones with less severe abnormalities, do survive. Among live-born children, about 1 in 155 (0.6 percent) has some major chromosomal abnormality. This number includes individuals with sex-chromosomal abnormalities, which are among the least severe chromosome abnormalities, but it does not include minor chromosomal variants of the sort discussed in Chapter 2 or chromosomal abnormalities that are too subtle to be detected through the light microscope.

NONDISJUNCTION

Any number of accidents can happen during meiosis that can cause an egg or a sperm to have two copies of a chromosome instead of only one; the same sort of errors can also lead to gametes that completely lack one particular chromosome. If a gamete carrying two copies of a chromosome participates in fertilization, the resulting zygote will be trisomic for that chromosome. Successful gametes that lack a particular chromosome will give rise to monosomic zygotes. Such abnormal gametes, which lead to monosomic or trisomic zygotes, result from failure of one of the pairs of chromosomes to be distributed normally to opposite poles during meiosis. The chromosomes that fail to separate (disjoin) properly are said to have undergone **nondisjunction**.

Since nondisjunction involves improper chromosome separation at anaphase during cell division, there are three instances in which it can occur. In meiosis, nondisjunction can occur in either the first meiotic division [Figure 6.1(*a*)] or during the second meiotic division [Figure 6.1(*b*)]. In either case, the upshot is the production of one or more gametes that carry an extra chromosome and one or more gametes that have a missing chromosome.

Nondisjunction can also occur during mitosis (see Figure 6.2), and in this case one daughter cell will be trisomic for the chromosome in question and the other daughter cell will be monosomic. If mitotic nondisjunction occurs in a cell of a developing embryo, the descendants of the trisomic or monosomic daughter cells will give rise to trisomic or monosomic clones in an otherwise normal individual. Thus, mitotic nondisjunction will lead to a chromosomal mosaic—an individual with two or more chromosomally distinct types of cells.

SEX-CHROMOSOMAL ABNORMALITIES

In human genetics, it is a general rule that among live-born children *chromosome abnormalities involving the autosomes have more severe phenotypic effects than those involving the sex chromosomes*. This generalization has a twofold basis. First, abnormal gene dosage of the Y chromosome has a milder effect than that of an autosome because, as emphasized in Chapter 5, the Y chromosome seems to carry few genes beyond those responsible for the development of maleness. Secondly, an abnormal number of X chromosomes is rendered less harmful than an abnormal number of autosomes because of dosage compensation involving the single active X principle (see Chapter 5). Indeed, trisomy of the sex chromosomes is in certain instances so mild phenotypically that affected individuals can lead normal or nearly normal lives.

On the other hand, X-chromosomal monosomy is a severe disorder, and complete lack of an X chromosome is **lethal** (incompatible with life) very early in embryonic development. This contrast between the milder effects of trisomy as compared with monosomy leads to another generalization that will be dis-

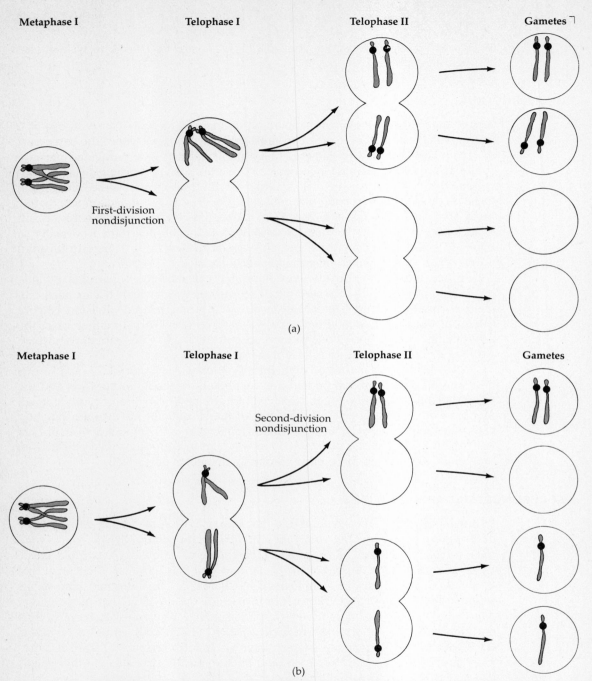

Metaphase I **Telophase I** **Telophase II** **Gametes**

First-division
nondisjunction

(a)

Metaphase I **Telophase I** **Telophase II** **Gametes**

Second-division
nondisjunction

(b)

Figure 6.1 Two types of meiotic nondisjunction. (*a*) First-division nondisjunction, in which homologous chromosomes fail to disjoin at anaphase I, leads ultimately to two gametes that have an extra copy of the chromosome in question and two gametes that lack the chromosome. (*b*) Second-division nondisjunction, in which sister chromatids fail to disjoin at anaphase II, leads to one gamete with an extra copy of the chromosome in question, one gamete lacking the chromosome, and two gametes that are normal. Although anaphase is not illustrated here, nondisjunction often results from a chromosome lagging behind at anaphase.

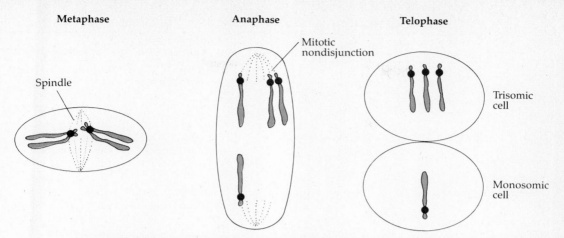

Figure 6.2 Mitotic nondisjunction occurs when sister chromatids of one member of a pair of homologous chromosomes fail to disjoin during anaphase, often due to anaphase lag. Of the resulting daughter cells, one will be trisomic for the chromosome in question and the other will be monosomic. Subsequent mitoses of the daughter cells result in a trisomic/monosomic chromosomal mosaic.

cussed further in the next chapter: generally speaking, *an increased dose of one or more genes has less severe phenotypic effects than a decreased dose of the same genes.*

TRISOMY X

The sex-chromosomal abnormalities found most frequently among live-born children are listed in Table 6.1. Among the most common is **trisomy X**. Women with this syndrome have 47 chromosomes, including three X chromosomes (XXX). (Technically speaking, the conventions of human karyotype nomenclature require that the total number of chromosomes be listed first, the sex-chromosome constitution next, if relevant, and other chromosome abnormalities following. Thus, trisomy X females would be denoted 47,XXX, but the abbreviation XXX is a useful shorthand for our purposes.) XXX females are physically quite normal and usually able to bear children, but some have menstrual irregularities and an early onset of menopause. Many of these individuals are mentally retarded.

Although the incidence of the XXX condition is about 1 per 950 newborn females, it is three times greater among the institutionalized mentally retarded. Mental ability of all kinds varies from individual to individual, and some XXX women are both mentally and physically normal. These women lead completely normal lives. They are born, grow to adulthood, bear children, grow old, and die, giving no hint of their extra X chromosome.

XYY AND CRIMINALITY

Another sex-chromosomal abnormality associated with few physical manifestations is the XYY condition—technically 47,XYY. Affected males have 47 chro-

Table 6.1   MOST FREQUENT SEX-CHROMOSOME ABNORMALITIES IN HUMANS

Sex-chromosome abnormality	Name of syndrome	Approximate frequency	Major symptoms
XXX (47,XXX)	Trisomy X "syndrome"	1 in 950 newborn females	Female; physically quite normal, but tendency toward mental retardation; fertile
XYY (47,XYY)	Double-Y "syndrome"	1 in 950 newborn males	Male; tend to be tall; behavioral effects highly controversial; fertile
XXY (47,XXY)	Klinefelter syndrome	1 in 1000 newborn males	Male; tend to be tall; sexual maturation absent, although some breast enlargement may occur; frequently associated with mental retardation; sterile
XO (45,X)	Turner syndrome	1 in 5000 newborn females	Female; short of stature, sometimes with "webbing" of skin between neck and shoulders; sexual maturation absent; mental abilities nearly normal; sterile

mosomes; they have the full chromosome complement of normal males plus a second Y chromosome. XYY males tend to be tall, averaging over 6 ft. Beyond this, distinctive physical symptoms are practically nonexistent, so it is questionable whether the term *syndrome* really applies to this case because there is no group of characteristic symptoms. Among newborn males, the incidence of the XYY condition is thought to be approximately 1 in 950.

The behavior of XYY males has been a matter of concern and controversy since 1965, when a Scottish study uncovered a high frequency of the XYY condition among imprisoned men convicted of violent crimes. More than 50 similar studies have now been carried out in prisons in Europe and the United States, and it has held true that the frequency of XYY men confined for crimes of violence is about 10 times greater than it is among a sample of newborn males. Many of these studies have concentrated on tall men; because XYY males tend to be tall, this will automatically create a bias, and the findings must therefore be interpreted with caution. Nevertheless, the higher than expected frequency of XYY males among tall criminals has suggested to some that XYY males may undergo impaired psychosocial development, perhaps caused by their height itself, which may produce social stresses tending to evoke violent, aggressive behavior, or perhaps caused by some subtle and invisible behavioral effect of the extra Y.

One Danish study of criminality among XYY males is especially important and revealing because it is free of many of the biases in other studies. The Danish research involved an examination of all available records on almost all men 28 to 32 years old whose height was 184 cm (72.4 in) or more and whose birthplace was Copenhagen. A total of 4139 sex-chromosomal determinations was made; among these there were 12 XYY (incidence 1 per 345), 16 XXY (incidence 1 per 259—the XXY sex-chromosomal constitution will be discussed in the next section), and 13 individuals with other types of chromosomal abnormality. The information examined included criminal records and types of crime, socioeconomic status of the men's parents, score on a Danish army intelligence test, and an index of educational level.

Parental socioeconomic status was lower among men with criminal convictions than among those without convictions, but no differences were found among XY, XYY, and XXY males. Thus, lower socioeconomic status is associated with cirminality but not with sex-chromosomal constitution.

Other Danish data pertaining to chromosomes and criminality are summarized in Table 6.2. Among XYY males, the rate of criminal convictions was 41.7 percent (5 out of 12), which is greater than the 9.3 percent found among XY males by a sufficient amount that the disparity cannot plausibly be attributed to chance variation. Based on these findings, there is indeed evidence for a greater rate of criminal convictions among XYY males than among tall XY males. Among XXY males there is also a somewhat greater incidence of criminal convictions (3 of 16) than found among XY males, but this difference is small enough to be due to chance.

The findings relative to height are summarized in the third column of Table 6.2. On the average, XYY males are taller than the XY males by about 3.7 cm (1.5 in), and XXY males are taller than the XY males by about 2.7 cm (1.1 in). Tallness alone is not related to criminality, however, because men with criminal records are somewhat shorter than their noncriminal counterparts.

One of the most important findings of the Danish study is summarized in the last two columns of Table 6.2. XYY and XXY males have significantly lower

Table 6.2 HEIGHT, CRIMINALITY, AND INTELLECTUAL FUNCTION AMONG XY, XYY, AND XXY DANISH MALES

	Number of males	Height, cm	Army intelligence test scores	Educational index
Average of all XY males	4096	187.1	43.7	1.6
XY males with no criminal record	3715	187.1	44.5	1.6
XY males with criminal record	381 (9.3%)	186.7	35.5	0.7
Average of all XYY males	12	190.8	29.7	0.6
XYY males with no criminal record	7	191.3	31.6	0.7
XYY males with criminal record	5 (41.7%)	190.2	27.0	0.5
Average of all XXY males	16	189.8	28.4	0.8

Source: Data from H. A. Witkin *et al.,* 1976, Science 193: 547–55.

scores on the Army intelligence test and a significantly lower index of educational level than do XY males. For each sex-chromosomal class, moreover, men with criminal records are substantially poorer in both measures of intellectual function than men without criminal records. Of course, this does not necessarily imply that men with lower levels of intellectual function actually commit more crimes; they may merely be more likely to get caught. In any case, the Danish data seem to imply that the high rate of criminality among XYY males is due to their moderately impaired mental function.

Some comments are in order about the nature of the crimes committed by the XYY males. Two of the XYY males were habitual criminals. One had more than 50 convictions for petty larceny, auto theft, burglary, embezzlement, and so on, but no record of violent or aggressive behavior. The other habitual criminal, who was also mentally retarded and had spent most of his life in institutions for the mentally retarded, had convictions for arson, theft, burglary, and embezzlement, but, once again, no convictions for aggressive crimes against people. The other three XYY males were petty criminals. Two had each been convicted twice for petty theft. The other had turned in a false report about a traffic accident and had started a small fire. Overall, *the crimes of the Danish XYY males were not crimes of violence and aggression; they were mainly crimes involving property.*

The Danish study illustrates the danger of preliminary scientific reports being sensationalized by the popular press. At about the time of the original Scottish discovery of XYY males among exceptionally violent criminals, a pathological killer sneaked into a dormitory in Chicago and brutally murdered several student nurses. A newspaper claimed that the man had been "born to kill" because he was XYY. This turned out to be untrue—the Chicago killer was, in fact, XY—but the false report was widely disseminated and believed. Several proposals were made for the mass screening of newborns for the detection of XYY males to provide them with special education and intensive psychological guidance to counteract their supposed "killer instinct."

It is good that the screening proposals were never carried out because in the atmosphere of the time they could have been harmful by creating a self-fulfilling prophecy. That is to say, if physicians and family are convinced that XYY males (or any other category of people) have severe, aggressive behavioral problems, then they may interact with the individuals in such a manner as to elicit the very behavior they expect, thereby fulfilling their own prophecy. In light of the Danish study, the rate of criminality is related more to level of intellectual function than to sex-chromosomal constitution.

Because XYY males have an extra Y chromosome, one would expect to find a high incidence of XXY and XYY karyotypes among their children. This is not the case, however. XYY males have chromosomally normal children (XX or XY) in the normal proportions. Evidently the extra Y chromosome is eliminated somehow during the formation of sperm. A similar situation is found among the offspring of XXX females, who have XY sons and XX daughters in the normal proportions and no elevated incidence of sex-chromosome abnormalities among their children.

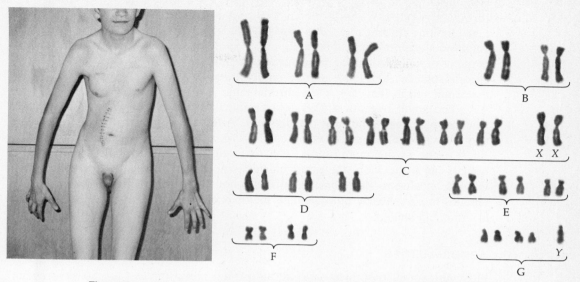

Figure 6.3 An individual with Klinefelter syndrome and the associated XXY karyotype.

XXY: KLINEFELTER SYNDROME

Although the XXX and XYY conditions have few and relatively mild physical manifestations, some sex-chromosomal abnormalities do have distinctive physical effects, chiefly affecting sexual development. Most frequent among these is the XXY condition (technically 47,XXY), which occurs in about 1 per 1000 newborn males. Males with the XXY chromosome constitution have a distinctive group of symptoms known as **Klinefelter syndrome** (Figure 6.3). Characteristic of the syndrome are very small testes but normal penis and scrotum. Although about half the patients have some degree of mental retardation, growth and physical development are, for the most part, quite normal, although affected males tend to be tall for their age (see Table 6.2). Puberty does not occur normally: the testes fail to enlarge, the voice remains rather high-pitched, and pubic and facial hair remains sparse. In about half the cases there is some enlargement of the breasts. Males with Klinefelter syndrome do not produce sperm and hence are sterile.

XO: TURNER SYNDROME

Another abnormality of the sex chromosomes, more rare than the XXY condition, is the XO condition. (The ''O'' used in the shorthand designation XO refers to the absence of the homologue of the X.) Individuals with this sex-chromosome constitution have 45 chromosomes altogether: 22 pairs of autosomes and a monosomic X. Thus, the XO chromosome constitution is technically designated 45,X. The frequency of XO among live-born females is about 1 per 5000, but this far under-represents the occurrence of the condition. As will be discussed later in this chapter, about 15 to 20 percent of spontaneously aborted fetuses that have a detectable chromosome abnormality are XO, making this one of the most common chromosomal abnormalities in spontaneous abortions. More than 1 percent of all recognized pregnancies in humans involves an embryo that is chromosomally XO, and approximately three-fourths of these are caused by abnormal sperm that lack a sex chromosome.

The small fraction of XO fetuses that survives develops into individuals who have **Turner syndrome** (Figure 6.4). Affected females are of short stature and often have a distinctive webbing of the skin between the neck and shoulders. There is no sexual maturation. The external sex organs remain childlike, the breasts fail to develop, the pubic hair does not grow. Internally, the ovaries are infantile or absent, and menstruation does not occur. Although the IQ of affected females is very nearly normal, they have specific defects in spatial abilities and arithmetical skills.

AUTOSOMAL ABNORMALITIES

Most autosomal trisomies—and all autosomal monosomies—are so severe that they are incompatible with life. The result of the chromosomal abnormality may be failure of blastocyst implantation or spontaneous abortion of an often grossly malformed embryo or fetus. Nevertheless, three autosomal trisomies are sometimes compatible with embryonic development and birth, although the trisomic children may be severely abnormal. The incidence of the three conditions among

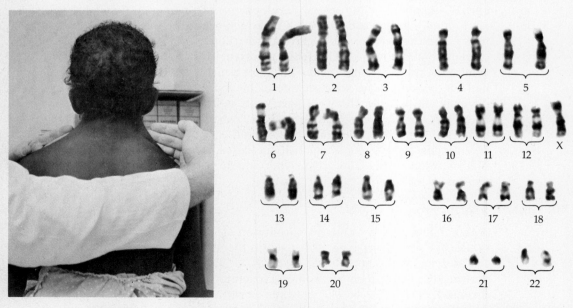

Figure 6.4 An individual with Turner syndrome and the associated XO (also symbolized 45,X) karyotype.

Table 6.3   MOST FREQUENT AUTOSOMAL TRISOMIES IN LIVE-BORN HUMANS

Trisomy	Name of syndrome	Approximate frequency
13 (47, + 13)	Patau syndrome	1 per 5000 live births
18 (47, + 18)	Edwards syndrome	1 per 6500 live births
21 (47, + 21)	Down syndrome	1 per 800 live births

live born children is summarized in Table 6.3. The most common autosomal trisomy among live-born children is **trisomy 21**, also known as **Down syndrome** after the physician who first described the condition over a century ago.

TRISOMY 21: DOWN SYNDROME

The cause of Down syndrome was for many years a complete mystery, but in 1959 it was discovered that affected children have 47 chromosomes (Figure 6.5), including three copies of chromosome 21. As noted in Table 6.3, the technically correct designation of the Down karyotype is 47, +21, the "+" preceding the "21" denoting an entire extra chromosome 21.

Children with Down syndrome are invariably mentally retarded. They are of short stature due to delayed maturation of the skeletal system, and their muscle tone is poor, leading to a characteristic facial appearance in older children and adults (Figure 6.6). Approximately 40 percent of children with Down syndrome have major heart defects. Many children with the syndrome also have epicanthus—a small fold of skin across the inner part of the eye. Altogether there are

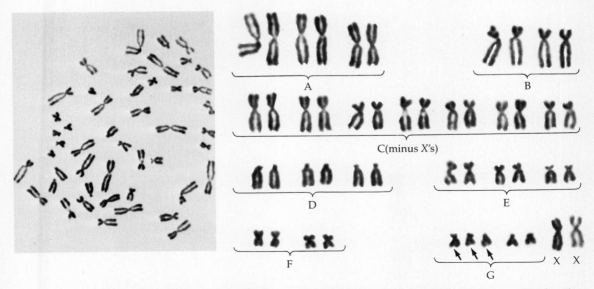

Figure 6.5   Metaphase spread and karyotype of a girl (XX) with Down syndrome showing the trisomy of chromosome 21 (arrows).

(a)

(b)

(c)

Figure 6.6 All the children in these photographs have Down syndrome. As is evident in (a) and (c), the facial features characteristic of Down syndrome are not always apparent in younger children.

10 or so major symptoms of the syndrome, but the condition is somewhat variable.

Down syndrome is very frequent in all racial groups (about 1 in 800 live births). The condition has probably been present in human populations since the ancient past, long before recorded history, perhaps even before our species, *Homo sapiens,* had evolved. There is, for example, a condition in chimpanzees that is very similar in symptoms to Down syndrome. This is also caused by trisomy, and the banding pattern of the chromosome responsible is similar to that of human chromosome 21.

Down syndrome does not usually run in families, and therefore most families that have had a child with Down syndrome have only a small risk of having a subsequent child with the same condition. About 97 percent of families with a trisomic-21 child are in this category. (However, as will be noted shortly, the risk of having a child with Down syndrome increases with the age of the mother.) A minority of families with an affected child has a high recurrence risk. In these cases one of the parents carries a chromosomal rearrangement that leads to a high risk of trisomy 21 in the children even though the parent is phenotypically normal. Such rearrangements will be discussed in Chapter 7. At present, we shall concentrate on the 97 percent; in these families the trisomic-21 children are the result of nondisjunction leading to a gamete with two copies of chromosome 21.

Actually, the abnormal gamete in the case of Down syndrome is usually the egg. Moreover, the risk of a pregnancy producing a child with Down syndrome increases dramatically with the age of the mother. This effect of maternal age is shown graphically in Figure 6.7. The risk increases with the mother's age to a high of about 6 percent(!) in women of age 45 and over. The **recurrence risk** of Down syndrome (i.e., the risk of a subsequent child having Down syndrome when one child is already affected) also increases with the mother's age; generally speaking, the risk of a woman having a second child with Down syndrome is about three times the risk of a woman of the same age who has not previously had an affected child.

Children with Down syndrome, as you might anticipate, have a shorter than normal life expectancy. Many have heart defects among their other multiple

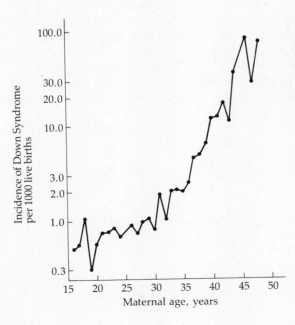

Figure 6.7 Rate of Down syndrome per 1000 live births related to maternal age. The graph is based on 330,859 live births (438 Down-syndrome births) occurring in Sweden from 1968 to 1970. The overall incidence of Down syndrome in this study was 1 per 755 live births.

physical abnormalities, and they are particularly susceptible to chest infections and bronchial pneumonia. In 1932 the average lifespan of children born with Down syndrome was nine years. But the life expectancy of affected individuals has been steadily increasing as antibiotics and mass immunizations have decreased the risk and severity of infections. The current life expectancy is about 50 years.

The increased life expectancy of individuals with Down syndrome has had a social consequence. Today more people with trisomy 21 are living than at any time in the past, and this creates a social cost because affected individuals are often utterly dependent on others for their welfare. If an affected child is institutionalized, the social cost may be paid through taxes. But virtually everyone acknowledges that many public institutions are understaffed by undertrained and underpaid personnel. Therefore, many physicians urge the parents to raise an affected child at home; indeed, there is a national organization of parents of Down syndrome children who do raise their children at home. The medical costs must then be borne by the family.

Unfortunately, there is also a psychological cost. Tragically, many people look upon the birth of an abnormal child as a stigma, a mark of Cain, and they inevitably feel remorse and guilt. The parents themselves must resist these emotions. We should long since have stopped believing with the Puritans that the birth of an abnormal child is retribution for past and hidden sins.

Concern for the welfare of society and the parents must not replace concern for the child. An affected child can benefit from special and intensive care and training, but often this is not delivered. Sometimes the child does not even receive the attention and stimulation required by every human being to remain alert and interested, regardless of ability. Public institutions are often too overburdened to provide the care, and the parents are not themselves trained to be able to provide it. Moreover, the retarded child is usually not capable of frolicking with other, normal children and learning from them. Indeed, small children can be especially cruel and heartless when they ridicule the physical abnormalities or mental deficiencies of others. All this spells out a very sad tale, not only for children with Down syndrome but also for countless others with various kinds of physical and mental abnormalities.

In grappling with the social issues raised by children with Down syndrome, it is all too easy to forget that the children are human beings as much in need of love, security, and affection as others. They have interests and personalities of their own, which have been poignantly described by Down-syndrome specialists Smith and Wilson:

> Children with Down syndrome usually take great pleasure in their surroundings, their families, their toys, their playmates. Happiness comes easily, and throughout life they usually maintain a childlike good humor. They are not burdened with the grown-up cares that come to most people with adolescence and adulthood. . . . Life is simpler and less complex. The emotions that others feel seem to be less intense for them. They are sometimes sad, happy, angry, or irritable, like everyone else, but their moods are generally not so profound and

they blow away more quickly. . . . A child with Down syndrome, though slow, is still very responsive to his environment, to those around him, and to the affection and encouragement he receives from others.*

TRISOMY 18: EDWARDS SYNDROME

Down syndrome is not the only autosomal trisomy compatible with the live birth of affected children, but it is by far the most frequent. Two other autosomal trisomies also occur in live-born children: trisomy 18 (i.e., 47, +18) and trisomy 13 (i.e., 47, +13). (See Table 6.3.)

Trisomy 18, also known as **Edwards syndrome**, is about eight times less frequent than Down syndrome; it affects about 1 in 6500 live-born children. As with trisomy 21, the incidence of trisomy 18 increases with the mother's age, but the increase is relatively small, far less than in Down syndrome. Children with Edwards syndrome have multiple congenital abnormalities, including severe mental and physical retardation (Figure 6.8). They have an elongated skull with low-set, malformed, sometimes pointed ears; their jaw and oral cavity are small. They carry their fingers in an abnormal position, with the second finger overlapping the third. Virtually all these children are born with heart defects. Approximately 65 percent of affected newborns are female, presumably because males with the syndrome are more likely to undergo spontaneous abortion. The life expectancy of affected children also indicates a greater severity in males: the life expectancy of males is about three months; in affected females it is about nine months.

TRISOMY 13: PATAU SYNDROME

Trisomy 13, also known as **Patau syndrome**, is as severe as trisomy 18. It occurs in about 1 in 5000 live births. As in other autosomal trisomies, affected children are severely retarded, both mentally and physically (Figure 6.9). Children with trisomy 13 have a small skull and eyes; the ears are often malformed, and deafness is common. Most have harelip and cleft palate. Many have malformed thumbs and extra digits. Nearly 70 percent have heart defects. Patau syndrome also has a slightly increased incidence with maternal age, but again the increase is far less than the age effect in Down syndrome. The sex ratio of affected children is about 1:1. Rarely do affected children survive more than three or four months after birth.

Most cases of trisomy 18 and trisomy 13 are sporadic. They are unpredictable and do not run in families, but there are a few high-risk families. Thus the situation insofar as occurrence and recurrence are concerned appears to be much like that in Down syndrome. The abnormalities caused by the extra chromosome in all three trisomies are major and multiple, physical and mental. Each trisomy is unique, though somewhat variable. Unlike the abnormalities caused by single-gene mutations, the abnormalities in the trisomies cannot be traced to some single

* Smith, D. W. and A. A. Wilson. 1973. The Child with Down's Syndrome (Mongolism), Saunders, Philadelpha.

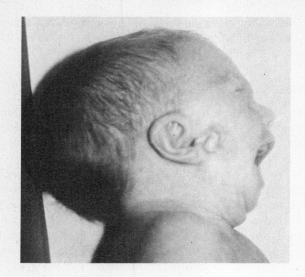

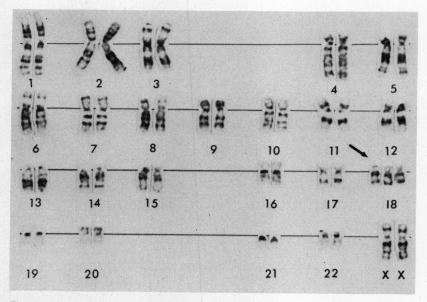

Figure 6.8 A child with Edwards syndrome (trisomy 18) and karyotype.

biochemical defect in the cells. Rather, the trisomy syndromes result from the cumulative action of abnormal gene dosage. Consequently there is at present no hope of effecting a "cure" for the trisomies.

AMNIOCENTESIS

Until the recent past nothing could be done to predict whether children would be born with incurable mental and physical defects. Now, in certain cases, it is

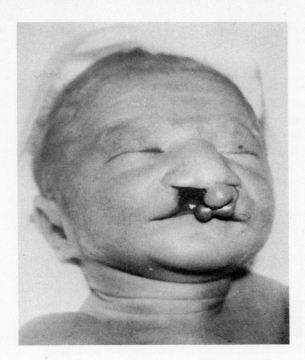

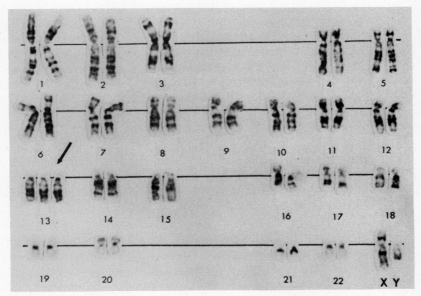

Figure 6.9 A child with Patau syndrome (trisomy 13) and karyotype.

possible to identify birth defects in the fetus while the fetus is still developing in the uterus. The principal method of procuring cells of the fetus suitable for laboratory study is known as **amniocentesis**.

The cells obtained in this procedure are not taken from the fetus itself, but they are genetically identical to those in the fetus. Every fetus is surrounded early in development by membranous sacs—sacs composed of cells that descend from the fertilized egg. The cluster of cells that reaches the uterus from the fallopian tube a few days after fertilization will develop into a fetus, but only certain cells in the cluster actually give rise to the body of the fetus. The rest divide and stick together as sheets of cells and move and fold so as to enshroud the fetus in the membranous sacs.

The innermost membrane, known as the **amnion**, is filled with a clear, watery fluid that cushions the fetus against mechanical shocks and prevents adhesion of parts of the fetus that might accidentally come into contact. Present in the amniotic fluid are cells that slough off from the fetus or from the amniotic membrane. These cells are, of course, genetically identical to those in the fetus.

At about the fourteenth to sixteenth week of pregnancy, the amnion becomes large enough to permit insertion of a hollow needle and removal of a sample of the amniotic fluid by piercing through the abdomen of the mother (Figure 6.10). A small amount of amniotic fluid is withdrawn, and the cells in this fluid are separated from the liquid and grown in cultures in the laboratory. The chromosomes and various biochemical or enzymatic capabilities of the cells can be examined, and in this way certain hereditary defects in the fetus can be diagnosed. If the cells are trisomic for chromosome 21, for example, this is indicative of a fetus with Down syndrome. All the known chromosomal defects can be prenatally diagnosed in this way. In addition, more than 100 other inherited disorders due to single mutant genes can be detected even though no visible chromosome abnormality is involved. For example, Tay-Sachs disease can be detected because the cultured amniotic cells lack the corresponding enzyme (see Chapter 4).

At major medical centers amniocentesis has become routine in high-risk pregnancies. These are pregnancies in which the fetus is known to have an especially high risk of having some detectable abnormality. In the case of Down syndrome, for example, the pregnancies at greatest risk are those in normal women over 35 or in the small proportion of families in which a chromosome rearrangement in one of the parents predisposes the fetus to high risk. In cases of recessively inherited conditions, the high-risk families are those in which both parents are carriers of recessive genes (the carriers of about 100 such genes can now be identified by appropriate studies of blood or other body fluids or cultured cells).

When amniocentesis is carried out and the embryo is found to be normal, then there is no problem and the parent can be put at ease with regard to the trait examined. When the fetus is affected, then an abortion may be considered. The decision is left to the parents, but in the present social climate many families choose abortion over the birth of a severely malformed and mentally retarded child.

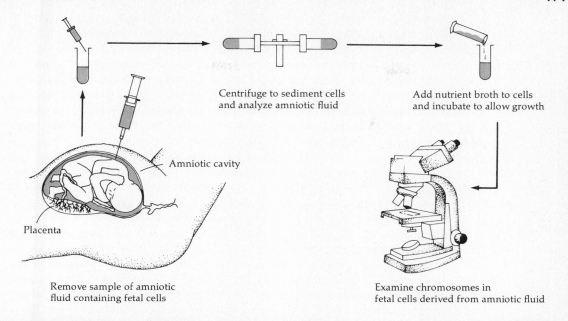

Centrifuge to sediment cells
and analyze amniotic fluid

Add nutrient broth to cells
and incubate to allow growth

Amniotic cavity

Placenta

Remove sample of amniotic
fluid containing fetal cells

Examine chromosomes in
fetal cells derived from amniotic fluid

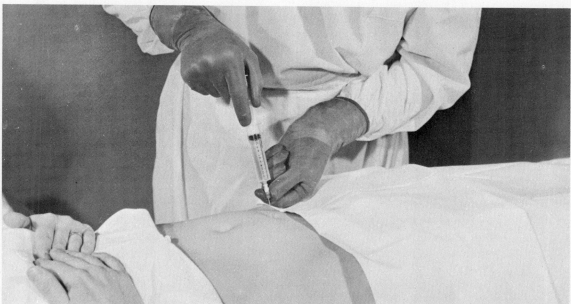

Figure 6.10 Outline of the procedure of amniocentesis in which cells from an unborn fetus are obtained and examined in the laboratory for signs of genetic disease or chromosome abnormality. The photo shows amniotic fluid being obtained from a pregnant woman.

The decision whether or not to abort is usually much more agonizing when the fetus is affected with a less severe abnormality. At the present time there are relatively few legal obstacles to abortion in the United States. Many thoughtful people have argued that this is the way it ought to be—that a woman should decide for herself, on whatever grounds she considers adequate, whether to nurture or to abort the fetus; society, they argue, has no right to force a woman to carry a fetus she does not want. Many equally thoughtful people feel differently. They argue that the legality of an activity does not make it morally right; they point out that slavery was legal for centuries, although we now perceive it to have been morally wrong. If abortion is wrong, the argument goes, then it is wrong for everyone, and the legality of it is a symptom of society's disrespect for life and not an absolution of the act. It would be inappropriate to discuss in further detail here the many sincerely held points of view concerning abortion. It is no simple matter, and great wisdom is called for.

An equally difficult matter—perhaps an even more explosive one—is the care of children born alive with severe, incurable, or lethal abnormalities. To what extent should extraordinary medical measures, often called "heroic" measures, be used to sustain and prolong the lives of such children? They cannot intentionally be killed, of course; infanticide is proscribed by the law, customs, and common morality of our society (although infanticide is accepted and practiced in some other societies). Plain decency requires these children to be fed and fondled and kept warm. But should their tortured lives be unnaturally prolonged? How long should a hopelessly deformed and doomed child be kept alive by medical gadgetry? And to what extent should heroic measures be used for children who are not so severely deformed? Should a child with a seriously defective heart due to Down syndrome receive heart surgery, for example? Some surgeons say no. They point out that a child otherwise normal would be completely well following successful surgery. But in cases such as Down syndrome, the patient will not be cured by surgery; the patient's life may be prolonged, but the major mental abnormalities will remain. And they argue that the well-being of the entire child and not just the heart should be of prime concern.

The other side of this question is perhaps best summarized by an admonition from the New Testament (Romans) that we who are strong ought to bear the infirmities of those who are weak and not live to please ourselves. The terrible dilemma applies not only to children, of course, but with equal force to the aged and to those hopelessly maimed by stroke, cancer, heart attack, or accident. The issues will be painful to resolve, and yet they should not be ignored. Almost everyone will have to decide, if not for a child or an aged parent or an injured spouse, then at least for himself or herself should the worst come to pass. So it is perhaps worth bearing in mind the ancient Chinese proverb that if you do not think about what is distant, you will discover sorrow near at hand.

DEATH BEFORE BIRTH

The fact is that most fetuses with major abnormalities of the autosomes do not survive to be born at all. Their development is so grossly abnormal that the pregnancy terminates early in a spontaneous abortion. The actual frequency of spontaneous abortions from all causes is a number virtually impossible to obtain. Since pregnancies are not usually diagnosed until they are two months along, abortions prior to this time are usually not detected. Very early abortions that result from failure of implantation or expulsion of the embryo within the first few weeks are often not recognized as abortions because the mother may be unaware of her pregnancy when the abortion occurs, and the abortion itself may cause her only minor discomfort. She may experience a particularly heavy menstrual discharge, for example, or perhaps she will miss a menstrual period, followed by a period with heavy flow—events that cause no particular wonder or alarm in most women because they are not uncommon occurrences. But some unknown fraction of these are actually very early abortions.

Of recognized pregnancies—those pregnancies that last long enough to be diagnosed—about 15 percent end in spontaneous abortion (expulsion of the fetus prior to the end of the third month of pregnancy) or miscarriage (expulsion of the fetus between the third and seventh month of pregnancy). The great majority of these abortions occurs before the fifth or sixth months. There is general agreement that at least as many abortions occur prior to recognition of the pregnancy as occur afterward, perhaps many more. Some studies have suggested that up to *half* of all fertilizations end in spontaneous abortions. Whatever the correct figure, spontaneous abortion in humans is an extremely common occurrence. Even the 15 percent of recognized pregnancies—nearly one out of six—represents a staggering proportion.

Table 6.4 is a composite of the results of many chromosome studies showing the chromosome constitutions that would be expected to occur among 100,000 recognized pregnancies. Results vary somewhat from study to study, so the numbers in the table should be regarded as approximate. Nevertheless, they are in the range actually found. As noted earlier, out of 100,000 recognized pregnancies, about 15,000 will undergo spontaneous abortion. These abortions represent a natural way of eliminating grossly malformed fetuses. Many of these fetuses carry genetic mutations that are incompatible with life; many others may have sustained environmentally induced damage or accidents during development. However, as the composite data in Table 6.4 indicate, approximately 50 percent of spontaneously aborted fetuses have major *chromosomal* abnormalities.

The most frequent class of abnormalities in trisomic abortuses is trisomy 16. Trisomics of D- and G-group chromosomes are also quite common; roughly another third of autosomal trisomic abortuses have trisomy of a D- or G-group chromosome. Included in this category are many fetuses with trisomy 21, the ones that do not survive until birth. About three-fourths of all fetuses with trisomy 21 are eliminated by spontaneous abortion and only one-fourth are born alive. To appreciate the significance of this finding, note that if all fetuses with trisomy 21 were born alive, about 1 in 200 live-born children would have Down syndrome! The remaining third of trisomic abortuses involve autosomes other than chromosome 16 or the D- or G-group chromosomes.

Two other types of chromosomal abnormalities account for a large fraction of chromosomally abnormal fetuses. One of the most common is the XO (i.e.,

Table 6.4 NUMBER AND TYPE OF CHROMOSOMAL ABNORMALITIES AMONG SPONTANEOUS ABORTUSES AND LIVE BIRTHS IN 100,000 HYPOTHETICAL RECOGNIZED PREGNANCIES*

		100,000 Recognized Pregnancies	
		15,000 spontaneous abortions 7,500 chromosomally abnormal	85,000 live births 550 chromosomally abnormal
Trisomy			
A:	1	0	0
	2	159	0
	3	53	0
B:	4	95	0
	5	0	0
C:	6–12	561	0
D:	13	128	17
	14	275	0
	15	318	0
E:	16	1229	0
	17	10	0
	18	223	13
F:	19–20	52	0
G:	21	350	113
	22	424	0
Sex chromosomes			
	XYY	4	46
	XXY	4	44
	XO	1350	8
	XXX	21	44
Translocations			
	Balanced	14	164
	Unbalanced	225	52
Polyploid			
	Triploid	1275	0
	Tetraploid	450	0
Other (mosaics, etc.)		280	49
Total		7500	550

* The numbers are approximate and vary somewhat from study to study. They are within the range found in most studies, though, and are close to the average values.

Source: Based on data in D. H. Carr and M. Gedeon, 1977, Population cytogenetics of human abortuses, in E. B. Hook and I. H. Porter (eds.), Population Cytogenetics: Studies in Humans, Academic Press, New York, pp. 1–9; and E. B. Hook and J. L. Hamerton, 1977; the frequency of chromosome abnormalities detected in consecutive newborn studies—differences between studies—results by sex and by severity of phenotypic involvement, in E. B. Hook and I. H. Porter (eds.), Population Cytogenetics: Studies in Humans, Academic Press, New York, pp. 63–79.

45,X) sex-chromosome constitution. Fetuses that are XO acccount for about 18 percent of chromosomally abnormal abortuses, but a small fraction of XO's—about 0.5 percent—do survive and have Turner syndrome. The high frequency of XO among abortuses is astonishing. Indeed, if all the XO fetuses were born alive, the incidence of the condition among live-born females would be about 3 percent!

Triploidy [69 chromosomes—see Figure 6.11(*a*)] and tetraploidy [92 chromosomes—see Figure 6.11(*b*)] are also relatively frequent among spontaneous abortions. Triploids alone account for about 17 percent of chromosomally abnormal fetuses. Triploid fetuses result from an abnormal meiosis in which one of the divisions fails to progress normally, so that two full sets of chromosomes become included in the same gamete; but they may also result from errors in fertilization, such as two sperm fertilizing the same egg or failure to eliminate one polar body.

To summarize the information in Table 6.4, about 15 percent of all recognized pregnancies terminate in spontaneous abortion, and about half of these abortions involve fetuses with major chromosomal abnormalities. Among the fetuses with chromosomal abnormalities, approximately 50 percent have autosomal trisomy, 18 percent are XO, and 23 percent are triploids or tetraploids.

The remaining aborted fetuses have multiple or more complex types of chromosomal abnormality. Included in the last category are fetuses that have **translocations**—chromosomes that have undergone an interchange of parts, a type of chromosomal abnormality that will be discussed in the next chapter. The last category of chromosomally abnormal fetuses also includes **chromosomal mosaics**—fetuses with chromosomal abnormalities in some tissues but not in others. Chromosomal mosaics are almost always the result of mitotic nondisjunction (see Figure 6.2) in the early mitotic divisions of the fertilized egg.

There is reason to believe that Table 6.4 significantly underestimates the true incidence of major chromosomal abnormalities among human zygotes. Chromosome studies of spontaneously aborted fetuses miss those instances in which a blastocyst fails to implant in the uterine wall or in which the embryo aborts so early that the pregnancy is unrecognized. For example, one category of chromosomal abnormality that is conspicuously absent among spontaneously aborted

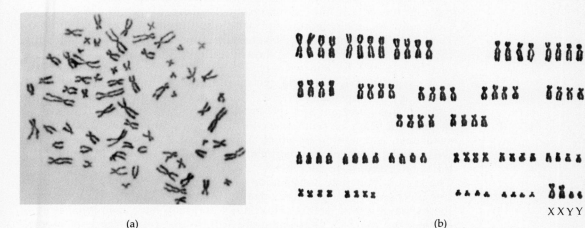

(a) (b)

XXYY

Figure 6.11 (*a*) Metaphase spread of a triploid human cell (69 chromosome). (*b*) Karyotype of a tetraploid human cell (92 chromosomes).

fetuses is **monosomy**—the absence of a single chromosome. XO fetuses are monosomic for the X, of course, and they are an exception because they are so common. But only very rarely is an autosomal monosomic found. The finding of fewer monosomics than trisomics is quite unexpected because monosomics arise from the loss of a chromosome during gamete formation, whereas trisomics result from the gain of a chromosome, and in virtually all organisms chromosome loss is more frequent than chromosome gain. The likely explanation is that many human zygotes are autosomal monosomics, but these are probably aborted so early that they are hardly ever found among abortions of recognized pregnancies.

SUMMARY

1. Abnormalities in chromosome number can involve entire sets of chromosomes. A cell such as an egg or sperm that has a single set of chromosomes is said to be **haploid**, a cell such as a normal somatic cell that carries two sets of chromosomes is called **diploid**, a cell with three sets of chromosomes is **triploid**, and one with four sets is **tetraploid**. Abnormalites in chromosome number can also involve individual chromosomes. An otherwise diploid cell that is missing one chromosome is said to be **monosomic** for the chromosome in question, and an otherwise diploid cell that has one extra chromosome is called **trisomic** for the chromosome in question.
2. Monosomic or trisomic zygotes are usually the result of nondisjunction during meiosis. **Nondisjunction** refers to an abnormal behavior of chromosomes during cell division that leads to the daughter cells having an extra copy (or a missing copy) of the chromosome instead of the normal number. Nondisjunction can occur during mitosis as well as meiosis. The result of nondisjunction in mitosis is a **chromosomal mosaic**—an individual who has two or more chromosomally distinct types of cells.
3. Generally speaking, monosomy or trisomy involving the sex chromosomes is phenotypically less severe than monosomy or trisomy involving the autosomes. In all cases, however, monosomy has more extreme phenotypic effects than trisomy. For the sex chromosomes, YO (i.e., 45,Y) zygotes invariably undergo spontaneous abortion so early that the pregnancy is unrecognized; XO (i.e., 45,X) zygotes usually undergo spontaneous abortion; and the small surviving fraction has **Turner syndrome**. The XXY chromosome constitution is associated with **Klinefelter syndrome**, a condition in which affected males are taller than normal, have moderate mental impairment, fail to undergo puberty, and sometimes have breast development.
4. The sex-chromosome constitution XXX (i.e., trisomy X) has no distinctive physical symptoms. Individuals who have trisomy X have a higher than normal risk of mental retardation and may have menstrual irregularities, but many XXX women are within the normal range of mental abilities and are fertile.
5. This sex-chromosome constitution XYY also has no distinctive physical symptoms, although XYY males tend to be taller than normal males.

Preliminary findings regarding XYY suggested a high rate of criminality, particularly involving crimes of violence against people. The concern over the criminality of XYY's was lessened somewhat by a later Danish study which found that while XYY males do have higher rates of criminal conviction than their XY counterparts, the higher rate of criminality generally involves petty crimes against property rather than violent crimes against people. Moreover, the high rates of criminal conviction may simply be due to the moderately impaired intellectual function of the XYY males. Both XXY and XXX individuals have normal (XX or XY) offspring rather than the high frequency of sex-chromosomal abnormalities that would be expected in theory.

6. Among the autosomes, all monosomies are incompatible with life and lead to failure of blastocyst implantation or to abortion prior to the diagnosis of pregnancy. Autosomal trisomies usually lead to abortion somewhat later during development, but three are compatible with live birth. **Trisomy 21** (also called **Down syndrome**) is the most frequent autosomal trisomy in live-born infants, affecting about 1 in 800 newborns. The incidence of Down syndrome increases dramatically with the mother's age to a maximum of about 6 percent in mothers aged 45 or older. In addition to other characteristic physical symptoms, children with Down syndrome are mentally retarded and many have heart defects. Nevertheless, many individuals with Down syndrome survive to adulthood.

7. **Trisomy 13 (Patau syndrome)** and **Trisomy 18 (Edwards syndrome)** are much more rare and more phenotypically severe than Down syndrome. Newborns with either of these syndromes usually die within the first few months of life, and virtually none survives for more than a year.

8. In the procedure of **amniocentesis**, cells of a developing fetus are obtained by piercing the amnion with a hollow needle inserted through the abdominal wall of a pregnant woman and drawing off a small amount of fetal-cell-containing amniotic fluid. The cells can then be cultured and studied in the laboratory. All major chromosomal abnormalities can be diagnosed before birth by means of amniocentesis, and over 100 genetic disorders due to single-gene defects can be detected in fetal cells. Ordinarily, however, amniocentesis is used only in high-risk pregnancies.

9. Chromosome studies of spontaneously aborted fetuses have revealed an astonishingly high frequency of major chromosomal abnormalities. On the average, among 100,000 hypothetical recognized pregnancies, about 15,000 will terminate in spontaneous abortion. Among these 15,000, about half will have major chromosomal abnormalities, including autosomal trisomy (50 percent of all chromosomally abnormal fetuses), X-chromosomal monosomy (18 percent), and polyploidy (23 percent triploids or tetraploids). Trisomy 16 is the single most frequent autosomal trisomy among spontaneously aborted fetuses, but D- and G-group trisomies are also relatively frequent. Indeed, about three-fourths of all trisomic-21 fetuses undergo spontaneous abortion. Because autosomal monosomics probably undergo spontaneous abortion prior to the recognition of pregnancy, chromosome studies of abortuses probably underestimate the true frequency of major chromosome abnormalities in human reproduction.

KEY WORDS

Amniocentesis	Lethal	Triploid
Diploid	Monosomic	Trisomic
Down syndrome	Nondisjunction	Turner syndrome
Edwards syndrome	Patau syndrome	47,XXX
Haploid	Recurrence risk	47,XYY
Klinefelter syndrome	Tetraploid	

PROBLEMS

6.1. What is the difference between *risk* and *recurrence risk*?

6.2. Why do many physicians routinely advise amniocentesis for pregnant women over 40?

6.3. Of the abnormal karyotypes discussed in this chapter, which two have the mildest phenotypic effects?

6.4. In what sense is Down syndrome an "inherited" disorder? In what sense is it not?

6.5. Amniocentesis always reveals the sex of a fetus (even though many parents request that the information be withheld from them). How is the sex revealed?

6.6. Two normal parents are both known to be carriers of the Tay-Sachs allele. The woman becomes pregnant and has amniocentesis performed. How can the physician tell where the fetus is affected, and what is the chance that a particular fetus will be affected?

6.7. How many Barr bodies would be found in individuals with the following karyotypes:

> (*a*) 45,X
>
> (*b*) 47,XXX
>
> (*c*) 47,XXY
>
> (*d*) 47,XYY

6.8. What type of nondisjunction is the most likely cause of XYY males?

6.9. If a 45,X female were fertile (all such individuals are sterile), what types of zygotes would be expected from a mating with a normal male, and in what frequencies? Which of the zygotes would undergo spontaneous abortion?

6.10. Using the data in Table 6.4, calculate the overall frequency of chromosome abnormalities among recognized pregnancies.

6.11. Using the data in Table 6.4, calculate the frequency of 45,X among live births. What would the frequency among live births be if none of the 45,X zygotes underwent spontaneous abortion?

6.12. A woman whose father was color blind marries a normal man and they have a son with Klinefelter syndrome who is also color blind. How can this happen?

FURTHER READING

Alfi, O. S., R. Chang, and S. P. Azen. 1980. Evidence for genetic control of nondisjunction in man. Am. J. Hum. Genet. 32: 477–83. Certain genes, when homozygous, seem to increase the rate of nondisjunction.

de Grouchy, J., C. Turleau, and C. Finaz. 1978. Chromosomal phylogeny of the primates. Ann. Rev. Genet. 12: 289–328. How human chromosomes came to be the way they are.

Edwards, R. G. 1983. Chromosome abnormalities in human embryos. Nature 303: 283. Discusses available information on chromosome abnormalities in very early embryos, from both *in vivo* and *in vitro* fertilization.

Emery, A. E. H. 1979. Elements of Medical Genetics. 5th ed. Churchill Livingstone, Edinburgh. A short but fine introduction to the medical aspects of human genetics.

Hamerton, J. L. 1971. Clinical Cytogenetics. Academic Press, New York. Advanced reference with detailed descriptions of trisomy syndromes.

Hook, E. B., and A. Lindsjo. 1978. Down syndrome in live births by single year maternal age interval in a Swedish study. Am. J. Hum. Genet. 30: 19–27. Maternal age effect in Down syndrome.

Hsu, T. C. 1979. Human and Mammalian Cytogenetics: An Historical Perspective. Springer-Verlag, New York. An entertaining book on human cytogenetics glittering with personal anecdotes of one who was there at the beginning.

Norwood, C. 1980. At Highest Risk: Environmental Hazards to Young and Unborn Children. McGraw-Hill, New York. Focuses on the dangers of drugs such as amphetamines, alcohol, and many others.

Smith, D. W., and A. A. Wilson. 1973. The Child with Down's Syndrome. W. B. Saunders, Philadelphia. A sympathetic account for parents, physicians, and other concerned persons.

Witkin, H. A., *et al.* 1976. Criminality in XYY and XXY men. Science 193: 547–55. An impeccably designed study.

chapter *7*

Abnormalities in Chromosome Structure

- Chromosome Breakage
- Duplications and Deficiencies
- Inversions
- Reciprocal Translocations
- Robertsonian Translocations
- Chromosomal Changes in Evolution
- Summary

In addition to abnormalities in chromosome number such as the examples discussed in the preceding chapter, there is a great variety of structural abnormalities in chromosomes. **Structural abnormalities** refer to chromosomes having an abnormal structure; examples include chromosomes that have a portion missing or a portion represented twice. They range in size from those that are so small as to be almost undetectable in the light microscope (and probably a great many are overlooked because they lie below the limit of this form of detection) to those that are so large as to be striking and obvious. They range in seriousness from those that have a few or no phenotypic effects to those that are almost as severe as monosomy or trisomy. Between these extremes are structural abnormalities that have no detectable effects on carriers but expose their children to great risk of severe abnormality.

Among abnormal types of chromosomes that have no known deleterious effects are the variants in the normal chromosome complement mentioned in Chapter 2—variants that are found in people with normal family histories, such as chromosomes with unusually prominent satellites, or variants in the length of the long arm of the Y chromosome. The word *abnormal* in reference to these variants should be taken to imply "atypical" rather than "harmful"; the important thing to remember about them is that the carriers and their offspring are phenotypically normal.

CHROMOSOME BREAKAGE

This chapter is about chromosome abnormalities that may have severe phenotypic effects in the carriers themselves or in some proportion of their offspring. These abnormal chromosomes originate from **chromosome breakage**. Chromosomes are fragile, and they sometimes break spontaneously. The incidence of chromosome breakage is greatly increased by exposure of the chromosomes to a wide variety of agents, including x-rays and certain chemicals (see Chapter 10). The effect of such agents is especially acute in cells that are actively undergoing division—embryonic cells, for example. Broken chromosomes also tend to "heal"—to undergo **restitution**. The broken ends behave as if they were "sticky" (they come together and fuse), and the restitution probably involves certain enzymes that aid in the repair process. Most of the time broken chromosomes restitute correctly, and the broken ends rejoin at the point of fracture. But sometimes they are not restituted correctly and chromosomal abnormalities result.

Fragments of chromosomes that have no centromere, which are called **acentric** chromosomes, are lost from daughter cells because they cannot be maneuvered to the poles during cell division. Generally, too, chromosomes that end up having two centromeres, which are called **dicentric** chromosomes, are lost rather quickly because there is a continual risk during cell division that the two centromeres will be pulled to opposite poles. In such a case the chromosome itself will form a bridge between the centromeres; the upshot is that either the chromosome forming the bridge will rupture somewhere in the middle and both poles will receive an abnormal, broken chromosome, or the chromosome forming the bridge will not rupture, in which case neither centromere may travel all the way to a pole where it can be included in a telophase nucleus and the entire dicentric chromosome will be lost. In short, *chromosomal abnormalities will usually persist only if they have exactly one centromere.*

Four kinds of structural abnormalities of chromosomes are of greatest importance in human genetics. Some cells have **deficiencies** (also called **deletions**), formed when a segment of a chromosome is broken off and lost. Some cells have **duplications**, in which a segment of a chromosome is present in more than the normal number of copies. The two other important abnormalities involve no loss or gain of chromosomal material; they involve instead the rearrangement of parts of chromosomes. In **inversions**, for example, a chromosome is broken in two places and the middle segment is flipped end for end before restitution. In **translocations**, a section of one chromosome is attached to another, nonhomologous chromosome. Inversions and translocations do not usually cause phenotypic abnormalities in the carriers, but they may, especially in the case of translocations, expose the offspring of carriers to substantial risk. In this chapter, each of these four types of structural abnormality will be briefly discussed.

In considering chromosome rearrangements, it is important to note whether the rearranged chromosomes still carry all of the genes they should in their proper dosage, although, of course, in an incorrect arrangement, or whether the rearranged chromosomes have lost or gained genes. A chromosome rearrangement that retains all the genes in proper dosage is said to be a **balanced** rearrangement. One that has lost or gained genes is said to be **unbalanced**.

DUPLICATIONS AND DEFICIENCIES

Figure 7.1 shows the origin of a deletion-bearing chromosome and a corresponding duplication-bearing chromosome. Figure 7.1(*a*) and (*b*) illustrates homologous chromosomes that have been broken in three places followed by repositioning of

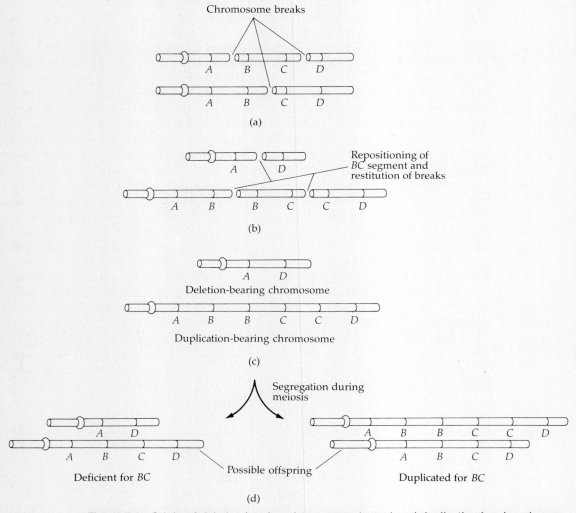

Figure 7.1   Origin of deletion-bearing chromosome (*c*, top) and duplication-bearing chromosome (*c*, bottom) from chromosome breaks indicated in (*a*) following restitution as in (*b*). Possible offspring of the individual in (*c*) are shown in (*d*).

the *BC* segment from one chromosome into the gap in the other. Restitution of the broken ends produces the chromosomes shown in Figure 7.1(*c*); one chromosome carries a deletion of the *BC* region, the other a duplication of the *BC* region. Taken together, the two chromosomes in Figure 7.1(*c*) represent a balanced chromosome rearrangement because all the genes are present in their proper dosage. However, during meiosis, the homologous chromosomes will segregate; half the resulting gametes will carry the deletion-bearing chromosome and half will carry the duplication-bearing chromosome. All of the resulting offspring will thus have an unbalanced chromosome rearrangement. That is to say, half of the offspring from the individual in Figure 7.1(*c*) will have a deficiency of *BC* [see Figure 7.1(*d*)] and half will have the corresponding duplication of *BC*.

The phenotypic effects of duplications and deficiencies depend on their size. Many children have been discovered who have partial deficiencies of any one of a number of chromosomes, including a deletion of part of the short arm of chromosomes 4, 5, or 18 or of part of the long arm of chromosomes 13, 18, or 21. Individuals with any of these deletions suffer from various physical abnormalities, and all have varying degrees of mental retardation.

The best-known human deletion is probably that of part of the short arm of chromosome 5 (Figure 7.2). Affected people have 46 chromosomes, but part of the short arm of one of the number 5 chromosomes is missing. The abnormalities caused by the deletion are comparatively minor, except for mental retardation. A distinctive feature of the syndrome is the peculiar meowing and catlike sound of the infant's cry, which results from a malformation of the larynx. The name of the syndrome—the **cat-cry** (or **cri-du-chat**) syndrome—derives from this. The majority of affected children are female, presumably because of increased fetal mortality of affected males. All affected children are physically and mentally retarded, and many have heart defects, a small skull with abnormally shaped ears, squinting or crossed eyes, and various other abnormalities. The condition is very rare.

The syndromes caused by deletion of part of the short arm or part of the long arm of chromosome 18 are also associated with mental and physical retar-

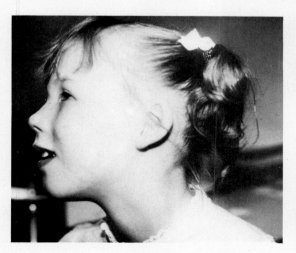

Figure 7.2 A child with the cat-cry syndrome (also called *cri-du-chat*) and a micrograph of the four B chromosomes. One chromosome has a deletion of part of the short arm (arrow), and banding studies reveal that the deletion is in chromosome 5.

dation. The physical abnormalities tend to be variable, however. Children with partial deletions of the short arm of chromosome 18 sometimes have ear and jaw malformations [Figure 7.3(*a*)]; those with partial deletions of the long arm of 18 frequently have rather severe eye and ear defects [Figure 7.3(*b*)]. Taking into account the size of the deleted segments in these and other deletion syndromes, the maximum length of chromosome that can be deleted and still be compatible with survival seems to be about 2 or 3 percent of the haploid autosomal complement. (The haploid autosomal complement refers to the sum of the lengths of all the autosomes in a normal gamete.)

Turning to duplications, and speaking in general terms, we find that small duplications tend to be less severe than deletions of similar sizes. For example, children who carry a duplication of the cat-cry segment of chromosome 5 are mentally retarded, but their physical abnormalities are very mild, some being completely normal in physical appearance. Also, the trisomies of chromosomes 13, 18, and 21 are essentially duplicated for these entire chromosomes; and, although trisomic children can survive, at least in some cases, the corresponding monosomics are invariably aborted spontaneously. Moreover, certain exceedingly rare individuals have **double trisomy**—simultaneous trisomy of chromosomes 18 and 21, for example. Since chromosomes 13, 18, and 21 constitute roughly 4, 3, and 2 percent of the length of a haploid set of autosomes, a duplication of 5 or 6 percent of the haploid autosomal complement can evidently be compatible with life, at least in a few instances.

INVERSIONS

Figure 7.4 depicts the events involved in the formation of an inversion. A chromosome that is broken in two places (*a*) undergoes a reversal of the broken

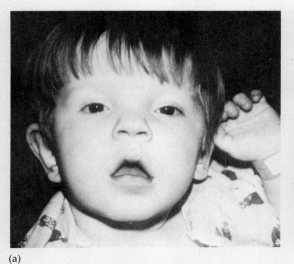

(a)   (b)

Figure 7.3 (*a*) Child with a deletion of part of the short arm of 18. (*b*) Child with a deletion of part of the long arm of 18.

Chromosome breaks

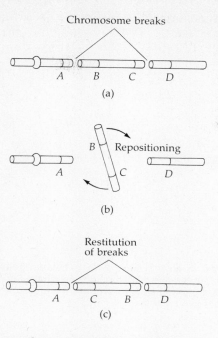

(a)

(b)

(c)

Inverted chromosome

(d)

Figure 7.4 Origin of a chromosome with a paracentric inversion (*d*) following the events of breakage and reunion in (*a*) through (*c*).

segment (*b*), followed by restitution of the broken ends (*c*), resulting in an inverted chromosome (*d*). The normal order of genes on this chromosome is *ABCD,* but the inverted chromosome has the order *ACBD.* The inversion in Figure 7.4(*d*) is a special type of inversion called a **paracentric inversion,** so called because the inverted segment does *not* include the centromere. Note that no genes are gained or lost because of the inversion; some genes merely occur in reverse of the normal order. Because inversions in human chromosomes seem to be relatively rare, the great majority of individuals who carry an inversion will be heterozygous for it; that is, the homologue of the inverted chromosome will have the normal gene sequence.

Heterozygous inversions can cause problems during meiosis. When synapsis occurs during prophase I, all regions of the inverted chromosome try to pair gene for gene with the corresponding regions of the normal chromosome. To accomplish this, the inverted region in one of the chromosomes must form a **loop,** as illustrated in Figure 7.5(*a*); this loop will permit gene-for-gene pairing with the homologue. Thus, except for a relatively small region around the breakpoints themselves, the normal and the inverted chromosomes can synapse all along their lengths.

Such inversion loops can readily be observed in fruit flies and other dipteran flies, which have a specialized type of giant chromosome in their salivary glands.

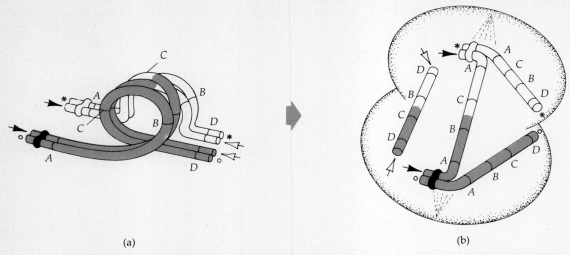

(a)   (b)

Figure 7.5 (*a*) Synapsis in an individual who is heterozygous for a paracentric inversion showing a crossover within the inversion loop. (*b*) Anaphase I configuration resulting from the crossover in (*a*). One of the chromatids involved in the crossover is a dicentric (solid arrows); the other is an acentric (open arrows).

Figure 7.6 shows an inversion loop (arrow) in salivary-gland chromosomes of the black fly *Eusimilium aureum*.

The consequence of a heterozygous inversion depends on whether or not a crossover occurs within the inversion loop. In the event that no crossover occurs in the inversion loop, the subsequent stages of meiosis occur without mishap; two of the resulting gametes receive the inverted chromosome, and the other two gametes receive its noninverted homologue.

When a crossover does occur within the inversion loop, trouble arises, as indicated in Figure 7.5. Part (*a*) shows the inversion loop with a crossover at the top of the loop, and part (*b*) shows the result of this configuration at anaphase I. The consequences of the crossover in Figure 7.5(*a*) can be deduced by tracing each chromatid in the configuration. For example, the chromatid whose tips are

Figure 7.6 Inversion loop (arrow) in giant salivary-gland chromosomes of the dipteran black fly, *Eusimilium aureum*.

marked with a small circle ends up with the normal gene sequence that it started out with because this chromatid was not involved in the crossover. Similarly, the chromatid whose tips are marked with an asterisk ends up with the inverted gene sequence that it started out with because this chromatid was also not involved in the crossover.

The tips of the crossover chromatids are marked with solid or open arrowheads. The one marked with solid arrowheads becomes a dicentric (two-centromere) chromosome because of the crossover. Conversely, the chromatid marked with the open arrowheads becomes an acentric (no-centromere) chromosome. Moreover, the dicentric chromatid is duplicated for A and deficient for D, whereas the acentric chromatid is deficient for A and duplicated for D.

Of the four possible gametes formed from the anaphase I cell in Figure 7.5(*b*), one will carry the normal chromosomes, one will carry the inverted chromosome, and two will have major chromosomal abnormalities as a result of the crossover. Indeed, the dicentric and the acentric often fail to be included in any telophase II nucleus, so the corresponding gametes completely lack the chromosome in question. In any event, *the chromatids involved in a crossover within the inversion loop of a heterozygous paracentric inversion become dicentric or acentric chromosomes.* Consequently, if an individual is heterozygous for a large inversion, so that crossing-over within the inversion loop will frequently occur, then there is substantial risk of major chromosomal abnormality among the offspring. In addition, among the phenotypically normal offspring (whether the inversion is large or small, and whether a crossover occurs or not), half will inherit the inverted chromosome and the other half will inherit its noninverted homologue.

The second principal type of inversion, which is called a **pericentric inversion** because it *does* include the centromere, is illustrated in Figure 7.7. When the breakpoints of a pericentric inversion are at appropriate distances from the centromere, the inversion can alter the physical appearance of the chromosome. The chromosome in Figure 7.7(*a*) is a metacentric chromosome, for example, but its inverted derivative in part (*b*) is a submetacentric. Some pericentric

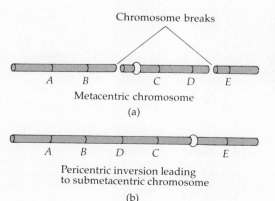

Chromosome breaks

A B C D E

Metacentric chromosome

(a)

A B D C E

Pericentric inversion leading
to submetacentric chromosome

(b)

Figure 7.7 Origin of a chromosome with a pericentric inversion (*b*) following chromosome breakage illustrated in (*a*). Note that the original chromosome is a metacentric but the inversion-bearing chromosome is a submetacentric.

inversions can therefore be detected microscopically because they change the relative position of the centromere.

As with heterozygous paracentric inversions, synapsis in meiosis produces an inversion loop [Figure 7.8(*a*)], and crossing-over within this loop results in abnormal chromatids [Figure 7.8(*b*)]. In the anaphase I configuration in Figure 7.8(*b*), it can be seen that one chromatid (its tips marked by small circles) is structurally normal and another (tips marked by asterisks) has the pericentric inversion; these chromatids are the ones not involved in the crossover. Of the two chromatids that were involved in the crossover, one (solid arrows) has a duplication of *A* and a deficiency of *D*, and the other (open arrows) has a deficiency of *A* and a duplication of *D*; both chromatids are **monocentric** (have a single centromere), in contrast to the situation with paracentric inversions outlined in Figure 7.5. In short, *the chromatids involved in a crossover within the inversion loop of a heterozygous pericentric inversion will carry duplications and deficiencies* and will lead to offspring who have major chromosomal abnormalities. Among the phenotypically normal offspring of an individual who is heterozygous for a pericentric inversion, half will receive the inverted chromosome and the other half will receive its noninverted homologue.

RECIPROCAL TRANSLOCATIONS

Reciprocal translocations involve an interchange of parts between nonhomologous chromosomes. The origin of a reciprocal translocation is illustrated in Figure 7.9. Notice that no chromosomal material is gained or lost because of the translocation. Notice also that each chromosome in the translocation must be monocentric.

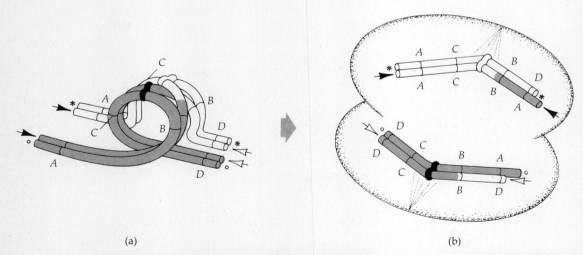

(a)   (b)

Figure 7.8 (*a*) Synapsis in an individual who is heterozygous for a pericentric inversion showing a crossover within the inversion loop. (*b*) Anaphase I configuration resulting from the crossover in (*a*). One of the chromatids involved in the crossover has a duplication of A and a deficiency of D (solid arrows); the other chromatid has a deficiency of A and a duplication of D (open arrows).

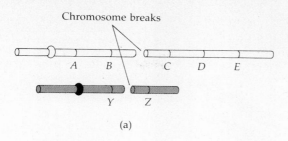

Chromosome breaks

(a)

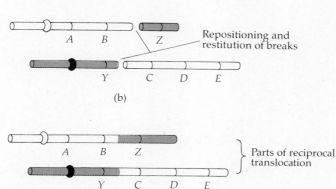

Repositioning and restitution of breaks

(b)

Parts of reciprocal translocation

(c)

Figure 7.9 Origin of a reciprocal translocation (*c*) from breaks in two nonhomologous chromosomes (*a*) followed by repositioning of broken ends and restitution of the breaks (*b*).

If the repositioning and restitution had joined *Y* with *B* and *Z* with *C*, the result would be a dicentric chromosome (*ABY*) and an acentric chromosome (*ZCDE*). Neither of these chromosomes could proceed normally through mitotic cell divisions and both would very quickly be lost, perhaps resulting in the death of the cell.

An individual who carries the parts of the reciprocal translocation in Figure 7.9(*c*) would also carry the normal homologue of both chromosomes (i.e., *ABCDE* and *YZ*). Such an individual is said to carry a **balanced translocation** because the individual carries all the normal genes in their proper dosage; only the order of genes has been changed. The composite data in Table 6.4 of Chapter 6 indicate that about 1 in 500 live-born children carries a balanced translocation.

Carriers of balanced reciprocal translocations, like carriers of inversions, are not themselves phenotypically abnormal, but they often produce eggs or sperm carrying duplications and deficiencies. Again like inversions, translocations can be passed intact from generation to generation, from carrier to carrier, without being detected. On the other hand, whereas abnormal gametes arise from heterozygous inversions only if a crossover occurs in the inverted region, abnormal gametes may arise from heterozygous translocations in the absence of crossing-over.

In an individual carrying a heterozygous translocation, synapsis involves four chromosomes instead of the normal two (see Figure 7.10). The types of gametes formed in a translocation carrier depend on which pairs of synapsed

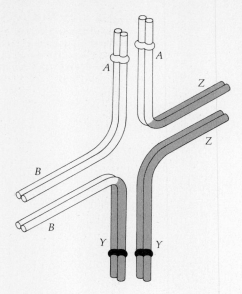

Figure 7.10 Structure formed during synapsis in an individual heterozygous for a reciprocal translocation. The chromosomes involved in the translocation are partly shaded and partly unshaded. Note that the structure has four centromeres; a normal bivalent has two centromeres.

chromosomes go together to the same pole during anaphase I. In Figure 7.10, various segments of the chromosomes have been labeled with letters for ease of reference. *AB* and *YZ* thus represent the normal chromosomes, and *AZ* and *YB* represent the parts of the reciprocal translocation. At anaphase I there are three possible modes of segregation, namely:

(1) *AB* & *YB* ⟷ *AZ* & *YZ*
(2) *AB* & *AZ* ⟷ *YB* & *YZ*
(3) *AZ* & *YB* ⟷ *AB* & *YZ*

where the double-headed arrow means "separate from." These modes of segregation determine the gametes formed by the translocation carrier, and these, in turn, determine the chromosome constitution of the zygotes formed in fertilization with a normal (i.e., *AB*- and *YZ*-bearing) gamete.

The possible outcomes of segregation and fertilization are illustrated in Figure 7.11, where the three modes of segregation are listed from left to right. The chromosome constitution of the zygotes is shown within the boxes, and it can be seen that there are many duplications and/or deficiencies resulting from gametes that receive one part of the reciprocal translocation but not the other. For example, the individual at the upper left has a duplication of *B* and a deficiency of *Z*, whereas the individual at the lower left has a duplication of *Z* and a deficiency of *B*.

Individuals who carry only part of a reciprocal translocation and thus have duplications and/or deficiencies are said to carry **unbalanced translocations**. When the interchanged parts of the reciprocal translocation are large, the individuals with unbalanced translocations will have large duplications or deficiencies and would be expected to undergo spontaneous abortion. Indeed, as reference to

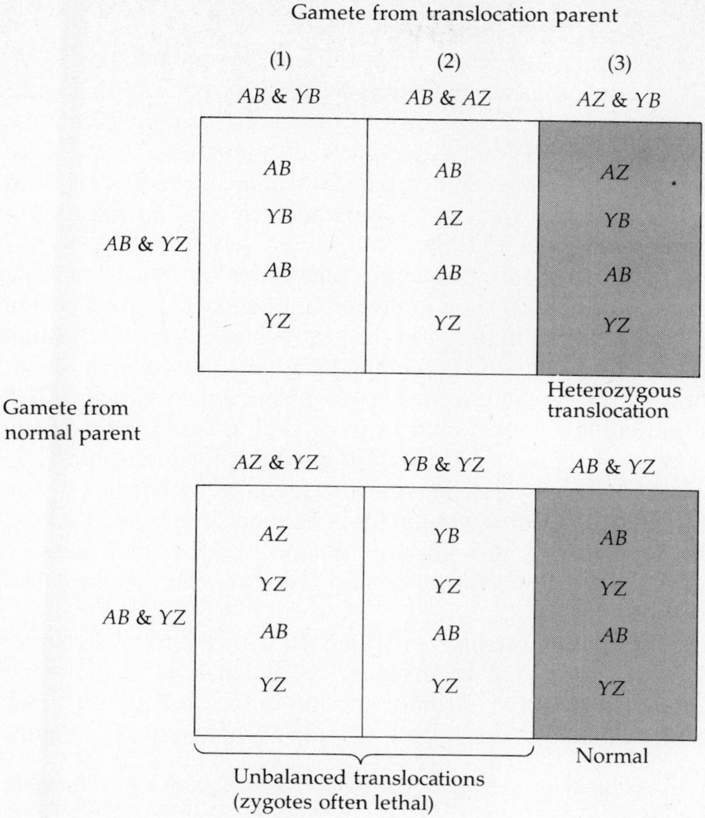

Gamete from translocation parent
(1)
(2)
(3)
AB & YB
AB & AZ
AZ & YB
AB
YB
AB
YZ
AB
AZ
AB
YZ
AZ
YB
AB
YZ
AB & YZ
Heterozygous translocation
Gamete from normal parent
AZ & YZ
YB & YZ
AB & YZ
AZ
YZ
AB
YZ
YB
YZ
AB
YZ
AB
YZ
AB
YZ
AB & YZ
Normal
Unbalanced translocations (zygotes often lethal)

Figure 7.11 Three modes of segregation of the translocation heterozygote in Figure 7.10. Modes (1) and (2) lead to unbalanced gametes. Mode (3) leads to balanced gametes, one of which carries both parts of the reciprocal translocation and one of which carries both normal chromosomes.

Table 6.4 in Chapter 6 will show, the frequency of unbalanced translocations among chromosomally abnormal fetuses in spontaneous abortions is about 3 percent. On the other hand, if the interchanged chromosomal parts are small, then certain of the unbalanced combinations may be compatible with live birth; it may also be noted in Table 6.4 that about 1 per 1600 live births carries an unbalanced translocation.

One mode of translocation segregation (mode 3 in Figure 7.11) yields balanced gametes leading to phenotypically normal offspring. In this mode of segregation, one class of gametes receives both parts of the reciprocal translocation, and the other class of gametes receives the structurally normal homologous chromosomes. The first type of gamete will lead to a phenotypically normal offspring that carries the translocation, and the second type will lead to a completely normal offspring having no chromosomal abnormality. Thus, with this third mode of segregation, all the offspring will be phenotypically normal, but half will be carriers of the reciprocal translocation.

Translocation carriers are therefore expected to produce a substantial fraction of unbalanced gametes and to have 50 percent translocation carriers among their phenotypically normal offspring.

ROBERTSONIAN TRANSLOCATIONS

A second type of translocation is sufficiently important in human genetics to be designated by a special term. A **Robertsonian translocation**, illustrated in Figure 7.12, involves the interchange of the long arm of one acrocentric chromosome with the short arm of a nonhomologous acrocentric chromosome [see Figure 7.12(*a*) and (*b*)], leading to a tiny metacentric chromosome consisting of both short arms along with a large metacentric or submetacentric chromosome consisting of both long arms [see Figure 7.12(*c*)].

The tiny metacentric chromosome frequently undergoes mitotic or meiotic nondisjunction and is lost, but its loss has no detectable phenotypic effect owing to the small amount of chromosomal material in this tiny metacentric. Such small metacentrics are sometimes discovered in screening programs, however. Indeed, in the Boston study of 13,751 consecutive newborns discussed in Chapter 2, 5 phenotypically normal individuals were found to have such a small metacentric chromosome, for an overall incidence of 1 per 2750 newborns. Nevertheless, in most cases the small metacentric is lost, leaving the large metacentric or submetacentric, which constitutes the Robertsonian translocation. In practical terms, a Robertsonian translocation results in a sort of "fusion" of the long arms of two acrocentrics, and for this reason a Robertsonian translocation is sometimes referred to as a **chromosomal fusion**.

Chapter 6 referred to certain kinships in which Down syndrome (trisomy 21) had a high risk of occurrence and recurrence. Such kinships account for about 3 percent of all cases of Down syndrome, and the reason for the high risk is the presence of a Robertsonian translocation involving chromosome 21. Figure

Chromosome breaks

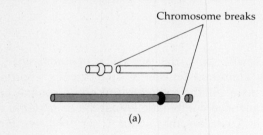

(a)

Figure 7.12 Origin of a Robertsonian translocation (*c*) from nonhomologous acrocentrics broken as shown in (*a*) with restitution as in (*b*). In most cases, the tiny metacentric chromosome consisting of both short arms undergoes nondisjunction and is lost.

Repositioning and restitution of breaks

(b)

This small fragment is usually lost

Robertsonian translocation

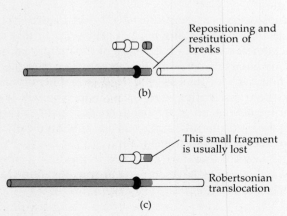

(c)

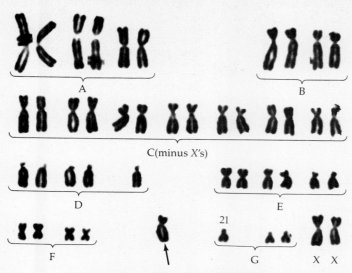

Figure 7.13 Karyotype of a phenotypically normal woman with a Robertsonian translocation involving chromosomes 15 and 21 (arrow). Note that there are only 45 chromosomes.

A

B

C(minus X's)

D

E

F

21

G

X X

7.13 shows the karyotype of a female carrying such a Robertsonian translocation, this one involving chromosomes 15 and 21; note that the individual carries one normal chromosome 15, one normal chromosome 21, and the Robertsonian translocation (indicated by the arrow). In this case, as in most similar cases, the small metacentric corresponding to the short arms of chromosomes 15 and 21 has been lost.

The woman with the karyotype in Figure 7.13, although she has only 45 chromosomes, is phenotypically normal. She has, however, a high risk of producing offspring with Down syndrome because of the various modes of segregation of the Robertsonian translocation. With a Robertsonian translocation, there are again three modes of segregation corresponding to those discussed earlier for reciprocal translocations. If we represent the Robertsonian translocation in Figure 7.13 by the symbol 15/21 and the normal homologues by the symbols 15 and 21, then the three modes of segregation can be represented as

(1) 15/21 & 21 $\longleftrightarrow$ 15
(2) 15/21 & 15 $\longleftrightarrow$ 21
(3)   15/21 $\longleftrightarrow$ 15 & 21

where, again, the symbol $\longleftrightarrow$ means "separate from" in anaphase I of meiosis.

The consequences of these modes of segregation with respect to the zygotes formed in subsequent fertilization with a normal gamete are illustrated in Figure 7.14. Modes 1 and 2 form trisomic zygotes and the corresponding monosomics. In many cases, these abnormal zygotes would undergo spontaneous abortion. In the case of a Robertsonian translocation involving chromosome 21, however, the trisomic-21 zygote can survive development and lead to a child with Down syndrome (upper left in Figure 7.14).

As with reciprocal translocations, the third mode of segregation produces gametes that carry either the Robertsonian translocation or the nontranslocated homologues; offspring from the translocation-bearing gametes will be phenotyp-

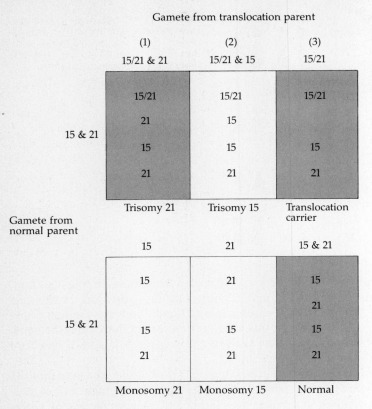

Gamete from translocation parent

	(1)	(2)	(3)
	15/21 & 21	15/21 & 15	15/21
15 & 21	15/21 21 15 21	15/21 15 15 21	15/21 15 21
	Trisomy 21	Trisomy 15	Translocation carrier

Gamete from normal parent

	15	21	15 & 21
15 & 21	15 15 21	21 15 21	15 21 15 21
	Monosomy 21	Monosomy 15	Normal

Figure 7.14 Three modes of segregation of Robertsonian translocation between chromosomes 15 and 21. Compare with Figure 7.11. Note that one of the possible offspring resulting from type (1) segregation will have Down syndrome.

ically normal carriers of the translocation (Figure 7.14, upper right), and offspring from the normal gametes will be phenotypically and chromosomally normal (Figure 7.14, lower right).

In summary, *a carrier of a Robertsonian translocation involving chromosome 21 will produce three categories of live-born offspring: (a) offspring with Down syndrome, (b) phenotypically normal carriers of the Robertsonian translocation, and (c) phenotypically and chromosomally normal offspring*. Moreover, the frequency of offspring in categories (*b*) and (*c*) will be equal because these arise as reciprocal products of the same mode of chromosome segregation. Strangely enough, and for reasons that are not yet understood, the frequency of gametes leading to offspring with Down syndrome depends on which parent is the carrier. The proportion of Down-syndrome children born to carrier mothers is high—15 to 20 percent or more; the proportion born to carrier fathers is lower—2 to 5 percent or less. These relatively high recurrence risks dictate that parents of Down children should be karyotyped.

CHROMOSOMAL CHANGES IN EVOLUTION

Chromosomal changes occur frequently enough during the course of evolution that related species rarely have identical chromosomes (unless the species are extremely closely related). For our purposes, **evolution** may be defined as cumulative change in the genetic makeup of a population, and a **species** may be defined as a group of organisms that can interbreed and produce fertile offspring. Chromosomal changes between related species are sufficiently common that some evolutionary biologists perceive chromosomal changes as playing a key role in the origin of new species.

Human beings are not exempt from the forces that govern the evolutionary process, and chromosomal changes have occurred in our own evolution. Comparison of human chromosomes with those of the chimpanzee reveals at least five recognizable pericentric inversions, for example. (Humans and chimpanzees are thought to have begun their evolution as separate species approximately 10 million years ago.) Remarkably, chimpanzees have 48 chromosomes, not 46, but chromosome banding studies have allowed the chromosomal correspondences between humans and chimpanzees to be deduced.

A detailed comparison between the karyotype of humans and chimpanzees is illustrated in Figure 7.15. In the chimpanzee, *Pan troglodytes,* no chromosome has a banding pattern like that of human chromosome 2. However, the chimp

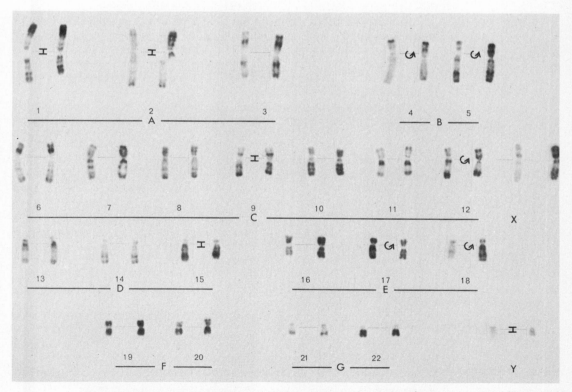

Figure 7.15 Comparison of human (left) and chimpanzee (right) chromosomes after R-banding. The curved arrow indicates a pericentric inversion, and the symbol I indicates altered chromosomal material in the region of the centromere. Note the two chimpanzee acrocentrics that have fused to form human chromosome 2.

has two acrocentric chromosomes not found in humans. One of these acrocentrics has a banding pattern like that of the short arm of human chromosome 2; the banding pattern of the other is like the long arm of human chromosome 2. The presence of these two acrocentrics is characteristic of nonhuman primates, so the inference is that human chromosome 2 arose as a fusion of the two acrocentrics during evolution. The fusion was not a typical sort of Robertsonian fusion, however. It involved the fusion of the short arms of the acrocentrics to form a dicentric chromosome, but the fusion was also accompanied by the inactivation of one of the centromeres by some unknown process. Modern humans arose from a population in which this fused chromosome had become homozygous.

SUMMARY

1. Chromosome breaks sometimes occur spontaneously and can be induced by certain agents such as x-rays. Broken ends of chromosomes tend to undergo **restitution** (rejoining), but they sometimes rejoin incorrectly, creating chromosomes with abnormal structures. Chromosomes with no centromere (**acentrics**) or with two centromeres (**dicentrics**) tend to be lost during cell division, so the only types of abnormal chromosomes that are perpetuated are those that have exactly one centromere.

2. A chromosome rearrangement may be balanced or unbalanced. A **balanced rearrangement** (such as an inversion) changes the arrangement but not the number of genes. An **unbalanced rearrangement** (such as a duplication) changes the number of genes. Unbalanced rearrangements upset the normal dosage of genes, and many of their phenotypic effects result from this upset in gene dosage.

3. A chromosome with a **duplication** carries two copies of certain genes. The phenotypic effects of a duplication depend on the size of the duplicated region and on the genes involved. Generally speaking, duplications of up to 5 or 6 percent of the haploid autosomal complement produce a variety of severe mental and physical abnormalities, but they are compatible with life. Small duplications may have virtually no phenotypic effects.

4. A chromosome with a **deletion (deficiency)** has certain genes missing. As with duplications, the phenotypic effects of a deficiency depend on the size of the deficiency and the genes involved. On the whole, however, deficiencies tend to have more harmful phenotypic effects than duplications of the same genes. In humans, the largest deficiencies that are compatible with life involve 2 to 3 percent of the haploid autosomal complement; larger deficiencies are almost always lethal. One of the best-known human deficiencies is a deficiency of part of the short arm of chromosome 5 (i.e., 5p–), which is associated with the **cri-du-chat** syndrome.

5. A chromosome with an **inversion** has part of its genes in the reverse of the normal order. Inversions that do not include the centromere are known as **paracentric** inversions; those that do include the centromere are known as **pericentric** inversions. Inversions are balanced rearrangements that have few or no phenotypic effects in the carriers. However, individuals who are heterozygous for inversions can produce unbalanced

gametes as a result of crossing-over within the inversion loop formed during synapsis of the inverted chromosome and its noninverted homologue. Crossing-over within the inversion loop of a heterozygous paracentric inversion produces a dicentric chromatid and an acentric chromatid; crossing-over within the inversion loop of a heterozygous pericentric inversion produces chromatids that have duplications and deficiencies.

6. A **reciprocal translocation** results from an interchange of parts between nonhomologous chromosomes. Reciprocal translocations are balanced rearrangements that usually have no detectable phenotypic effects in carriers. Heterozygotes for reciprocal translocations produce a substantial frequency of unbalanced gametes because of the ways in which the reciprocal translocation can segregate with its untranslocated normal homologues during meiosis. In two out of three possible modes of segregation, only unbalanced gametes are formed; these gametes will typically lead to spontaneous abortion of the fetus, especially if the translocated chromosomal segments are large. The third mode of segregation results in gametes carrying either both parts of the reciprocal translocation or both normal homologues. The first class of gametes gives rise to phenotypically normal carriers of the translocation; the second class gives rise to phenotypically and chromosomally normal offspring.

7. A **Robertsonian translocation** is a chromosome consisting of the long arms of two nonhomologous acrocentric chromosomes. The short arms of the chromosomes are typically lost without causing phenotypic harm. Robertsonian translocations involving chromosome 21 are particularly important in human genetics because one class of unbalanced gametes carries what amounts to a duplication of the long arm of chromosome 21, which leads to offspring who have Down syndrome. About 3 percent of all Down-syndrome births involve a parent with such a translocation, and in these cases the recurrence risk is high.

KEY WORDS

Acentric chromosome	Dicentric chromosome	Restitution
Balanced rearrangement	Duplication	Robertsonian translocation
Cat-cry syndrome	Inversion	Structural abnormality
Chromosomal fusion	Inversion loop	Translocation
Chromosome breakage	Paracentric inversion	Unbalanced rearrangement
Deficiency	Pericentric inversion	
Deletion	Reciprocal translocation	

PROBLEMS

7.1. Distinguish between a balanced chromosome rearrangement and an unbalanced one.

7.2. Why are acentric chromosomes not of great importance in most contexts in genetics?

7.3. What type of chromosomal abnormality discussed in this chapter changes the chromosome number among carriers?

7.4. A ring chromosome is a circular chromosome having a single centromere. For a metacentric chromosome broken in two places, where would the breaks have to occur to be able to form a ring chromosome, and how would the restitution have to occur?

7.5. A phenotypically normal woman with a history of repeated miscarriages has 46 chromosomes, but two of them have abnormal banding patterns (see illustration).

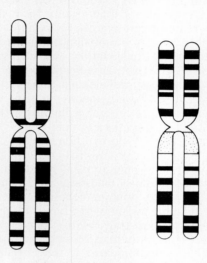

Can you identify the abnormal chromosomes, the type of chromosome abnormality involved, and the origin of the abnormality?

7.6. What type of chromosome abnormality is illustrated by the chromosome 7 shown here?

7.7. The C-group chromosome shown in the illustration has, among other abnormalities, a misplaced centromere. The individual carrying this chromosome is phenotypically normal and has 45 normal chromosomes in addition to the one shown. Identify the abnormal chromosome and the nature of the abnormality.

7.8. What type of abnormality is illustrated by the chromosome 2 shown here?

7.9. Sketch the inversion loop that would be formed by synapsis between a chromosome having the gene sequence *A B C G F E D H* with its normal homologue *a b c d e f g h*. If *C* (*c*) represents the centromere, what kind of inversion is this? What kinds of chromatids would result from crossing over within the inversion loop?

7.10. If the three modes of segregation of a Robertsonian translocation are equally likely, what is the probability that a carrier will have a chromosomally normal offspring?

7.11. A phenotypically normal male and two of his phenotypically normal offspring have 47 chromosomes. The extra chromosome is a tiny metacentric. The male's cousin has a 14/21 Robertsonian translocation. What is a likely explanation of the tiny metacentric?

7.12. A phenotypically normal woman has three children. One is phenotypically normal, one has trisomy 13, and one has trisomy 21. The woman and her phenotypically normal child are both found to have 45 chromosomes, one of which is shown in the illustration. Identify the abnormal chromosome and the chromosomal origin of the offspring. Would the woman have any chance of producing a chromosomally normal offspring?

7.13. In the case described in Problem 12, how many chromosomes would the trisomy-13 child have? How many chromosomes would the trisomy-21 have?

FURTHER READING

Borgaonkar, D. S. 1977. 2d. ed. Chromosomal Variation in Man. Alan R. Liss, New York. A catalogue of chromosomal variants and abnormalities.

de Grouchy, J., and C. Turleau. 1977. Clinical Atlas of Human Chromosomes. John Wiley and Sons, New York. A compendium of chromosomal abnormalities.

Gilbert, F. 1983. Chromosome aberrations and oncogenes. Nature 303: 475. A theory of cancer formation implicating specific genes and chromosomal aberrations.

Hamerton, J. L. 1971. General Cytogenetics. Academic Press, New York. An advanced text on chromosomes and chromosome behavior in humans.

Jacobs, P. A. 1981. Mutation rates of structural chromosome rearrangements in man. Am. J. Hum. Genet. 33: 44–51. The overall rate of occurrence of new detectable structural abnormalities is about one per 1000 gametes.

Niebuhr, E. 1978. The *cri du chat* syndrome. Human Genet. 44: 227–75. A review.

Rowley, J. D. 1980. Chromosome abnormalities in human leukemia. Ann. Rev. Genet. 14: 17–39. Discusses the variety of leukemia-related chromosomal abnormalities.

Sandler, L., and F. Hecht. 1973. Genetic effects of aneuploidy. Am. J. Hum. Genet. 25: 332–39. Discusses the effects of duplications and deficiencies in terms of their size.

Therman, E. 1980. Human Chromosomes: Structure, Behavior, Effects. Spring-Verlag, New York. Excellent discussion of human chromosomal abnormalities.

Vogel, F., and A. G. Motulsky. 1979. Human Genetics: Problems and Approaches. Springer-Verlag, New York. An excellent advanced-level survey of human genetics.

Williamson, B. 1982. Gene therapy. Nature 298: 416–18. Prospects for gene therapy and related ethical concerns.

chapter 8

The Double Helix

The emphasis in previous chapters was primarily on cells and their inheritance because cells are alive. They are the smallest living parts of living things. But cells are composed of smaller parts—nucleus, chromosomes, mitochondria, ribosomes, and so on—and the smaller parts are composed of still smaller parts, their atoms and molecules. A cell is thus a highly organized tiny sackful of chemicals. The life of a cell is basically chemical, and therefore living things, like people, are basically chemical. This statement should not be taken to mean that such human activities as dancing and writing poetry can or should be reduced completely to chemical equations, but it does imply that these and other activities are rooted in the body's chemical processes. Of course, life would be dreary indeed if human behavior were to be interpreted solely in terms of chemical equations. There are no chemical equations to express the beauty, romance, irony, and excitement of being alive, nor are there equations for life's ugliness and tragedy, its heartbreaks and disappointments. The meaning of life is surely to be found in spheres beyond mere chemistry. On the other hand, chemical processes are an essential part of being alive, and to ignore them is to be ignorant of one of life's most fascinating dimensions.

A CHEMICAL PRELUDE

Nowadays we learn as early as elementary school that all physical objects in the universe, including living things, are composed of **atoms**, those ubiquitous entities whose principal constituents are protons and neutrons in the atom's nucleus and electrons whirling in space around the nucleus. We also learn that certain atoms tend to be chemically combined or bound together by sharing a pair of electrons, a sharing that creates a strong type of chemical bond known as a **covalent bond**. Each atom involved in a covalent bond shares one electron with the other atom, and the shared pair of electrons whirls in space around both atomic nuclei. This ability of atoms to share electrons binds the atoms together.

A good example of covalent bonding is found in the water molecule. (A **molecule** is a group of atoms bound together chemically in a specific physical arrangement.) A hydrogen atom, represented by the letter H, has only one electron available for sharing; an oxygen atom, symbolized by O, can share two electrons. A mutual accommodation between hydrogen and oxygen is reached by sharing electrons. Two hydrogen atoms share their single electrons with one oxygen atom, and the oxygen atom shares one of its electrons with each of the hydrogens. The shared electrons travel around both atomic nuclei, but much of their time is spent between the nuclei. The sharing of a pair of electrons is represented by a line between the symbols for the atoms. H—O—H therefore symbolizes water.

The water molecule is actually rather V-shaped, with the two hydrogen nuclei near the tips of the arms and the oxygen nucleus near the point of the base. Although the molecule as a whole is electrically neutral, the shared electrons tend to remain nearer the oxygen nucleus than the hydrogen nuclei. Therefore the tips of the V's arms have a slight positive charge whereas the base has a slight negative charge; this makes possible a second and rather weak kind of chemical bond between different molecules because negative and positive charges tend to attract each other like tiny magnets. These weak bonds are called **hydrogen bonds**, and they play a key role in such important biological molecules as DNA. Figure 8.1 shows the relationships between covalent bonds and hydrogen bonds, using water molecules as an example.

The four most abundant elements in the human body are oxygen (O), carbon (C), hydrogen (H), and nitrogen (N); these constitute 96 percent of the total weight. Another 3.95 percent is accounted for by the seven elements calcium (Ca), phosphorus (P), potassium (K), sulfur (S), chlorine (Cl), sodium (Na), and magnesium (Mg). The remaining 0.05 percent is made up of traces of such elements as manganese (Mn), iron (Fe), copper (Cu), zinc (Zn), and a few others. The key element in this list is carbon.

The unique features of the chemistry of living things can largely be attributed

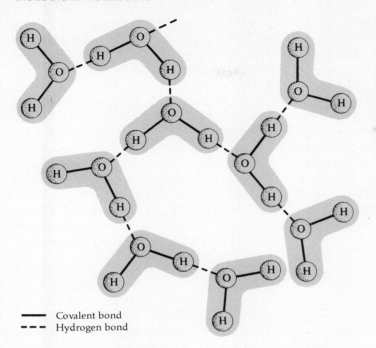

Figure 8.1 Relationship between strong covalent bonds and weak hydrogen bonds in water. Each V-shaped water molecule has the chemical formula H_2O, and the covalent bonds in the molecule are represented by solid lines. The oxygen atom in water has a slight negative charge, whereas the hydrogen atoms have a slight positive charge. Since positive and negative charges tend to attract, adjacent water molecules tend to orient with the oxygen of one aligned toward a hydrogen of the other; the weak bonds thus formed are called hydrogen bonds and are represented here by dashed lines.

——— Covalent bond
- - - Hydrogen bond

to the wonderful abilities of carbon atoms. Every carbon atom has four electrons available for sharing and thus can form four covalent bonds. More importantly, carbon atoms will bond covalently with each other so they can form long chains of carbon atoms, perhaps with side branches here and there; they can even form closed rings, sometimes with another atom such as nitrogen or oxygen as part of the ring. The combining ability of carbon makes the number of different kinds of molecules that can be formed virtually infinite.

BIOLOGICAL MOLECULES

The almost limitless variety of molecules possible in living tissues would at first glance seem quite enough to confuse and bedevil any attempt to make sense of it. But things haven't turned out that way. In much the manner that a child learns to make sense out of a complex world by lumping things into categories based on their common features ("This is a dog—whether big or small or shaggy or smooth, I can tell it's a dog"; or "That's a table—whether square or round or squat or tall, I can tell it's a table"), the biochemist has learned to make sense out of large assortments of complex molecules by classifying them suitably.

Four classes of large molecules are of particular importance in living things; these are the **nucleic acids**, **proteins**, **carbohydrates**, and **lipids**. Nucleic acids are important in cells largely because one type of nucleic acid, DNA, is the genetic material—genes are DNA—and because another type of nucleic acid, RNA, plays a major role in enabling the cell to implement the genetic instructions in DNA.

Proteins, carbohydrates, and lipids are principal constituents of cells by weight. Live cells are about 20 percent protein, 3 percent lipid, 1 percent carbohydrate, and 65 percent water. These numbers are averages; each tissue, such as bone or brain or muscle or blood, has a somewhat different composition.

DNA (deoxyribonucleic acid), RNA (ribonucleic acid), proteins, carbohydrates, and lipids are each composed of smaller molecules joined together; these components are the basic units or building blocks of the different classes of large molecules. DNA molecules are composed of four different units called **deoxyribonucleotides** (or just "nucleotides"); RNA molecules are mostly made up of four units called **ribonucleotides**. Proteins are composed of approximately 20 in different building blocks known as **amino acids**; carbohydrates are molecules constructed from smaller **sugars**. There are several different classes of lipids, including fats, steroids (testosterone is an example), waxes, oils, and certain alcohols. In addition to the principal classes of large molecules—nucleic acids, proteins, carbohydrates, and lipids—cells contain combinations of them, such as nucleoproteins (which are partly nucleic acid and partly protein), lipoproteins (part lipid, part protein), and so on.

DNA is found in the cell as a long, skinny, threadlike molecule. RNA is also long and threadlike, but some types fold back upon themselves to form precise three-dimensional configurations. A principal function of RNA is to receive and carry the genetic information from DNA in the nucleus into the cytoplasm where particular proteins are produced that correspond to the genetic information. Each gene carries information instructing the cell to hook together chemically a certain specific sequence of amino acids, forming the protein coded for by that gene.

Carbohydrates and lipids have several important functions. Various fats and oils along with starch, a carbohydrate, are primary food-storage molecules in plants. Fats and the carbohydrate glycogen, which is chemically similar to starch, are the main food-storage molecules in animals. The covalent bonds in these molecules store energy needed by the cell, energy released when the bonds are cleaved. That fats and starches have a high energy content is no secret to dieters, who measure the energy content of foods in terms of calories and who know that potatoes, breads, cream, peanut butter, and countless other temptations are the kiss goodbye to hopes of losing weight. Carbohydrates are also a component of cell membranes; whether a person has blood group A, B, AB, or O depends on a particular kind of carbohydrate present on the membranes of the red blood cells, for example. Lipids, too, are a major constituent of cell membranes; their disinclination to disperse in water is thereby put to good use, for otherwise the cell would dissolve away. The nervous system is especially rich in lipids—the brain is almost 15 percent lipid, for example—but the exact role of lipids in the function of the nervous system is not well understood. Many hormones—those ubiquitous chemical messengers of the body—are members of the steroid class of lipids.

The real workhorses of the cell are proteins. Proteins are notable for their diversity. Some aggregate into long skinny structures such as microtubules; some aggregate into large flat sheets; some fold and knot up in a characteristic way to form holes or traps into which smaller molecules can fit. Two general classes of proteins may be distinguished: **structural proteins**, which serve as structural support in the cell, and **enzymes**, which accelerate specific chemical reactions. Some enzymes split other molecules apart. An example is hexosaminidase A, an enzyme that occurs in nerve cells. This enzyme latches onto a certain carbohydrate-lipid molecule (called ganglioside GM_2) and cleaves a specific covalent bond between two of the sugars in the carbohydrate part of the molecule. This is one step in the breakdown of GM_2. The absence of hexosaminidase A or its failure to function leads to an accumulation of GM_2 ganglioside in nerve cells throughout the body, and this in turn leads to blindness and severe mental retardation. Tay-Sachs disease, discussed in Chapter 4, results from an inherited defect in the gene for hexosaminidase A.

Other enzymes put atoms or molecules together; they encourage the formation of new covalent bonds. Carbohydrates and lipids are manufactured by specific enzymes that work in assembly-line fashion to hook together their various subunits. Testosterone, for example, is produced step by step from cholesterol by a series of different enzymes that works sequentially, one kind of enzyme for each step. It is important to remember that enzymes do not make impossible events occur. They only "hurry" or *catalyze* chemical reactions that, given time, would occur anyway. Enzymes, by facilitating the breakup and rearrangement of chemical bonds in other molecules, make reactions happen in split seconds that otherwise might require centuries.

Proteins are so important in the cell that much cellular machinery is concerned with translating the genetic information of DNA into specific proteins corresponding to the information. The genetic information in a region of DNA is copied to produce a specific molecule of RNA. The RNA then proceeds from the nucleus into the cytoplasm, where the information it contains is translated into a corresponding specific sequence of amino acids in a newly produced protein molecule. The relationship among DNA, RNA, and protein is roughly analogous to the following example: Suppose you had a machine that could translate the symbols on an ancient scroll. The scroll is so precious that it cannot be removed from the library (nucleus), and the translating machine is so bulky that it must stay outside (in the cytoplasm). The obvious thing to do would be to go into the library and transcribe or copy the scroll, symbol for symbol, and then bring the copy (the RNA) to the cytoplasm for translation. In this analogy the scroll corresponds to a gene, actually a segment of a DNA molecule, and the translation of the scroll into another language corresponds to the production of a molecule of a new protein using the chemical machinery in the cytoplasm. (The details of the process of protein synthesis will be examined in Chapter 9.)

DNA STRUCTURE: A CLOSER LOOK

The principal subject of this chapter is the structure and replication of **DNA (deoxyribonucleic acid)**. The subject was briefly introduced in Chapter 1 in connection with the organization of DNA in chromosomes (see Figures 1.14 and 1.15) and again in Chapter 2 in connection with recombination during meiosis (see Figure 2.13). Here we are interested in a more detailed examination of the structure of DNA, but it will be best to begin with a conventional sort of analogy that highlights some of the principal features of the theadlike molecule. This analogy, which depicts DNA as a ladderlike structure, is illustrated in Figure 8.2, and several important points are to be noted:

1. A DNA molecule consists of two strands. In Figure 8.2 (*left*), the molecule is depicted as a sort of ladderlike structure with long sidepieces and rungs across the middle. Using the ladder analogy, each DNA strand in the molecule is a "half-ladder" consisting of one of the sidepieces with half-rungs jutting off toward the middle. The complete ladder is formed by the side-by-side alignment of these half-ladders. As will become clear shortly, the chemical bonds *within* the half-ladders are strong covalent bonds, but the chemical bonds *between* the half-ladders that hold them together are weak hydrogen bonds.

2. Each DNA strand is composed of a linear string of constituents called **nucleotides**. These are illustrated in Figure 8.2 as little ladder parts, each consisting of a section of sidepiece with a half-rung jutting off. Four nucleotides are found in DNA, which were symbolized in Chapter 1 as A (crosshatched circle in Figure 8.2), T (open circle), G (crosshatched square), and C (open square).

3. The nucleotides in the individual DNA strands that face each other across the way are **complementary**; where one strand carries an A, the other

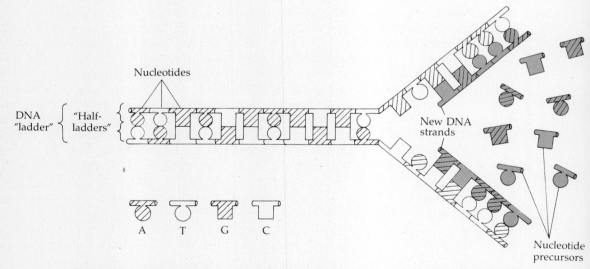

Figure 8.2 Ladder analogy of DNA structure. The fundamental units are nucleotides (A, T, G, and C); each is represented as a short section of sidepiece and an attached half-rung (square or circle). Nucleotides are attached by their sidepieces to form a half-ladder, and the double-stranded DNA molecule is composed of two such half-ladders (left). Note that A pairs only with T, and G pairs only with C. Replication is illustrated at the right, where the original DNA strands have become separated and each serves as a template for the synthesis of a complementary strand (shaded) by the one-by-one addition of nucleotides.

across the way carries a T; and where one strand carries a G, the other across the way carries a C. This pairing of complementary nucleotides accounts for the configuration of facing half-rungs illustrated in Figure 8.2: a hatched circle (A) is always paired with an open circle (T), and a hatched square (G) is always paired with an open square (C).

4. In the process of DNA replication, illustrated at the right in Figure 8.2, the individual strands of a DNA molecule open out, and new complementary strands (represented as shaded half-ladders) are synthesized using each original strand as a **template** (a sort of "blueprint"). During the synthesis of the new DNA strands, nucleotide precursors that occur free in the cell's nucleus are added one by one to the growing strands according to the nucleotide pairing rules of A with T and G with C.

5. In any specific gene, only one DNA strand carries the actual genetic information in its nucleotide sequence; this strand is called the **sense** strand of the gene in question; we may, for purposes of illustration, assume that the upper strand in Figure 8.2 is the sense strand. The other DNA strand in the gene, which carries the complementary nucleotides of the sense strand, is known as the **antisense** strand. In the case of Figure 8.2, the antisense strand is the lower one. During DNA replication when the sense and antisense strands separate, the old sense strand serves as a template for the synthesis of a new antisense strand, and the old antisense strand serves as a template for the synthesis of a new sense strand. Consequently, the mode of DNA replication illustrated in Figure 8.2 results in two double-stranded DNA molecules, each identical in nucleotide sequence to the original unreplicated molecule.

A more detailed view of two "ladder rungs" is shown in Figure 8.3. Part (*a*) illustrates an AT nucleotide pair; part (*b*) illustrates a GC nucleotide pair. It is to be noted that each nucleotide consists of three parts:

1. A **phosphate group** consisting of an atom of phosphorus (P) attached to three atoms of oxygen, which is indicated in Figure 8.3 as an encircled P.

2. A molecule of the sugar **deoxyribose**, which basically has the shape of a flattened ring composed of four carbon atoms and one oxygen atom; a fifth carbon atom pokes up from the ring and is attached to the phosphate group. In the sort of representation in Figure 8.3, each vertex in the rings represents an atom of carbon, but the symbol C is not written to avoid cluttering the illustration. The deoxyribose sugar has hydrogen atoms attached to the carbons at various places, but again these are omitted for the sake of simplicity. In addition, the carbon atom attached to the phosphate group is depicted in Figure 8.3 as a kink in the line jutting up from the deoxyribose ring.

3. A molecule of a constituent called a **base**, which is attached to the side of the sugar ring opposite the phosphate group. Four bases are found in the nucleotides that make up DNA. Each base is composed of a carbon-nitrogen (N) ring to which other atoms are attached. In the cases of **thymine** (T) and **cytosine** (C), the carbon-nitrogen ring is a single ring; these single-ring bases are known collectively as **pyrimidines**. In the cases

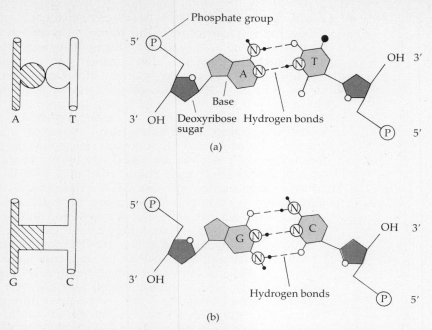

Figure 8.3 Ladder analogy and more detailed chemical structures of (a) A-T base pair and (b) G-C base pair. The base pairing is facilitated by hydrogen bonds that can form between certain atoms on the bases (broken lines). A-T base pairs form two hydrogen bonds; G-C base pairs form three hydrogen bonds. The large solid dot on the thymine represents a methyl (—CH₃) group.

of **adenine** (A) and **guanine** (G), the carbon-nitrogen ring is actually a fused double ring; these double-ring bases are referred to as **purines**. The unit consisting of a sugar, a base, and a phosphate group constitutes a **nucleotide**, and the particular base determines which nucleotide we are dealing with:

Deoxyadenosine phosphate (A) carries the base **adenine**

Deoxythymidine phosphate (T) carries the base **thymine**

Deoxyguanosine phosphate (G) carries the base **guanine**

Deoxycytidine phosphate (C) carries the base **cytosine**

Figure 8.3 also illustrates two other important features of DNA structure. First, each nucleotide has an orientation, a "top" and a "bottom." One end of the nucleotide is marked by the —OH group sticking off the sugar; this is called the **3′ (three-prime) end** of the nucleotide, and it is so named because of the way biochemists number the carbon atoms in the sugar ring. The opposite end of the nucleotide is marked by the phosphate group; this is called the **5′ (five-prime) end** of the nucleotide.

The second important feature of DNA structure illustrated in Figure 8.3 is the role of hydrogen bonds in holding the nucleotides in partner DNA strands together. In Figure 8.3 the hydrogen bonds are indicated by broken lines, and it can be seen that AT pairs form two hydrogen bonds whereas GC pairs form three. The hydrogen bonds come about because the facing configurations of atoms on the bases permit certain atoms to share hydrogen atoms (small solid dots). In the case of adenine and thymine, for example, one hydrogen bond is formed by a nitrogen and an oxygen sharing a hydrogen, and the other is formed by two nitrogens sharing a hydrogen. Because of the configurations of atoms in the bases, hydrogen bonds can form only between A and T and only between G and C. It is this feature of DNA that establishes the nucleotide pairing rules of A with T and G with C, and it is perhaps the single most important aspect of DNA structure.

How the nucleotides are attached to one another to form a complete molecule of DNA is illustrated in Figure 8.4. In essence, the 5′ end of the nucleotide below becomes attached to the 3′ end of the one above, which, when repeated with many nucleotides, forms a long alternating sugar-phosphate-sugar-phosphate-sugar-phosphate ''backbone'' with the bases jutting off. The partner DNA strand has a similar arrangement, but note that the orientation of the partner strand is reversed relative to the original strand. In Figure 8.4 the left-hand strand runs 3′ to 5′ from bottom to top, whereas the right-hand strand runs 5′ to 3′. This opposite polarity of the DNA strands is referred to by saying that the strands are **antiparallel**. The molecule in Figure 8.4 could be written as

5′-TACT-3′
3′-ATGA-5′

or, completely equivalently, as

5′-AGTA-3′
3′-TCAT-5′

Figure 8.4 illustrates a very short DNA molecule consisting of only four nucleotide pairs. Most naturally occurring DNA molecules are very much longer. The single, circular DNA molecule that forms the ''chromosome'' of the bacterium *Escherichia coli* contains about 4 million nucleotide pairs, and the DNA molecule in human chromosome 1 contains about 285 million nucleotide pairs. Moreover—and this is a fundamentally important point—there is no limitation whatever on the order or sequence of nucleotide pairs along the DNA. Since there are four possible nucleotide pairs (i.e., AT, TA, GC, and CG), a single nucleotide pair has 4 possibilities, two pairs would have $4 \times 4 = 16$ possible sequences, three pairs would have $4 \times 4 \times 4 = 4^3 = 64$ possible sequences, and so on. A DNA molecule as short as 100 nucleotide pairs would have 4^{100} (approximately 1.6×10^{60}) possible sequences, which is a number so large as to be beyond human comprehension. The genetic diversity among human beings and, indeed, the diversity of life forms on earth can be attributed in large part to the immense information-carrying capabilities of DNA.

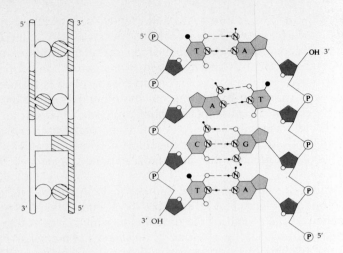

THREE-DIMENSIONAL STRUCTURES OF DNA

One of the most important discoveries that preceded a detailed understanding of DNA structure was that DNA contains an equal number of A and T nucleotides and an equal number of G and C nucleotides. In human DNA, for example, 31 percent of the nucleotides are deoxyadenosine phosphate (A), 31 percent are deoxythymidine phosphate (T), 19 percent are deoxyguanosine phosphate (G), and 19 percent are deoxycytidine phosphate (C). Thus, relative to their amounts, A = T and G = C. In the fruit fly, *Drosophila melanogaster,* to take another example, the nucleotides are 27.5 percent A, 27.5 percent T, 22.5 percent G, and 22.5 percent C, so, here again, A = T and G = C.

The equality of A with T and G with C in DNA was just an inexplicable curiosity until the early 1950s, when Maurice Wilkins and Rosalind Franklin at Kings College in London discovered through their x-ray diffraction studies that a complete molecule of DNA consists of two strands. (X-ray diffraction is a technique in which x-rays are transmitted through a crystal of some substance, and an atomic view of the substance is inferred from the pattern of reinforcement or interference of the transmitted x-rays.) As this information about two strands in DNA became available to James Watson and Francis Crick at Cambridge University, they began to try to construct a tinker-toy-like model of DNA. In one of their first models, the bases attached to the nucleotides stuck off toward the outside of the molecule instead of pointing inward. But they could not get such a porcupine model to work; every version was incompatible with details of the x-ray diffraction data.

If the bases point toward the interior of the molecule, how could they fit together without requiring too much space or too little space or otherwise violating the laws of physics and chemistry? According to Watson, who later wrote *The Double Helix,* a fascinating account of the discovery of the structure of DNA, he began to play with little metal cutouts in the shapes of the nucleotides to see how they could fit together. The x-ray data implied that the distance between the DNA strands was the same everywhere, which ruled out the pairing of purines (i.e., A-A, A-G, or G-G) because such pairs would be too large; it also ruled out the pairing of pyrimidines (i.e., T-T, T-C, or C-C) because such pairs would be too small. Evidently, each pair must involve one purine and one pyrimidine—but which ones?

One happy day Watson discovered that his nucleotide cutouts could be fit together as illustrated at the right side of Figure 8.4. With the appropriate orientation of the nucleotides, adenine will pair with thymine and guanine will pair with cytosine. Watson realized that this was a key discovery for several reasons. First, it explained why, relative to their amounts, A = T and G = C. Since A and T are paired, each A must have a partner T and each T a partner A, so their amounts must be equal. Likewise for G and C. Second, it explained how the proper distance between the DNA strands is preserved: a large purine base (A or G) is always paired with a small pyrimidine base (T or C, respectively), thus giving the same spacing between the DNA strands. Third, the pairing of A with T and G with C explained how the DNA strands in a molecule are held together; the answer lies in the formation of hydrogen bonds between the bases. Although each individual hydrogen bond is relatively weak, in a long DNA molecule consisting of hundreds of thousands of nucleotide pairs, their cumulative effect will be substantial and will stabilize the two-stranded structure under most conditions found in cells.

Once the possibilities of A-T and G-C base pairing were realized, Watson and Crick set out to construct a physical model of DNA structure. The model they finally devised is shown at the bottom of Figure 8.5. Here the sugar-phosphate backbones of the individual DNA strands can be seen to wind around one another on the outside of the DNA molecule, reminiscent of the winding of the red and white stripes on a barber pole. (The paths of the backbones can be traced easily by following the black spheres representing the phosphate atoms.) The paired bases are flat and are stacked on top of one another like a roll of coins, and the backbones wind around the outside of the roll.

The three-dimensional DNA duplex in Figure 8.5 is called a **double helix**— "double" because of the two DNA strands it contains, and "helix" because the strands wind around one another in the form of a right-handed helix. The configuration of DNA illustrated in Figure 8.5 is the one originally proposed by Watson and Crick. It is known as the **B form** of DNA, and the dimensions of the molecule are well known: the diameter of the molecule is about 20 Å; each backbone makes a complete turn around the helix every 34 Å; and there are 10 base pairs in every turn; thus the thickness of each base pair is 3.4 Å.

The space-filling model of the DNA duplex at the bottom of Figure 8.5 blends into several alternative representations that highlight certain features of its structure. Immediately above the space-filling model is a representation in which the sugar-phosphate backbones are denoted as ribbons and the base pairs as thick horizontal lines. In the middle region of Figure 8.5, two base pairs have been tilted sideways (with apologies to the twisted and contorted covalent bonds) to illustrate the hydrogen bonding between A and T and between G and C (indicated by the short vertical lines between the atoms involved). At the top of

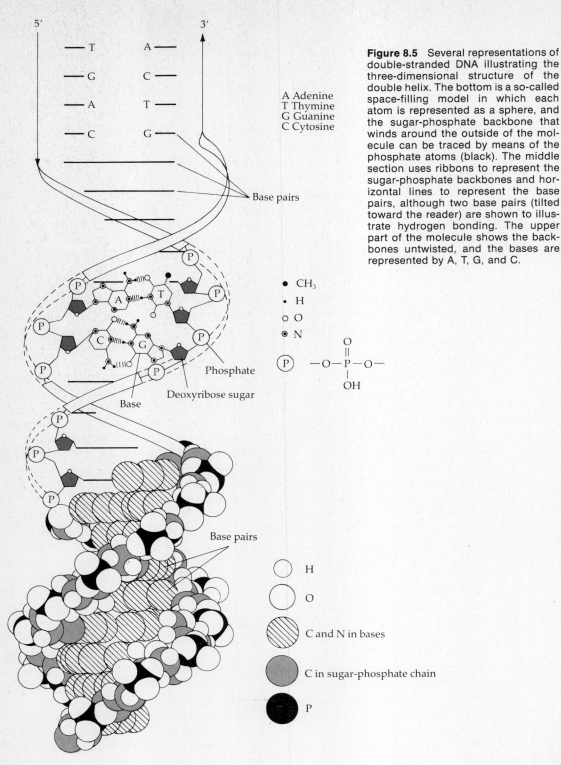

5′   3′

— T   A —
— G   C —
— A   T —
— C   G —

A Adenine
T Thymine
G Guanine
C Cytosine

Base pairs

P

CH₃ → CH_3 ● CH₃
• H
○ O
◉ N

P $-O-\overset{\overset{\displaystyle O}{\|}}{\underset{\underset{\displaystyle OH}{|}}{P}}-O-$

Phosphate

Deoxyribose sugar

Base

Base pairs

○ H
○ O
▨ C and N in bases
⬤ C in sugar-phosphate chain
⚫ P

Figure 8.5 Several representations of double-stranded DNA illustrating the three-dimensional structure of the double helix. The bottom is a so-called space-filling model in which each atom is represented as a sphere, and the sugar-phosphate backbone that winds around the outside of the molecule can be traced by means of the phosphate atoms (black). The middle section uses ribbons to represent the sugar-phosphate backbones and horizontal lines to represent the base pairs, although two base pairs (tilted toward the reader) are shown to illustrate hydrogen bonding. The upper part of the molecule shows the backbones untwisted, and the bases are represented by A, T, G, and C.

the figure, the backbones of the double helix have been untwisted and the possible base pairs are symbolized by the first letter of each base; the 3′ end of each DNA strand is indicated by the arrow (upper right) and the 5′ end by the blunt end of the arrow (upper left).

In the years since 1953, when Watson and Crick first proposed their structure for B-form DNA, their proposal has been confirmed in virtually every important detail. The structure at the bottom of Figure 8.5 appears to be the usual conformation of duplex DNA in the living cell. However, research by x-ray crystallographers has revealed that more than 20 variations of right-handed helical DNA can also occur. These variant forms involve ones in which the bases are tilted at different angles or stacked more tightly than in the B form. Indeed, it has recently been discovered that under certain conditions duplex DNA can even form a left-handed helix, which has come to be known as the **Z form** of DNA because the sugar-phosphate backbone zigzags instead of forming a smooth coil.

The contrast between Z (left-handed) DNA and B (right-handed) DNA is illustrated in Figure 8.6, where the sugar-phosphate backbones are outlined in black. It should be emphasized that any DNA molecule can potentially assume either the B or the Z configuration depending on conditions in the cell. Indeed, in a long molecule, some stretches may be in the Z configuration and others in the B configuration, and these can switch back and forth. However, the Z and B structures are markedly different, and a protein or other molecule that has an

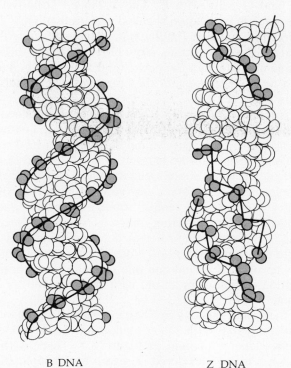

Figure 8.6 Contrasting three-dimensional structures of B DNA and Z DNA. Note that the sugar-phosphate backbone in Z DNA coils to the left and that it zigzags instead of being a smooth coil.

B DNA   Z DNA

appropriate molecular configuration to attach to B DNA might not be able to attach to Z DNA, and vice versa. Consequently, the configuration of DNA may be important in such genetic processes as the regulation of gene expression. The precise role played in cells by Z DNA is at present unclear, but it is clear that Z DNA is widespread. For example, the interband regions of dipteran giant chromosomes (Figure 7.6) are especially rich in Z DNA. Whatever the role of Z DNA may be, the predominant B form of DNA and all the variant forms, including Z, still obey the fundamental A-T and G-C base-pairing rules.

DNA REPLICATION

Watson and Crick's double-stranded DNA structure immediately suggested a manner in which DNA might replicate. The two strands of the original molecule might separate, with each existing strand serving as a template (or pattern) for the synthesis of a new DNA strand complementary in sequence to the template (see Figure 8.2).

In broad outline this conception of DNA replication has proven correct, and some of the details are illustrated in Figure 8.7, which shows the replication of a DNA sequence corresponding to part of a human hemoglobin gene. In the first step in replication, the strands in the original molecule shown in part (*a*) separate [part (*b*)]. Then an enzyme called **DNA polymerase** attaches to each free single-stranded template and creates new strands (shaded) according to the rules that A pairs with T and G pairs with C. The Y-shaped structure in part (*c*) corresponding to this stage in replication is called a **replication fork**. The enzyme works in only one direction, however, adding new nucleotides only to the 3' end of a growing strand (i.e., the new strands grow in length in the direction of the arrowheads). Consequently, although the new strand at the top in part (*c*) would start at the rightmost tip of the template, the bottom strand would start at the base of the replication fork.

In part (*d*) the parental DNA strand is shown as having opened up further, and synthesis of the new strands continues. The top strand can continue straightaway to the left, because this represents strand elongation of its 3' end, but the bottom strand has to start farther to the left and proceed toward the right. That is to say, one of the daughter DNA strands (the top one here) can be synthesized as one continuous molecule, but the other new strand (the bottom one) must be synthesized in smaller segments (in reality about a thousand nucleotides in length), which are fused together by other enzymes when they meet. The end result of the process is the production of two daughter molecules identical in nucleotide sequence to the original [part (*e*)]. Note that in each daughter molecule one of the strands is an original strand and one is a newly synthesized strand. This mode of replication, which pairs an old with a new strand, is called **semi-conservative replication**.

On the whole, the ladder analogy in Figure 8.2 gives a relatively accurate description of the overall process of DNA replication, but it does gloss over many of the molecular details apparent in Figure 8.7. There is, however, an additional matter that has to be considered in regard to the replication of DNA in a eukaryotic chromosome. Human chromosome 1, to take an example, contains

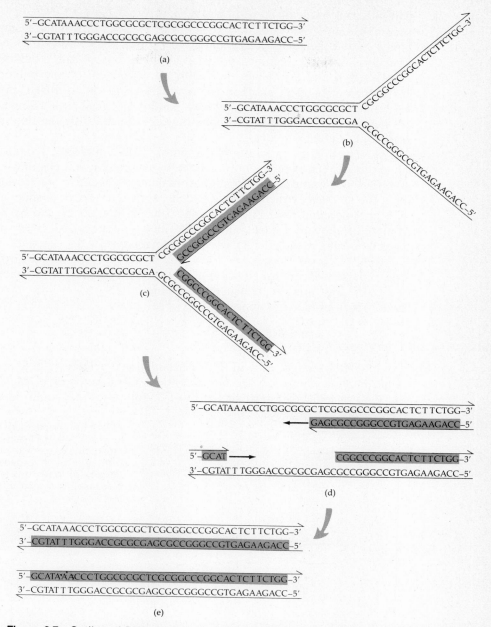

Figure 8.7 Outline of DNA replication involving opening out of the original double-stranded molecule and synthesis of new strands (shaded) using the original strands as templates.

about 285 million nucleotide pairs of DNA. At a rate of DNA synthesis of 40 nucleotides per second (the rate for DNA polymerase), replication of a 285-million-nucleotide length of DNA, if it occurred from one end straight to the other, would require nearly 3 months! This is absurd, of course; replication of chromosome 1 occurs in a matter of hours, not months. The way this feat is accomplished is that replication is initiated by helix unwinding at many interior points of the molecule. Replication and further helix unwinding occur in both directions from these points, producing the sort of replication bubbles shown in Figure 8.8(*a*) and (*b*). When the rightward movement of one bubble meets the leftward movement of another, the new DNA strands are fused, producing the larger bubble in Figure 8.8(*c*). In short, hundreds of regions of a long DNA duplex are being replicated simultaneously, thus greatly reducing the time required for replication to be completed.

Figure 8.9(*a*) is an autoradiograph of a DNA molecule from a human cell caught in the act of replication. To obtain such pictures, cells undergoing DNA synthesis are bathed in a solution containing radioactive nucleotides. After a time, the DNA is extracted from the cells, spread onto a glass slide, and overlaid with photographic film. The dark spots on the film are caused by radioactivity from the radioactive nucleotides in the newly synthesized DNA strands. The black line underneath the DNA molecule represents 50 μm (i.e., 500,000 Å). Figure 8.9(*b*) is an interpretative drawing of the molecule in part (*a*). The black lines represent DNA strands synthesized in the presence of radioactive nucleotides; the gray lines represent nonradioactive DNA. Note the two replication forks, which proceed in opposite directions.

RESTRICTION MAPPING OF DNA

Any long DNA molecule will have certain short nucleotide sequences repeated within it. For example, if the DNA contains equal amounts of each nucleotide, then any specific two-base sequence (TG, for instance) would be expected to

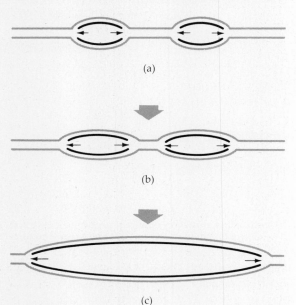

(a)

(b)

(c)

Figure 8.8 Replication bubbles in eukaryotes. A long DNA molecule begins replication at many interior points (*a*), with replication proceeding bidirectionally (*b*). When two replication bubbles meet, they fuse and form a larger bubble (*c*).

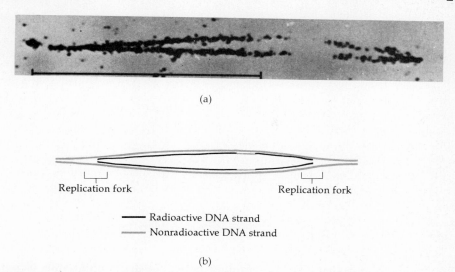

(a)

Replication fork   Replication fork

—— Radioactive DNA strand
~~~~ Nonradioactive DNA strand

(b)

Figure 8.9 (a) Autoradiograph of a DNA molecule in a human cell caught in the act of replication. The black line represents 50μm (0.002 in). (b) Interpretative drawing of the molecule in (a). Note similarity with Figure 8.8(c).

occur on the average once every $4 \times 4 = 4^2 = 16$ nucleotides. This would be the average distance between TG sites, of course, but there would be wide variation: some TG sites may be nearly adjacent; others may be separated by a hundred or more nucleotides. Similarly, any specific four-base sequence (AGCT, for example), would be expected to occur on an average of once every $4^4 = 256$ nucleotides, again with wide variation around this average. Continuing the same reasoning, a specific six-base sequence (for instance, GAATTC) would be expected to occur with an average interval between sites of $4^6 = 4096$ nucleotides.

As it happens, there is a large number of enzymes called **restriction enzymes** that are able to recognize and attach to specific, short nucleotide sequences in double-stranded DNA and to cleave both strands at these sites. The nucleotide sequence recognized and cleaved by the restriction enzyme is called the enzyme's **restriction site**. For example, the restriction enzyme *Alu*I recognizes and cuts double-stranded DNA wherever the sequence $\frac{5'\text{-AGCT-}3'}{3'\text{-TCGA-}5'}$ may occur, so this four-base sequence is *Alu*I's restriction site. These enzymes are normally found in certain bacterial cells, and they seem to play a role in the bacteria's defense against invading DNA, such as the DNA of viruses that attack bacterial cells.

Because of their specific sites of action in DNA, restriction enzymes provide a principal tool for the analysis of DNA sequence. Several widely used restriction enzymes and their sites of action are illustrated in Figure 8.10 (more than 100 restriction enzymes are commercially available). Some enzymes, such as *Alu*I, *Hha*I, and *Hae*III, attack four-base restriction sites; others, such as *Eco*RI, *Bam*HI, and *Xho*I, attack six-base restriction sites. Some restriction enzymes (e.g., *Alu*I) cut both DNA strands at the same position; others (e.g., *Eco*RI) make staggered symmetrical cuts in the strands. Details aside, the important

*Alu*I   5′–AG|C T–3′
       3′–TC|GA–5′

*Hha*I   5′–G CG|C–3′
       3′–C |GC G–5′

*Hae*III   5′–GG|CC–3′
       3′–CC|GG–5′

*Eco*RI   5′–G|AATT C–3′
       3′–C TTAA|G–5′

*Bam*HI   5′–G|GATC C–3′
       3′–C CTAG|G–5′

*Xho*I   5′–C|TCGA G–3′
       3′–G AGCT|C–5′

Figure 8.10 Restriction enzymes and their restriction sites in DNA. The solid line represents the position at which the enzyme cleaves the DNA strands.

point is that a restriction enzyme will cleave both strands of a double-stranded molecule at positions of the corresponding restriction sites and will therefore cut the molecule into smaller segments. For example, if a DNA molecule is 3000 nucleotide-pairs in length and contains *Alu*I sequences (i.e., $\frac{5'\text{-AGCT-}3'}{3'\text{-TCGA-}5'}$) at positions 1000 and 2000 along the molecule, then digestion of this molecule with *Alu*I enzyme will produce three DNA fragments, each 1000 nucleotide-pairs in length.

When a piece of double-stranded DNA is cut by a restriction enzyme, the resulting fragments (called **restriction fragments**) can be separated according to size by placing the fragments in slots at the edge of a slab of a jellylike material (usually **acrylamide** or **agarose**) and subjecting the gel to an electric current for several hours (Figure 8.11). This procedure, which is called **electrophoresis**, sep-

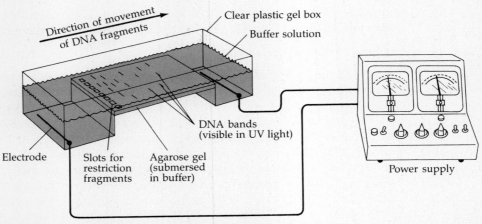

Figure 8.11 Typical setup for electrophoresis of DNA fragments.

arates the DNA fragments by size because smaller fragments move more rapidly in response to the electric field. Usually the DNA is combined with a dye that fluoresces in response to ultraviolet light. After electrophoresis, the several size classes of DNA are visible as discrete bands when viewed with ultraviolet light (see Figures 8.11 and 8.12).

The fragments produced by one restriction enzyme can be removed from the gel and digested with a different restriction enzyme. The size of the fragments produced by the second enzyme reveals the positions of the recognition sequence of the second enzyme relative to those of the first. For example, if an *Alu*I restriction fragment is cut into two fragments of equal size by *Eco*RI, then we know that somewhere on the DNA molecule there is an *Eco*RI site flanked by two equally distant *Alu*I sites. When such studies are carried out systematically with a large number of different restriction enzymes, the relative positions of all the restriction sites in a particular piece of DNA can be determined. The resulting diagram showing the locations of the restriction sites is known as a **restriction map**. Figure 8.13 is a restriction map of a segment of mouse DNA that includes a hemoglobin gene.

Generating a restriction map is often a preliminary to the ultimate level of DNA study—**DNA sequencing**, which refers to the determination of the actual nucleotide sequence in a molecule. Restriction enzymes are important tools in DNA sequencing because they provide fragments of convenient length with

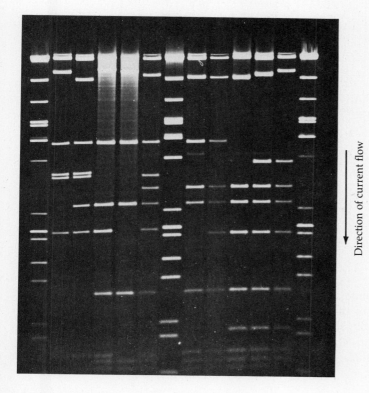

Figure 8.12 DNA fragments produced by restriction-enzyme digestion of cloned DNA molecules from yeast.

Direction of current flow

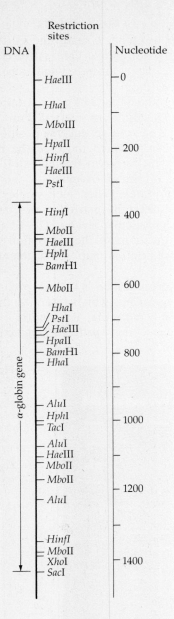

Figure 8.13 Restriction map of mouse α-globin gene showing position of restriction sites for several restriction enzymes.

known terminal sequences. The detailed methods involved in DNA sequencing are beyond the scope of this book, but the important point is that modern DNA sequencing is technically straightforward, accurate, and rapid—so much so that molecules of more than 50,000 nucleotides have been sequenced, and sequence information is accumulating much more quickly than its significance can be fully comprehended.

RECOMBINANT DNA

Restriction analysis and DNA sequencing both require relatively large amounts of a specific DNA fragment. Suitable amounts are usually obtained by means of **recombinant DNA**, which refers to DNA molecules that have been artificially created by chemically splicing together DNA sequences from different organisms. One widely used method of recombinant DNA involves bacterial plasmids. **Plasmids** (see Figure 8.14) are small, ring-shaped molecules of DNA often found inside bacterial cells in addition to the cells' own DNA and often present in multiple copies (i.e., 30 to 35 per cell). Many plasmids carry genes for antibiotic resistance and therefore are important to the cells that carry them.

Plasmids are transmitted from cell generation to cell generation via the processes of DNA replication and cell division. As the bacterial DNA replicates, the plasmid DNA replicates, too, and when the bacterial cell divides, each daughter cell receives the plasmid DNA along with the bacterial DNA. When a plasmid carrying a DNA fragment of interest replicates inside a living bacterial cell, the fragment of interest is replicated also. When the bacterial cell divides, all its offspring will receive the artificially created plasmid. Within a large clone of bacteria descended from a single, plasmid-carrying cell, there may be 10^{15} or more replicas of the plasmid, each carrying the DNA fragment of interest. All that is required to recover the DNA fragment is to purify the plasmid DNA from the clone.

Some of the steps involved in creating recombinant DNA are illustrated in Figure 8.15. Part (*a*) shows a gene of interest (black line) along with plasmid DNA (gray ring), and restriction sites flanking the gene and in the plasmid are indicated. When these DNA molecules are cleaved with the appropriate restriction enzyme, the fragments in part (*b*) are obtained. (Note that the plasmid ring has been opened at its restriction site.) The fragments in (*b*) are then mixed

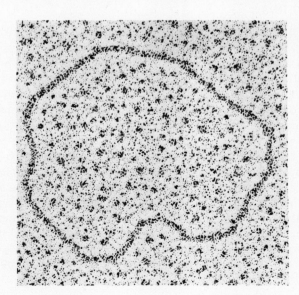

Figure 8.14 Electron micrograph of a bacterial plasmid.

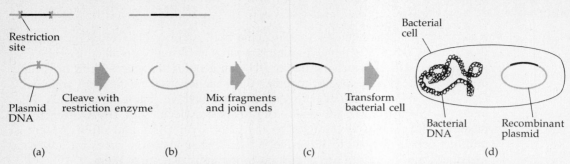

Figure 8.15 Commonly used method for production of recombinant DNA, which involves cleavage of a plasmid and the DNA of interest with a restriction enzyme, joining together of the fragment with the opened plasmid ring, and introduction of the recombinant plasmid into a bacterial cell.

together and combined with an enzyme that joins the free ends of DNA molecules, and this joining will insert the gene of interest into the plasmid ring [see part (c)]. The next step (d) is called **transformation**, and in this step the plasmid is reintroduced into a bacterial cell. When such a newly infected cell is provided with nutrients for growth, it begins to grow and divide, producing a clone of cells carrying an identical recombinant-DNA plasmid. Because transformed cells form such a clone from which the recombinant-DNA plasmid can be isolated, the procedure for producing recombinant DNA molecules is often called **DNA cloning**. Because bacterial cells are small and divide rapidly, DNA cloning can provide virtually unlimited quantities of any particular DNA fragment of interest from any organism.

APPLICATIONS OF RECOMBINANT DNA

The development of the technology of DNA cloning has caused a revolution in genetics because it has opened the way for a direct biochemical analysis of genes and of the manner in which the expression of genes is controlled. With traditional approaches, information about a gene's structure and function had to be inferred from studies of the ultimate product coded by the gene (usually a protein, such as hemoglobin) or from the phenotypic effects caused by mutations in the gene. A vast amount of important information about gene structure and function has been obtained from such traditional approaches, but studies of the DNA itself are often more direct and to the point. Indeed, one new approach to genetics— an approach still in its infancy—involves the chemical alteration of cloned DNA fragments to produce precisely defined changes in nucleotide sequence. When these altered fragments are introduced back into the cell, the phenotypic effects of the engineered mutation can be determined. This new approach stands genetics on its head. In the past, geneticists tried to understand the chemical nature of a mutation by studies of its phenotypic effects; with engineered mutations, geneticists can now study the phenotypic effects of mutations whose chemical nature is already known.

In addition to causing a revolution in genetics and molecular biology, the technology of recombinant DNA has caused a revolution in pharmacology—so much so that a number of specialized genetic-engineering companies are thriving, and many huge pharmaceutical and chemical companies have eagerly expanded their research departments to encompass the new technology. The promise of recombinant DNA is, of course, that bacteria carrying an appropriate DNA fragment might produce the gene product corresponding to the fragment. If this product was of medical or other importance, the bacteria could be used to produce it in abundance and relatively inexpensively.

The promise of recombinant DNA has already been achieved in several noteworthy instances. Bacterial production of human insulin is a prime example. In the past, the insulin used in the treatment of diabetes had to be extracted from animal pancreases, and it was scarce and expensive. In addition, many diabetics developed allergic reactions to the animal form of the protein. Recombinant DNA techniques have now been used to create bacterial strains that produce human insulin in large amounts, and this insulin has been used clinically with no reported adverse effects.

Insulin production is one of the cardinal practical achievements of recombinant DNA, but other successes are human interferon (a protein thought to be important in the body's defense against viral infection) and human growth hormone (a protein needed for successful treatment of children with a rare disease called pituitary dwarfism; it is otherwise obtainable in active form only from the pea-sized pituitary gland of human cadavers). These examples of the practical uses of recombinant DNA are only the early successes. Many other examples, some equally dramatic, are sure to follow.

The blessings of recombinant DNA are not entirely unmixed, however. From the very beginnings of the research, some people, including a few scientists themselves, have had a vague sense of unease. In some cases, these feelings were motivated by a general philosophical belief that one ought not to tamper with living things, and these feelings were enhanced by sensationalist press reports about scientists "creating new forms of life" with allusions to Doomsday Bugs and Frankensteins. For others, the source of anxiety was not the creatures that might be unleashed by design; it was the potential catastrophes that might be triggered by accident. These people pointed out that *E. coli* (then and still a popular bacterium for DNA cloning) is a normal resident of the intestinal tract of all humans and other mammals. If the wrong genes were cloned in *E. coli* by accident (an often-used hypothetical example involved the genes for tumor induction from human cancer-causing viruses), and if the *E. coli* clone carrying these genes escaped from the laboratory, the escaped clone might be able to invade the intestine and replace the normal bacteria. In such an event, an epidemic of awesome proportions might ensue. On the other hand, almost everyone agreed that the dangers of recombinant DNA were potential, not imminent. Nevertheless, critics of the procedures argued that the risks had not been properly assessed. They urged caution, circumspection.

Sympathetic to the need for caution, a group of 11 prominent and influential scientists banded together in the summer of 1974 and urged a worldwide moratorium on several lines of research involving recombinant DNA that were judged to be potentially hazardous. Four scientists in the group were themselves directly involved in the research they agreed to suspend. The moratorium on the potentially harmful research was followed by an international conference held February 24 to 27, 1975, at the rustic, former YMCA camp of Asilomar on the California coast. The conference included 86 American and 53 foreign scientists from 15 countries, and its purpose was to discuss the research and try to reach agreement on how to minimize the potential dangers.

The conferees at Asilomar, who included attorneys and public health officials as well as scientists, decided that there were certain experiments for which the risks outweighed the benefits. Among these were experiments to clone DNA from known disease-causing organisms or from organisms that produce toxins or venoms. The Asilomar group urged that such experiments be banned. The group also decided that other experiments could be classified according to their degree of risk and that each degree of risk could be minimized with appropriate levels of physical and biological containment of the recombinant clones.

Physical containment refers to the laboratory setting of the research—to such issues as whether it is open to unauthorized access and whether proper facilities for working with potentially dangerous organisms are available and used. The highest degree of physical containment applies to laboratories in which the organisms are maintained and manipulated in glove boxes and not directly handled by anyone. Such facilities were originally developed for the study of such extremely pathogenic organisms as the ones that cause smallpox, anthrax, or botulism. **Biological containment** refers to the organism that harbors the recombinant DNA. The lowest level of biological containment involves the normal laboratory strain of *E. coli*; the highest level involves strains carrying certain mutations that render them so weak and feeble and so nutritionally fastidious that they are virtually unable to survive outside of the laboratory. The conferees at Asilomar decided that each allowable experiment could be carried out with levels of physical and biological containment appropriate to its degree of risk.

The Asilomar conference was followed by others, both in the United States and abroad, with the result that the Asilomar recommendations, somewhat modified, are now embodied in regulations that embrace all government-supported research. Other countries have followed suit with similar regulations. These regulations are periodically reexamined and revised (although some types of experiments are still forbidden), and there is now widespread agreement that the risks of DNA cloning have been appropriately weighed and minimized.

Perhaps the greatest risk of recombinant DNA research has been insufficiently discussed. It is not the risk that some harmful bacteria will escape from the laboratory and cause an epidemic. It involves the manner in which the products of such clones are to be used. Drug abuse is one of the most serious problems faced by today's society; the main offenders are tobacco, alcohol, and tranquilizers. Will the miracle products of DNA cloning be used more responsibly than other drugs? Brain biochemists have recently discovered entire classes of mind-bending small-protein hormones; some obliterate pain and others cause feelings of elation and well-being. These small proteins will be easy to produce in large quantity with DNA cloning. How are such drugs to be used? It is hoped that they will be used responsibly, but one is naggingly reminded of Aldous Huxley's *Island,* in which the inhabitants willingly maintain themselves in a state of drug-induced euphoria, oblivious to their political and spiritual oppression.

SUMMARY

1. DNA occurs in cells as a double-stranded molecule, and many of the important structural characteristics of DNA are adequately illustrated by a ladder analogy. In this analogy, double-stranded DNA is represented as a ladder consisting of two long sidepieces connected at intervals by rungs. Each strand in the double-stranded molecule is represented as a sidepiece with half-rungs jutting off, and the ladder is completed by bringing corresponding half-rungs on the two sidepieces into contact. When DNA replicates, the two half-ladders separate, and each one is used as a **template** (pattern) for the step-by-step creation of a new partner half-ladder.

2. Actually, each DNA strand consists of a linear sequence of chemical constituents called **nucleotides**. A nucleotide is composed of three parts: (*a*) a sugar ring (deoxyribose sugar in the case of DNA), (*b*) a phosphate group attached to the $5'$ carbon of the sugar ring, and (*c*) one of four chemical groups called **bases** attached to the other side of the sugar ring. Four bases are found in DNA: **adenine, guanine, thymine,** and **cytosine.** The first two of these are known collectively as **purines**, the second two as **pyrimidines**. These four bases determine the four nucleotides that occur in DNA: **deoxyadenosine phosphate** (symbolized **A**), which carries adenine; **deoxyguanosine phosphate** (**G**), which carries guanine; **deoxythymidine phosphate** (**T**), which carries thymine; and **deoxycytidine phosphate** (**C**), which carries cytosine.

3. Nucleotides are linked together by means of the phosphate group of one becoming attached to the $3'$ carbon of the sugar ring of the next nucleotide in line. The alternating sugar-phosphate-sugar-phosphate pattern so produced provides the backbone of the DNA strand, and each sugar carries a base that juts off the backbone. (These protruding bases correspond to the half-rungs in the ladder analogy.) One end of the backbone of a DNA strand terminates with a phosphate group; this is called the **$5'$ end** of the strand. The other end of the strand terminates in a free hydroxyl (—OH) on the $3'$ carbon of the sugar ring; this end is called the **$3'$ end** of the strand.

4. Double-stranded DNA is formed when two DNA strands come together so that their bases meet in the middle (forming the ladder). The strands are brought together by weak chemical bonds called **hydrogen bonds** that form between corresponding bases in the two strands. However, adenine will form hydrogen bonds only with thymine, and guanine will form hydrogen bonds only with cytosine. Because of these base-pairing restrictions, the nucleotides in the two strands of double-stranded DNA are **complementary** in the sense that wherever one strand carries an A the other across the way will carry a T, and wherever one strand carries a G the other across the way will carry a C. In addition, hydrogen bonds

can form between the bases only if the two strands are **antiparallel**; that is, the strands must run in opposite directions so that the 5′ nucleotide of one strand is paired with the 3′ nucleotide of the other strand.

5. The normal, three-dimensional configuration of double-stranded DNA is a **double helix**, in which the sugar-phosphate backbones wind around the outside of the molecule in a right-handed helix, and the paired bases are stacked on top of one another in the interior. This double-helical configuration is 20 Å in diameter. Each backbone makes a complete turn around the helix every 34Å, and there are 10 base pairs per complete turn. This right-handed helix is known as the **B form** of DNA, but many alternative configurations of DNA are also known. Many of these involve minor alterations in the double helix in which the bases are somewhat tilted or more tightly stacked. However, DNA can also coil in a left-handed configuration called **Z DNA**. DNA inside living cells can probably switch back and forth between the B conformation and the Z conformation, but the biological role of these alternative DNA configurations is at present unknown.

6. DNA replication begins with the unwinding of part of the double-stranded molecule to produce a Y-shaped **replication fork**. The single-stranded regions are then used as templates for the synthesis of new partner strands by the enzyme **DNA polymerase**, which adds successive nucleotides to the 3′ end to progressively elongate the new strands. Because DNA polymerase can add to only the 3′ end of a growing strand, only one new strand (the one growing in the 5′ to 3′ direction) can be synthesized continuously. The other new strand is synthesized in relatively small fragments that are attached to one another when they meet. Long DNA molecules in eukaryotes initiate replication at many points along their length, and the **replication bubbles** so produced have two replication forks. When DNA replication is completed, each daughter molecule consists of one newly synthesized strand and one original strand; this mode of replication is called **semiconservative**.

7. **Restriction enzymes** are enzymes that attach to particular nucleotide sequences of double-stranded DNA. These recognition sequences are called **restriction sites**, and a restriction enzyme will cleave both strands of the DNA molecule at positions of its restriction site. An example of a restriction enzyme is *Eco*RI, whose restriction site is 5′-GAATTC-3′ / 3′-CTTAAG-5′. Study of DNA fragments produced by a group of different restriction enzymes can be used to construct a restriction map of the DNA molecule in question. A **restriction map** is a diagram showing the relative positions of restriction sites along a DNA molecule. Restriction fragments of convenient length can then be used to determine the nucleotide sequence of the DNA.

8. **Recombinant DNA** (or **DNA cloning**) refers to DNA molecules that have been artificially created by chemically splicing together DNA fragments from different organisms. One method for producing recombinant DNA involves the use of bacterial **plasmids**, which are small, ring-shaped molecules of DNA found naturally inside certain bacterial cells. The method uses a restriction enzyme to separate the gene of interest from flanking sequences and also to open the plasmid ring. These fragments are then mixed together and an enzyme added to splice the gene of

interest into the plasmid. The recombinant plasmid can then be reintroduced into a living cell by means of **transformation**, and the resulting transformed cell can be allowed to divide repeatedly to create a large, genetically identical population (a **clone**). Many replicas of the original DNA fragment can be recovered from the clone by isolating the plasmid DNA.

9. Uses of recombinant DNA range from the production of large quantities of a particular DNA sequence for use in restriction mapping or DNA sequencing to practical applications in the creation of bacterial clones that produce gene products of medical importance, such as insulin, growth hormone, and interferon. In the early years of recombinant DNA research, some anxiety arose about its potential hazards. It is now widely agreed that the potential risks have been minimized by appropriate standards of **physical containment** and **biological containment**. Less fully discussed is whether the miracle drugs that will come out of future recombinant DNA research will be used wisely and responsibly.

KEY WORDS

Adenine	Double helix	Replication fork
Antiparallel	Electrophoresis	Restriction enzyme
Antisense strand	Enzyme	Restriction fragment
B form of DNA	5′ end	Restriction map
Base	Guanine	Restriction site
Biological containment	Hydrogen bond	Semiconservative replication
Complementary DNA strands	Nucleotide	Sense strand
Covalent bond	Phosphate	Template
Cytosine	Physical containment	3′ end
Deoxyribonucleic acid	Plasmid	Thymine
Deoxyribose	Purine	Transformation
DNA polymerase	Pyrimidine	Z form of DNA
	Recombinant DNA	

PROBLEMS

8.1. Why is it important that the hydrogen bonds holding DNA strands together are relatively weak bonds, in contrast to covalent bonds, which are relatively strong?

8.2. In the context of recombinant DNA, what is transformation?

8.3. What is the distinction between physical containment and biological containment in the context of recombinant DNA?

8.4. How many different DNA sequences exactly three nucleotides in length are possible?

8.5. A DNA strand has the sequence 5′-GCATTACGAATGC-3′. What is the sequence of its partner strand?

8.6. A DNA strand consists of alternating G's and C's. What is the sequence of its partner strand?

8.7. A certain DNA strand has the nucleotide composition 23% A, 32% T, 19% G, and

26% C. What is the nucleotide composition of the partner strand? Considering both strands together, what is the nucleotide composition?

8.8. In a normal double-stranded DNA molecule, considering both strands, what is the ratio of A's to T's? Of G's to C's? Of purines (A's and G's) to pyrimidines (T's and C's)?

8.9. A DNA molecule has 17 occurrences of the sequence 5'-GGCC-3' along one strand. How many occurrences of the same sequence occur along the other strand?

8.10. A DNA strand has the sequence 5'-ATGGCGCTA-3'. When this strand was produced by DNA polymerase, which end, the 5' end or the 3' end, was synthesized first?

8.11. DNA polymerase will use bromodeoxyuridine (B) in DNA synthesis, incorporating it in place of thymidine. If the DNA strand 5'-TATAGGCAAATTG-3' is replicated in the presence of bromodeoxyuridine, what will be the sequence of the partner strand?

8.12. Restriction enzyme *Bam*HI cleaves double-stranded DNA molecules at the sequence 5'-GGATCC-3'. How many DNA fragments would be produced by *Bam*HI digestion of the following molecule?

5'-ATTGGATCCCTGAGATCCGATGGATCCGGATCTTA-3'
3'-TAACCTAGGGACTCTAGGCTACCTAGGCCTAGAAT-5'

8.13. Restriction enzyme *Xho*II has restriction site 5'-RGATCY-3', where R means any purine (A or G) and Y means any pyrimidine (T or C). If the DNA molecule in the previous problem were digested with *Xho*II, how many fragments would be produced?

8.14. In how many ways could the following DNA fragments be spliced together, and what would be the sequence of the resulting molecule?

5'-ATGC-3'　　　5'-GCTA-3'
3'-TACG-5'　　　3'-CGAT-5'

FURTHER READING

Abelson, J. 1980. A revolution in biology. Science 209: 1319–21. A brief overview of the potentials of recombinant DNA.

Bauer, W. R., F. H. C. Crick, and J. H. White. 1980 (July). Supercoiled DNA. Scientific American 243: 118–33. In many forms of DNA the double helix itself forms a higher-order helix.

Berg, P. 1981. Dissections and reconstructions of genes and chromosomes. Science 213: 296–303. Nobel-prize lecture by one of the originators of recombinant DNA.

Dickerson, R. E., et al. 1982. The anatomy of A-, B-, and Z-DNA. Science 216: 475–85. A more detailed look at the structures.

Gilbert, W., and L. Villa-Komaroff. 1980 (April). Useful proteins from recombinant bacteria. Scientific American 242: 74–94. Why and how recombinant DNA has produced an ongoing revolution in pharmacology.

Hopwood, D. A. 1981 (September). The genetic programming of industrial microorganisms. Scientific American 245: 90–102. Recombinant DNA techniques in making certain useful microbes still more useful.

Johnson, I. S. 1983. Human insulin from recombinant DNA technology. Science 219: 632–37. A remarkable achievement of modern science.

Kornberg, R. D., and A. Klug. 1980 (February). The nucleosome. Scientific American 244: 52–64. About the discovery of the elementary subunit of chromatin structure.

Lilley, D. M. J., and J. F. Pardon. 1979. Structure and function of chromatin. Ann. Rev. Genet. 13: 197–233. Details of chromatin are reviewed.

Motulsky, A. G. 1983. Impact of genetic manipulation on society and medicine. Science 219: 135–40. Genetic manipulation in a broad perspective.

Novick, R. P. 1980 (December). Plasmids. Scientific American 243: 102–27. A good overview of plasmids and their way of "life."

Richards, J. (ed.). 1978. Recombinant DNA. Academic Press, New York. An informal account of the early years of the recombinant DNA debate.

Sanger, F. 1981. Determination of nucleotide sequences in DNA. Science 214: 1205–10. Nobel-prize lecture by the inventor of one sequencing technique.

Science, February 11, 1983. (Vol. 219). An entire issue devoted to the practical applications of recombinant DNA and other molecular techniques.

Seidel, G. E., Jr. 1981. Superovulation and embryo transfer in cattle. Science 211: 351–58. Modern reproductive management in cattle husbandry.

Watson, J. D. 1976. 3d. ed. Molecular Biology of the Gene. W. A. Benjamin, Menlo Park, Calif. A classical textbook well written and superbly produced.

———. 1980. The Double Helix. W. W. Norton, New York. An entertaining account of the discovery of DNA structure with commentary, reviews, and original papers.

chapter **9**

Gene Expression: Nature Is Blind and Reads Braille

Genetic information is carried in the sequence of nucleotides in DNA, but proteins and a few kinds of RNA are the day-to-day molecular workhorses in the cell. Protein molecules have a prodigious variety of shapes, which provides them with extraordinary versatility in the functions they can carry out. Each type of protein molecule assumes a specific shape that enables it to carry out its specific function in the cell. The genetic information for the production of proteins is not found in the shape of DNA, however; it is found in the sequence of nucleotides. In this sense genetic information in DNA is like a set of instructions for making a carpenter's tools. The instructions are verbal and structurally monotonous—a string of letters and words on paper. The tools made from these instructions—mallets, awls, clamps, planes, chisels, and so on—are structurally very different because of their diverse functions. The "tools" in the cell are its proteins. Much of the biochemical machinery in cells is involved in converting the genetic information in DNA into the structure of proteins. How this reading and processing of genetic information occurs is the subject of this chapter.

THE STRUCTURE OF PROTEINS

Protein molecules have complex, convoluted, three-dimensional structures, as illustrated in Figure 9.1, depicting the oxygen-carrying protein **hemoglobin** found in red blood cells. Every kind of protein molecule has a three-dimensional structure that is different from every other one. This variety of shapes is what allows proteins to carry out so many different functions in the cell. Fortunately, proteins have an underlying simplicity that permits certain fundamentals to be comprehended, even though their three-dimensional structures are complex. Proteins are composed of linear strings of building blocks known as **amino acids**. There are only twenty commonly occurring types of amino acids, although a typical protein may have a hundred or more of these hooked together. Amino acid structures (and their one-letter and three-letter abbreviations such as F or phe for phenylalanine) are shown in Figure 9.2.

Some of the structures are relatively complex, but each amino acid can be

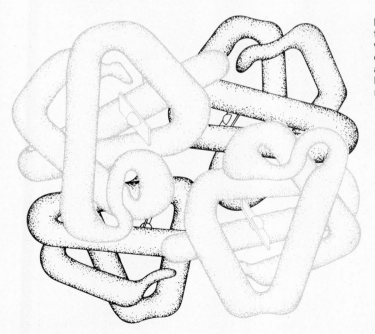

Figure 9.1 Three-dimensional structure of adult hemoglobin molecule, which contains two each of two types of polypeptide chains, called alpha (α) and beta (β). The heme groups that bind with oxygen are shown as small rectangles.

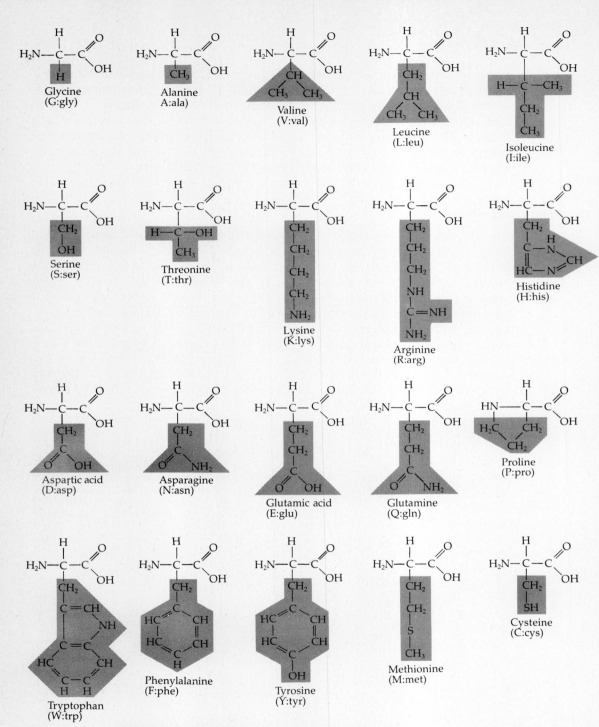

Figure 9.2 Structures of the 20 commonly occurring amino acids, laid out as in Figure 9.3. Note particularly the various types of R groups. Proline is an exceptional amino acid in that its R group is attached back to its own amino group. Each amino acid has two abbreviations, one letter and three letters.

represented in the manner shown in Figure 9.3. Some teachers use the many colors and shapes of toddlers' plastic pop beads to illustrate how diverse amino acids can connect together to form long chains. Like pop beads, which have a knob at one end and a hole at the other, one end of any amino acid is different from the other. One end of the amino acid carries an N, and this defines the **amino end** of the amino acid (so called because —NH₂ is an amino group). The other end carries a C, and this defines the **carboxyl end** of the amino acid (because —COOH is a carboxyl group). The part denoted R (for "radical") and shaded in Figure 9.3 corresponds to the shaded parts of the amino acids in Figure 9.2. The R group is different for each amino acid and gives to each amino acid its unique chemical and physical properties.

Although the role of amino acids as the building blocks of proteins is of greatest concern in this chapter, some amino acids are important in their own right. For example, phenylalanine is the amino acid whose presence in excess in body fluids causes **phenylketonuria** (or **PKU**), a disease inherited as an autosomal recessive and characterized by severe mental retardation and light skin pigmentation and hair color (see Figure 9.4). The disease results from a defect in an enzyme that converts phenylalanine into another amino acid, tyrosine, and it affects about 1 in 10,000 newborn children. Such inherited defects in body chemistry (**metabolism**) are often called **inborn errors of metabolism**. (Other examples of inborn errors, such as Tay-Sachs disease and Lesch-Nyhan syndrome, were discussed in previous chapters.)

Because children affected with PKU are unable to convert phenylalanine to tyrosine, toxic amounts of phenylalanine accumulate in the body fluids, causing, among other effects, severe brain damage. At the same time, the abnormally low levels of tyrosine in these children result in a reduced amount of pigmentation because the pigment molecule melanin is produced from tyrosine by a series of chemical alterations. If children with PKU are diagnosed soon enough after birth, they can be placed on a specially concocted diet low in phenylalanine. The child is allowed only as much phenylalanine as can be used by the body in manufacturing proteins, and phenylalanine is prevented from accumulating in the body. By this special dietary treatment the detrimental effects of excess phenylalanine on mental development are almost completely avoided. Several years after birth, when the nervous system is fully developed, the affected child can often return to a normal or nearly normal diet with no ill effects.

Dietary treatment of PKU represents one of the earliest successes of sci-

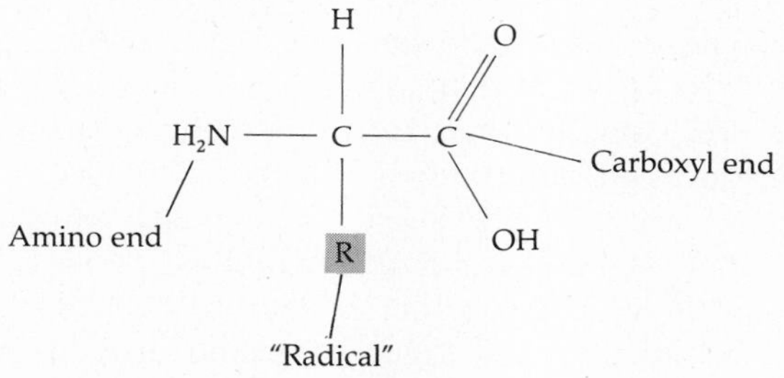

Figure 9.3 Generalized structure of an amino acid having a central carbon attached to an amino group, a carboxyl group, and a "radical" (shaded). The structure of the "radical" differs from amino acid to amino acid and gives each its unique chemical characteristics.

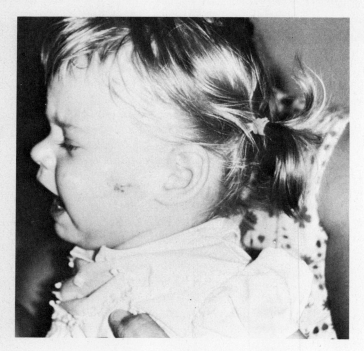

Figure 9.4 A 13-month-old child with phenylketonuria.

entific medicine, and in most parts of the United States today all newborns are routinely examined for PKU. The success with PKU has not been entirely unmixed, however. We now know that the women who have PKU genetically but were saved from mental retardation by the dietary treatment have a very high frequency of mentally retarded offspring, even when the father is normal and the child does *not* have PKU. This type of mental retardation evidently results from abnormal phenylalanine metabolism in the mother's uterus. A completely successful dietary treatment of this condition has yet to be developed, but attempts to control it are being made by putting the mother back on a low-phenylalanine diet early in pregnancy or even prior to conception.

Earlier we noted that proteins consist of hundreds of amino acids hooked together in the manner of pop beads. The details of the amino acid attachment are illustrated for three amino acids in Figure 9.5. In this figure, each amino acid is arranged as in Figure 9.3. The bonds connecting them (heavy lines) are called **peptide bonds**. The N—C—C—N—C—C—N—C—C backbone of the protein is highlighted in boldface type, and it can be seen that the amino end of the protein is at the left, the carboxyl end at the right. During protein synthesis, the amino acids are added to the chain one by one, each new amino acid being added to the carboxyl end of the growing chain. In the lower part of Figure 9.5 two simpler versions of the chemical structure are shown, one based on the three-letter abbreviations and the other on single-letter abbreviations.

A chain of amino acids hooked together as shown in Figure 9.5 is properly called a **polypeptide**. The number and kinds of amino acids in such a chain

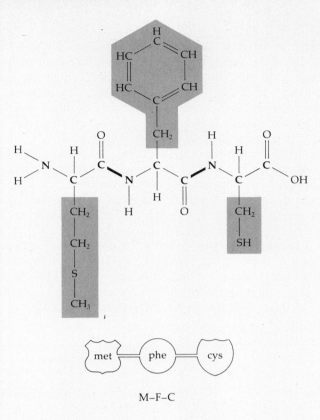

Figure 9.5 Polypeptides consist of amino acids hooked together by means of peptide bonds (heavy lines). Just three amino acids are shown here.

M–F–C

determine its three-dimensional structure because the "R" groups of amino acids along the chain interact by means of hydrogen bonds and other chemical forces to coil and twist the chain into a relatively stable configuration. Sometimes two or more different polypeptide chains must come together to form the biologically active protein molecule. For example, a molecule of hemoglobin (Figure 9.1) contains four polypeptide chains. Two of the chains are known as **α chains**; these are products of the α-globin genes. The other two chains are **β chains**, which are products of the β-globin gene. (The mutation causing sickle cell anemia is in the β-globin gene.) The α and β polypeptides come together as they do because of chemical interactions determined by their precise configuration of amino acids.

The three-dimensional structure of a polypeptide chain is thus an automatic consequence of its amino acid sequence, so the question of how the genetic information in DNA becomes expressed in the three-dimensional structure of a protein can be rephrased as the much simpler problem of how the information in DNA becomes expressed in the amino acid sequence of a polypeptide chain. This process involves two rather distinct steps: **transcription**, which refers to the synthesis of an RNA molecule using the genetic information in DNA, and **translation**, which refers to the manufacture of a polypeptide chain from the information in the RNA molecule.

TRANSCRIPTION

The molecular structure of RNA is very much like that of DNA, with three important exceptions. First, the sugar in RNA is ribose rather than deoxyribose [Figure 9.6(a)]. Second, most RNA is made up of the bases adenine, guanine, cytosine, and **uracil** instead of adenine, guanine, cytosine, and **thymine**. Uracil (U) fills the role of thymine (T); and, since the structures of uracil and thymine are very similar [see Figure 9.6(b)], uracil undergoes base pairing with adenine in RNA just as thymine pairs with adenine in DNA. Third, most RNA in cells is single stranded, unlike DNA, which is normally double stranded. Although RNA is a single-stranded molecule, it can fold back on itself to form hairpinlike structures in which the pairing of complementary bases can occur.

The base sequence along an RNA molecule is specified by the base sequence along a corresponding length of DNA. The transfer of information from the base sequence of DNA to a base sequence of RNA is accomplished in the process of **transcription**. In this process, an enzyme called **RNA polymerase** binds with the "sense" strand of DNA at a particular region of the DNA that carries a base sequence called a **promoter** sequence. Once attached to the DNA strand, the polymerase proceeds in one direction along the DNA molecule and uses the DNA as a template for the synthesis of RNA. As the polymerase travels along the

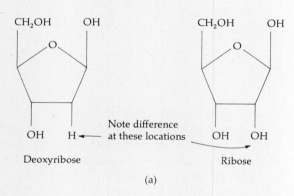

Figure 9.6 Two major differences between DNA and RNA. (a) The sugar deoxyribose in DNA is replaced with ribose in RNA. (b) The base thymine (T) in DNA is replaced with uracil (U) in RNA.

DNA it manufactures an RNA molecule exactly complementary in base sequence to the sense DNA strand (the template) by adding **ribonucleotide** (phosphate-ribose-base) units one by one to the growing RNA chain (Figure 9.7). Like DNA polymerase, RNA polymerase can add successive units only to the 3′ end of a growing strand, so the enzyme can travel in only one direction. A DNA strand with the sequence 3′-ATGCATGC-5′ would be transcribed as the RNA sequence 5′-UACGUACG-3′. At some point farther along the DNA strand a "stop" signal for transcription is encountered by the polymerase, and at this point transcription ceases and the newly synthesized RNA strand is released. As a polymerase molecule moves off the promoter at the beginning of transcription, a second polymerase can bind and initiate transcription, proceeding on the heels of the first. As soon as the second polymerase vacates the promoter sequence, a third polymerase can bind, and then a fourth, a fifth, and so on. In this way many RNA molecules can be produced in close order by polymerase molecules moving in single file along a length of DNA (Figure 9.8).

The essential feature of transcription is that it is the process whereby RNA molecules are produced from the genetic information in the sense strand of DNA and that the nucleotide sequence of an RNA transcript is complementary to the nucleotide sequence of the DNA strand from which it was transcribed. Transcription is an essential step in gene action. Indeed, transcription provides a suitably modified definition of a gene that is useful for some purposes: a *gene* is a sequence of nucleotides in DNA that is transcribed into an RNA molecule. Since the DNA in the nucleus stays in the nucleus, whereas RNA can travel

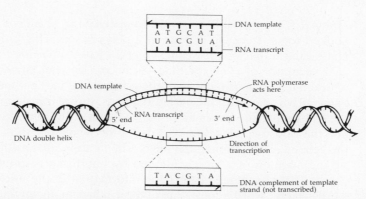

Figure 9.7 Transcription is the process whereby an RNA molecule is produced that is complementary in base sequence to a section of one of the DNA strands (the "template" strand) in a DNA double helix. The diagram shows a DNA double helix, partly unwound, with transcription occurring from the DNA template in the unwound region. (The sugar-phosphate backbone of the RNA transcript is denoted by the inside line.) The upper box shows in more detail a six-base segment of the template and transcript; note the base pairing (A with T and G with C), although, in RNA, U replaces T. The lower box shows the base sequence of the nontemplate DNA strand for the same six-base segment as detailed in the upper box. Note here, first, that the base sequence is complementary to the base sequence of the DNA template strand, and, second, that the base sequence is the same as that found in the RNA transcript from the template strand (except, of course, that in RNA, U replaces T). The process of transcription is carried out by a special enzyme—RNA polymerase.

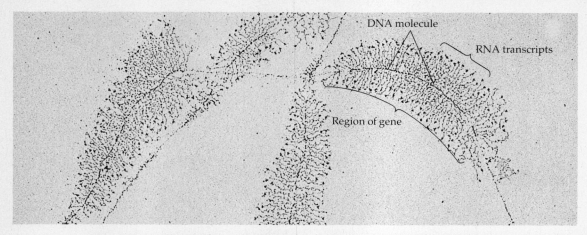

Figure 9.8 A DNA molecule caught in the act of transcription. The DNA shown codes for rRNA in the spotted newt *Triturus viridescens*. Note that many RNA molecules, or transcripts, are produced simultaneously from each distinct length of DNA because many polymerases move in tandem along the DNA.

from nucleus to cytoplasm, transcription of the genetic information of DNA into RNA provides the means by which genetic information travels from nucleus to cytoplasm.

TRANSLATION

Many RNA transcripts are used to specify the sequence of amino acids to be hooked together in the making of a polypeptide. This process—using the information in the nucleotide sequence of an RNA molecule to determine the amino acid sequence of a polypeptide—is known as **translation**, and RNA molecules whose nucleotide sequences become translated into polypeptides are called **messenger RNAs** (or **mRNAs**). What happens during translation is that the 5' end of the RNA molecule becomes associated with a ribosome particle in the cytoplasm and the ribosome slowly moves down the whole length of the mRNA molecule, the genetic information in the mRNA being translated as the ribosome goes along by the addition of amino acids one by one to an ever-lengthening polypeptide chain. The ribosome "reads" the information in the mRNA molecule word by word—each "word" consisting of three adjacent nucleotides in the mRNA. Each such sequence of three nucleotides triggers the addition of a specific amino acid to the growing polypeptide chain.

The translation process involves two other types of RNA that are not mRNA. That is to say, these two types of RNA are not translated into polypeptides but carry out different functions in the cell. **Ribosomal RNA (rRNA)**, as its name implies, is associated with the ribosomes. These tiny organelles, found by the thousands in the cytoplasm of most cells, are the sites of polypeptide synthesis. Translation of mRNA molecules takes place on the ribosomes. As one might expect of structures able to perform such a prodigious task as polypeptide synthesis, ribosomes are very complex little organelles, each containing one copy of each of three distinct RNA molecules and more than 50 different kinds of proteins.

Ribosomes are but one major part of the translational machinery; transfer RNAs constitute the other major part. **Transfer RNA (tRNA)** molecules are the molecules that actually "translate" the nucleotide sequence of mRNA into the amino acid sequence of a polypeptide. One might say that the ribosome is the site of translation, the reader, whereas the tRNAs provide the dictionary. During translation, the nucleotides in the mRNA are read from one end to the other in groups of three, each group called a **codon** and each codon specifying a particular amino acid to be inserted into the polypeptide chain at the corresponding position (see Figure 9.9). The sequences 5′-UUU-3′, 5′-UAU-3′, 5′-AUG-3′, and 5′-CCC-3′ are all examples of codons. Specific transfer RNAs recognize the codons and bring the correct amino acids into line, and each amino acid is then linked onto the growing polypeptide chain.

Transfer RNAs, of which there are dozens of different kinds, contain 70 to 80 nucleotides. Prior to participation in translation, the tRNAs undergo extensive modification in the cytoplasm through the addition of a few nucleotides to one end of the molecule and the chemical alteration and modification of certain bases within the molecule. The fully tailored tRNA is able to fold back on itself in a sort of cloverleaf configuration to allow base pairing to occur between certain complementary sequences.

This structure is shown schematically in Figure 9.10(*a*), although the three-

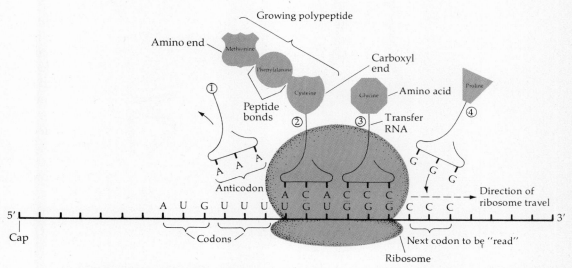

Figure 9.9 An mRNA molecule at one instant during translation. The tRNA for phenylalanine (1) has just been released from the ribosome. The cysteine of the growing polypeptide will now be detached from its tRNA (2), and a peptide bond will form with the glycine attached to its tRNA (3). The ribosome will then shift to the right one codon, tRNA (2) will be released as tRNA (3) (carrying the polypeptide) comes to occupy the left-hand position, and the proline tRNA (4) will come to its place at the right-hand position.

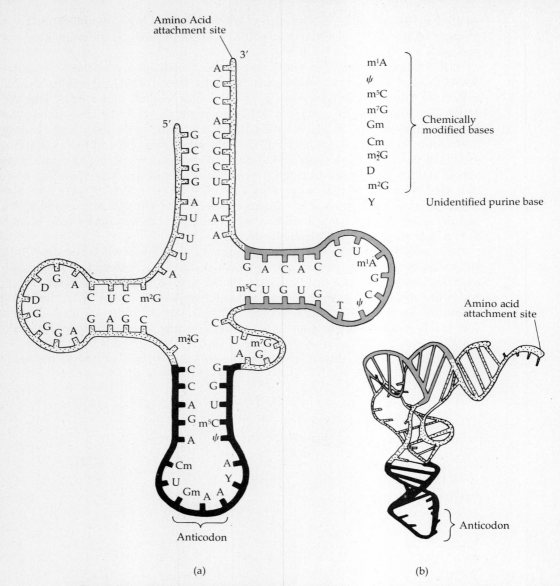

Figure 9.10 Structure of phenylalanine tRNA from yeast cells. (*a*) The ribonucleotide sequence in a sort of "cloverleaf" configuration; note a number of chemically modified bases, the black loop containing the anticodon, and the gray loop. (*b*) The same tRNA molecule in the three-dimensional form it assumes, where the bars indicate hydrogen-bonded base pairs. The gray and black regions correspond to those in part (*a*).

dimensional configuration of the molecule is slightly more complex [Figure 9.10(*b*)]. All known tRNAs have this same overall structure: four base-paired regions and three loops of unpaired bases. The business end of the molecule is the bottom loop in the figure, containing a specific sequence of three nucleotides known as the **anticodon**. By base pairing, the anticodon is able to recognize the corresponding three-base codon in the mRNA. Each of the many kinds of tRNA has a unique anticodon that distinguishes it from all other tRNAs.

The other end of a tRNA molecule ready to participate in translation has an amino acid attached to it. Transfer RNAs with the same anticodon always become coupled with the same amino acid, and this coupling is carried out by a group of cellular enzymes, each enzyme specific for one amino acid. Thus, in translation, a codon in the mRNA specifies an anticodon according to the rules of base pairing, and the tRNA bearing this anticodon is arranged in advance to carry the correct amino acid at the other end. In this way a particular three-base codon in mRNA specifies a particular amino acid.

All in all, well over a hundred different molecules are required to translate any mRNA molecule correctly. The many constituents of ribosomes are necessary, as are the tRNAs, each charged with its proper amino acid. The first step in translation is the recognition of mRNA by the ribosome. This involves recognition of the mRNA **cap**, which refers to an unusual nucleotide in an unusual orientation that is added to the 5′ end of the mRNA before the RNA leaves the nucleus (see Figure 9.9). (Such caps are found only in eukaryotic mRNAs, not in prokaryotes.) Once the ribosome is attached to the cap it proceeds in the 3′ direction along the mRNA until it encounters the codon 5′-AUG-3′, which corresponds to the amino acid methionine and is the actual "start" signal for polypeptide synthesis.

When the 5′-AUG-3′ codon is encountered, methionine is established as the first amino acid in the polypeptide. Then the second codon is read and its corresponding amino acid is attached to the methionine. The third codon is read and its corresponding amino acid is attached to the second amino acid. Then the fourth codon is read, and the fifth, and sixth, and so on, each additional codon causing another amino acid to be attached to the growing polypeptide chain. This process continues until a "stop" signal is encountered in the mRNA; the "stop" signal can be any one of the codons 5′-UAA-3′, 5′-UAG-3′, or 5′-UGA-3′, and such a codon causes the finished polypeptide to be released from the ribosome. The methionine with which translation started is often clipped off the finished polypeptide by an enzyme. Translation always occurs in the 5′ to 3′ direction along the mRNA, and polypeptide synthesis always proceeds from the amino end of the polypeptide to the carboxyl end and progresses at the rate of about 15 amino acids per second. A molecule of, say, hemoglobin (the β chain) is completed in roughly 10 seconds. Moreover, several ribosomes can move simultaneously in tandem along an mRNA molecule (see Figure 9.11).

THE GENETIC CODE

The detailed correspondence between codons in mRNA and the amino acids they stand for is known as the **genetic code**; this is shown in Table 9.1. Note first that all 20 amino acids are signaled by at least one codon, most by several codons. 5′-AUG-3′ (coding for methionine) is the start signal for translation, and 5′-AUG-3′ codons also serve to insert methionine at interior positions in the polypeptide chain. Codons 5′-UAA-3′, 5′-UAG-3′, and 5′-UGA-3′ correspond to stops in translation.

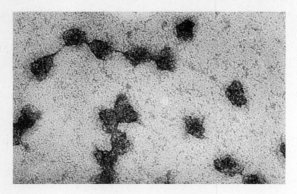

Figure 9.11 Electron micrograph of an mRNA molecule being translated simultaneously by several ribosomes moving along it in tandem. This mRNA and group of ribosomes, taken from a rabbit cell, is active in hemoglobin synthesis.

The codon assignments in the code were worked out by a variety of techniques; one of the most straightforward techniques involved the use of artificially synthesized mRNAs of known nucleotide sequence. Such mRNAs are translated in a test tube by an appropriate mixture of ribosomes, tRNAs, amino acids, and so on, and the amino acid sequence of the polypeptide produced can be compared with the known sequence of bases in the mRNA. (By treating the ribosomes with large amounts of magnesium, translation can be made to begin with any codon and not just with 5'-AUG-3'.) An mRNA sequence 5'-UUUUUUUUUUUU-····-UUU-3', for example, will be translated as phe-phe-phe-····-phe (that is to say, a long string of phenylalanines); thus 5'-UUU-3' must be a codon for phenylalanine.

Most amino acids are denoted by more than one codon—that is, the genetic

Table 9.1   THE GENETIC CODE

		U			C			A			G		
							Second nucleotide in codon						
U	U	UUU	phe	F	UCU	ser	S	UAU	tyr	Y	UGU	cys	C
		UUC	phe		UCC	ser		UAC	tyr		UGC	cys	
		UUA	leu	L	UCA	ser		UAA	(stop)		UGA	(stop)	
		UUG	leu		UCG	ser		UAG	(stop)		UGG	trp	W
	C	CUU	leu	L	CCU	pro	P	CAU	his	H	CGU	arg	R
		CUC	leu		CCC	pro		CAC	his		CGC	arg	
		CUA	leu		CCA	pro		CAA	gln	Q	CGA	arg	
		CUG	leu		CCG	pro		CAG	gln		CGG	arg	
	A	AUU	ile	I	ACU	thr	T	AAU	asn	N	AGU	ser	S
		AUC	ile		ACC	thr		AAC	asn		AGC	ser	
		AUA	ile		ACA	thr		AAA	lys	K	AGA	arg	R
		AUG*	met	M	ACG	thr		AAG	lys		AGG	arg	
	G	GUU	val	V	GCU	ala	A	GAU	asp	D	GGU	gly	G
		GUC	val		GCC	ala		GAC	asp		GGC	gly	
		GUA	val		GCA	ala		GAA	glu	E	GGA	gly	
		GUG	val		GCG	ala		GAG	glu		GGG	gly	

First nucleotide in codon (5' end) — left margin label

Note: The table shows the correspondence (the genetic code) between three-base codons in mRNA and the amino acid inserted into the polypeptide. Amino acids are represented by two types of abbreviations: a three-letter and a single-letter abbreviation. The asterisk (*) on the codon AUG (methionine) denotes that this codon is the "start" signal for protein synthesis. Codons UAA, UAG, and UGA are termination (stop) signals.

code has synonyms. Notice that most synonyms are identical in their first two nucleotides and differ only in the third position. This comes about because the anticodons in tRNAs do not base-pair with complete fidelity in the third position. Many anticodons can distinguish U or C from A or G in the third codon position, but they cannot distinguish U from C or A from G in this position. The weak fidelity of base pairing in the third-codon position is evidently due partly to the structure of the anticodon itself and to how the tRNA adheres to the ribosome during translation.

At one time it was thought that the genetic code was **universal**, that the same code was used in all organisms from bacteria to slime molds to peach trees to marmosets to humans. This has turned out to be very nearly true. All *organisms* so far studied do use the same genetic code, but interesting exceptions are found in the small number of proteins produced in mitochondria by their own translational apparatus. For example, human mitochondria use the code in Table 9.1 with the following exceptions: 5'-AUA-3' codes for methionine instead of isoleucine; 5'-UGA-3' codes for tryptophan instead of stop; and 5'-AGA-3' and 5'-AGG-3' are stop codons instead of codons for arginine. These exceptions are matters of detail, however, and they support the view that all living things are fundamentally related and that the genetic code is ancient.

THE MECHANICS OF TRANSLATION

The genetic code specifies the correspondence between a segment of mRNA and the polypeptide produced. As an example, consider the mRNA sequence 5'-AUGUUUUGUGGGCCC-3'. The first five codons in the mRNA are therefore 5'-AUG-3', 5'-UUU-3', 5'-UGU-3', 5'-GGG-3', and 5'-CCC-3'. Ordinarily, AUG would not be the leading end of the mRNA sequence; there would be some prior recognition sequence terminated by a cap important in initiating translation, but this sequence would not be translated. Initially the mRNA and a ribosome join together along with methionine(met)-carrying tRNA (at the 5'-AUG-3' codon) and phenylalanine(phe)-carrying tRNA (at the 5'-UUU-3' codon next in line). The structure can be represented schematically like this:

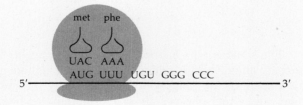

In this diagram the tRNAs are drawn as simple hairpins, omitting the side loops. Also, the anticodons are shown to be exact base-pair complements of the codons. This is not strictly correct, especially with respect to the nucleotide that pairs in the third-codon position, which is often one of the chemically modified bases mentioned previously. Thus the diagram takes some liberties in this regard.

When the mRNA-ribosome-tRNA complex is established as in the diagram above, four things happen almost simultaneously. First, the met (methionine) is cleaved from its tRNA. Second, the methionine is connected chemically to the next amino acid, in this case phe (phenylalanine), by a peptide bond of the type indicated by the heavy lines in Figure 9.5. Third, the tRNA for methionine is released from the ribosome and becomes free in the cell, there to become re-charged with another molecule of methionine by its associated enzyme. Fourth and finally, the ribosome shifts along the mRNA to the next three-base codon, in this case 5'-UGU-3'. We then have:

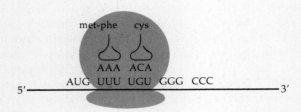

Then the same things happen: the met-phe is clipped off the tRNA for phenylalanine and attached to cys (cysteine), the tRNA for phenylalanine falls free, and the ribosome moves on to the next codon—5'-GGG-3' in the example:

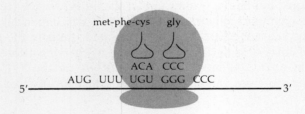

The same events happen over and over again. With each codon translated, a specific amino acid is added to the ever-lengthening polypeptide. The first five amino acids of the polypeptide produced by the mRNA in the diagrams above would be met-phe-cys-gly-pro (methionine-phenylalanine-cysteine-glycine-proline), corresponding to the codons 5'-AUG-3', 5'-UUU-3', 5'-UGU-3', 5'-GGG-3', and 5'-CCC-3'. The string of amino acids grows until one of the triplets 5'-UAA-3', 5'-UAG-3', or 5'-UGA-3' is encountered in the mRNA. These, you recall, mean "stop." At this point the polypeptide attached to the last tRNA remaining on the ribosome is clipped off, and the polypeptide, which began folding into its three-dimensional form as it was being synthesized, becomes free to carry out its function in the cytoplasm, perhaps in conjunction with other polypeptides.

In overall summary, transcription of DNA results in the production of an RNA code-bearing tape (the mRNA). This code-bearing tape is translated into a string of amino acids by means of transfer RNAs, which recognize each successive codon in the mRNA and bring the correct amino acid into line. These remarkable processes both depend on the hydrogen bonding that can occur between bases in paired nucleic acids (DNA with RNA in the case of transcription, mRNA with tRNA in translation).

At least one scientist involved in several key discoveries in the early years of molecular biology has expressed disappointment that the processes were not more mysterious. " 'Terrible complexifier' that I was—always two steps ahead of reality—[I] dreamed of something much more grandiose than a plain code-bearing tape. What I did not want to acknowledge is that nature is blind and reads Braille."*

RNA SPLICING: INTRONS AND EXONS

Transcription (synthesis of RNA complementary to DNA) and translation (synthesis of polypeptide corresponding to RNA) are two fundamental processes in gene expression. In prokaryotes these are the only two major steps, and the processes are coupled. That is to say, translation of the 5' end of an RNA can be occurring as the 3' end of the same RNA is still being synthesized along the DNA. This coupling between transcription and translation is possible in prokaryotes because prokaryotic cells lack a nucleus and nuclear envelope, so the transcriptional machinery and the translational machinery are in close physical proximity.

The situation is very different in eukaryotes because the nuclear envelope separates the cell into compartments—the nucleus and the cytoplasm—and the transcriptional machinery is in the nucleus whereas the translational machinery is in the cytoplasm. This separation provides an opportunity for the RNA produced in transcription to be altered prior to its export to the cytoplasm for use in translation. Such alterations are known as **RNA processing**, and in eukaryotes RNA processing represents a third major step in gene expression that occurs between transcription and translation. One aspect of RNA processing has already been mentioned—the addition of the 5' cap to the RNA transcript. The 3' end of mRNAs is also modified during processing in the nucleus, namely by the addition of 50 to 100 consecutive A's to the 3' end, forming what is often called the **3' poly-A tail** of the mRNA.

Undoubtedly the most dramatic aspect of eukaryotic RNA processing involves the removal of entire segments of the original transcript by means of a splicing process. The effect of RNA splicing in eukaryotes can be appreciated most readily by contrasting the situations in eukaryotes and prokaryotes. In prokaryotes, RNA transcripts are translated straightaway from one end to the

* Chargaff, E. 1978. Heraclitean Fire. Rockefeller University Press, New York, p. 98.

other. Adjacent amino acids in the polypeptide are therefore coded by adjacent codons in the RNA, and these derive from adjacent triplets of nucleotides in the DNA. This sort of correspondence is summarized by saying that the molecules are **colinear**. That is, the RNA is colinear with the DNA, and the polypeptide is colinear with the RNA. The effect of RNA splicing in eukaryotes is that colinearity breaks down. Adjacent amino acids in a polypeptide need not necessarily correspond to adjacent triplets of nucleotides in the DNA.

Because the RNA produced in transcription is modified by RNA processing before it is exported to the cytoplasm to act as mRNA, these two RNAs can be very different in nucleotide sequence, and it will be necessary to distinguish them. The RNA molecule produced by transcription is called the **primary transcript** to distinguish it from the fully processed mRNA that is used in translation. It will also be necessary to refer to that part of a DNA or RNA molecule that actually codes for amino acids, which is called a **coding region**. A fully processed eukaryotic mRNA has just one coding region, which corresponds to the one polypeptide that can be synthesized from such an mRNA. However, in the corresponding DNA, this coding region is often found to be split into smaller coding regions, sometimes a dozen or more, and these are not adjacent in the DNA. Bringing these split coding regions into juxtaposition is one of the primary functions of RNA splicing.

A typical sort of eukaryotic genes-in-pieces structure is illustrated in Figure 9.12, which shows the mouse α-globin gene. (The human gene is virtually identical.) Across the top is arrayed the nucleotide sequence of the antisense DNA strand with its 5′ end at the left and its 3′ end at the right. The antisense strand is often depicted instead of the sense strand because it facilitates comparison with an RNA transcript of the sense strand; the transcript will be identical in

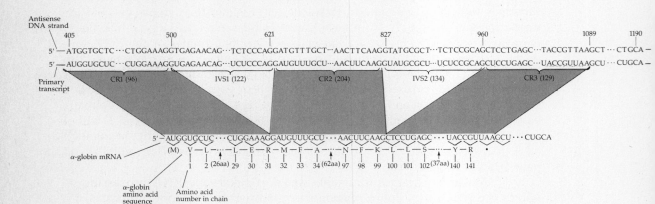

Figure 9.12 Intervening sequences (introns) in the mouse α-globin gene. Sequences corresponding to IVS1 and IVS2 do appear in the α-globin RNA transcript, but they are removed by a highly precise splicing process that brings the coding regions (CR1, CR2, and CR3) into juxtaposition in the processed mRNA. The human globin genes also have three exons (coding regions) and two introns.

base sequence to the antisense strand except that T in DNA is replaced with U in RNA. There are 787 nucleotides in the region of interest (the numbering is as in the original report), but only the key ones are shown; the rest are represented by ellipses (. . .).

The second line in Figure 9.12 shows the nucleotide sequence of the primary transcript of the mouse α-globin gene; it corresponds nucleotide-by-nucleotide with the DNA antisense strand, so the primary transcript and the DNA are colinear. The third line shows the nucleotide sequence of the α-globin mRNA, and it can be seen that two large segments of 122 nucleotides and 134 nucleotides have been spliced out of the primary transcript during the course of RNA processing. These spliced-out regions have come to be known as **intervening sequences** or **introns**. What remains in the mRNA are sequences corresponding to three coding regions, which are called **exons**.

The bottom line in the figure shows the amino acid sequence of mouse α-globin, using the single-letter amino acid abbreviations in Table 9.1. The initial methionine (M), later cleaved off, is evident at the left, and the terminating codon UAA is evident at the right. Although the polypeptide is colinear with the mRNA, it is by no means colinear with the DNA. This lack of colinearity occurs because of the RNA splicing that eliminates the intervening sequences. For example, amino acid 31 is coded by nucleotides 498–499 and 622, which are by no means adjacent nucleotides; and amino acids 99 and 100 are coded by nucleotides 824–825–826 and nucleotides 961–962–963, so adjacent amino acids do not correspond to adjacent triplets in the DNA.

In short, while the DNA and its primary transcript are colinear, and the mRNA and the polypeptide are colinear, the mRNA and the primary transcript are *not* colinear because of the removal of intervening sequences from the primary transcript by splicing. Consequently, knowledge of the nucleotide sequence of the DNA in a eukaryotic gene is not sufficient to predict the amino acid sequence that will be produced by this gene because there is no way to know merely by examining the nucleotide sequence which parts of the gene correspond to introns and which parts correspond to exons.

Much remains to be learned about the possible functions of intervening sequences and RNA splicing. Speculations range widely from hypotheses which propose that the positions and lengths of intervening sequences play an essential role in regulating the timing or level of gene expression to hypotheses at the other extreme which propose that intervening sequences are molecular remnants of the way in which genes evolve and have no present functions at all. Whatever the final answer may turn out to be, the important point for present purposes is that split genes are apparently typical of eukaryotes and that RNA processing, particularly RNA splicing, is an essential step in gene expression along with transcription and translation.

REGULATION OF GENE EXPRESSION

Transcription, RNA processing, and translation summarize the process of gene **expression**, but they do not convey how gene expression is regulated. Human beings are estimated to have about 200 different types of cells—liver cells, spleen cells, white blood cells, various types of cells in the brain and nervous system, and so on. These cells have virtually identical genetic constitutions, yet they express different groups of genes, or the same genes to different extents. Just as a pianist can produce many different-sounding chords by depressing different groups of keys, the same genetic constitution can produce many different types of cells by expressing different groups of genes. The ability of genes to be expressed only in certain cell types, only at certain times, or only at certain levels is known as gene **regulation**, and there are many levels (**control points**) at which regulation can occur; these control points correspond to the various processes involved in gene expression.

Theoretically, genes could be regulated by changing the DNA itself—by changing its structure from, say, the B form to the Z form or vice versa, or by chemically altering the DNA to make it more or less accessible to enzymes such as RNA polymerase. Genes can also be regulated at the level of transcription by proteins that interact with DNA and so make it more or less accessible to RNA polymerase. The next step in gene expression is RNA processing, and here too there are many opportunities for regulation via rates of processing or transport of RNAs from the nucleus. Translation is yet another potential control point for regulation, as certain mRNAs are "masked" in the cytoplasm and are unable to be translated until they are "unmasked." After translation there are yet more control points of gene expression, as enzymes can be modified, activated, or inhibited by their interactions with other proteins or with appropriate small molecules.

Examples of regulation at all these control points are known. Thus, there is not a single, universal process of regulation. There are many possible control points for genes, and different genes can therefore be regulated in different ways. Moreover, a population does not evolve in such a way that its regulatory processes will be elegant, clever, or simple (though many such processes are). Populations evolve to survive and they are opportunistic. If a cumbersome system of regulation should occur by chance and work tolerably well, there will be further evolution for refinement and effectiveness, but not necessarily for simplicity. On the whole, regulation seems to encompass a hodgepodge of diverse, ad hoc processes, each of which is retained because it happens to do the job. Some of these processes are beautiful in their simplicity; others are bewilderingly complex and poorly understood. In the rest of this chapter we shall provide several examples of gene regulation, more as a sampler than as a thorough review of all the possibilities.

HEMOGLOBIN REGULATION

Precise regulation of the expression of the genes for hemoglobin—the oxygen-carrying protein in the blood—occurs during human development. Humans do not have a single type of hemoglobin; we have several. As illustrated in Figure 9.1, every hemoglobin molecule consists of four polypeptide chains. In the molecules that constitute the bulk of hemoglobin in the adult, two of these chains are α chains and two are β chains. The majority of adult hemoglobin can therefore be represented by the polypeptide formula $\alpha_2\beta_2$. The α chain is the product of the α-globin gene, but most normal humans have two copies of this gene; that is to say, the α-globin gene is duplicated (see Chapter 7). The β polypeptide is the product of the β-globin gene. (The sickle cell anemia mutation affects the β chain.) Adults also have a third hemoglobin chain, the δ (**delta**) chain, which resembles the β chain but is encoded by the δ-globin gene. The δ-globin gene is not as strongly expressed as the β-globin gene, however; adult hemoglobin consists of 97 percent $\alpha_2\beta_2$ but only 3 percent $\alpha_2\delta_2$.

Each stage of human development has its characteristic types of hemoglobin, presumably because each type of hemoglobin has its own oxygen-carrying capability, and the genes expressed at any one developmental stage are appropriate to the availability of oxygen at that stage. The earliest embryos produce two other hemoglobin chains (Figure 9.13)—The ζ (**zeta**) chain, which is α-like,

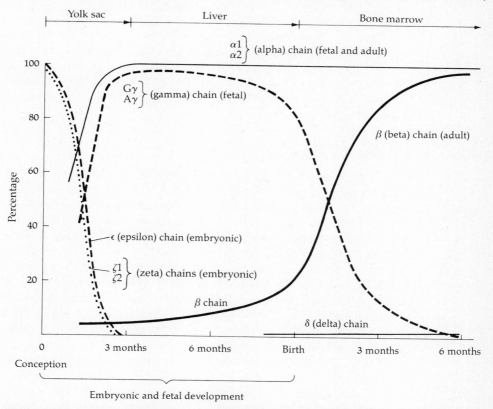

Figure 9.13 Graph showing regulation of the production of hemoglobin polypeptide chains and the site of synthesis during human development. Each type of polypeptide chain arises as the product of a different gene. The vertical axis denotes the proportion of hemoglobin molecules containing the various chains. Since each hemoglobin molecule contains four chains, the proportions at any age add up to more than 100 percent. Note the ε and ζ chains, found only in embryos, the fetal γ chain, and the small quantity of δ chain. By six months, a baby has essentially the adult complement of hemoglobin—97 percent $\alpha_2\beta_2$ and 3 percent $\alpha_2\delta_2$.

and the ϵ (**epsilon**) chain, which is β-like. There are actually two ζ-globin genes (a duplication like the α-globin situation), and in earliest embryonic development the predominant hemoglobin has the polypeptide formula $\zeta_2\epsilon_2$. Later embryos and fetuses begin a hemoglobin transition by initiating expressing of the α-globin genes and still another duplicate pair—the $^G\gamma$ (G-gamma) and $^A\gamma$ (A-gamma) pair—in liver cells. The γ chains are identical polypeptides except at amino acid position 136, where $^G\gamma$ carries glycine but $^A\gamma$ carries alanine. Since the γ chains are β-like, late embryos have three types of hemoglobin—namely, $\zeta_2\epsilon_2$, $\alpha_2\epsilon_2$, and $\zeta_2\gamma_2$. By the end of embryonic development, the embryonic globins ζ and ϵ have all but disappeared, and the fetal hemoglobin consists largely of $\alpha_2\gamma_2$ (see Figure 9.13).

At about the fifth month of development, there begins a gradual decline in the quantity of the γ chain, and expression of the β chain begins as the focus of hemoglobin synthesis switches from the liver to the bone marrow. The decrease in γ chains is almost exactly compensated for by the increase in β chains. The δ chain is not produced until just before birth, and it is present in only small amounts. At birth most of the hemoglobin is $\alpha_2\gamma_2$, with the rest largely $\alpha_2\beta_2$ except for a trace of $\alpha_2\delta_2$. The exchange of β for γ then occurs more rapidly; at about two months, the proportions of $\alpha_2\beta_2$ and $\alpha_2\gamma_2$ have been reversed from what they were at birth. The production of β instead of γ continues, and slightly more δ is produced, until at six months the child has essentially its adult complement of hemoglobin.

REGULATION BY DNA MODIFICATION

The regulation of expression of the human globin genes in Figure 9.13 is a preeminent example of regulation because it involves genes being turned on (α, β) or off (ζ, ϵ, γ) at different times in development or in different tissues (fetal liver, adult bone marrow). How this regulation occurs is not understood in detail, but recent evidence suggests that it may involve the chemical modification of certain strategically located cytosines in the DNA. In mammalian DNA, 2 to 7 percent of the cytosine bases in DNA are modified by an enzyme that creates **5-methyl cytosine** by the additon of a methyl group ($—CH_3$) to the base. (5-methyl cytosine has a methyl group at the same relative position as does thymine—see Figure 8.3.) Most of the 5-methyl cytosine bases are found in the dinucleotide sequence 5'-CG-3', but lower levels of methylation of cytosines in other sequences can also occur.

Figure 9.14 shows the organization of globin genes in human DNA. There are two clusters of such genes: the cluster of β-like genes located on chromosome 11 and the cluster of α-like genes on chromosome 16. The entire β-like cluster occupies about 55,000 base pairs of DNA (expressed as 55 kb, where the symbol kb represents a **kilobase**, a length of DNA equal to 1000 base pairs); the α-like cluster occupies about 25 kb. Note that there are two functional α genes ($\alpha1$ and $\alpha2$), two functional ζ genes ($\zeta1$ and $\zeta2$), and two functional γ genes ($^G\gamma$ and $^A\gamma$). The direction of transcription is indicated, and it is of interest that the order of genes in both clusters is the same as their order of expression during development.

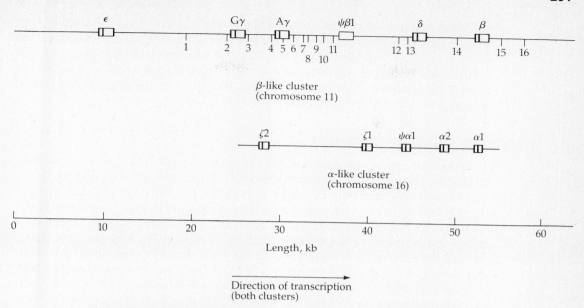

Figure 9.14 Organization of human β-globin-like gene cluster (chromosome 11) and α-globin-like gene cluster (chromosome 16). Note the β pseudogene (ψβ1) and the α pseudogene (ψα1). The numbers represent possible sites of cytosine modification.

In addition to the functional globin genes, the clusters have two known pseudogenes—one α pseudogene (ψα1) and one β pseudogene (ψβ1). **Pseudogenes** are genes that have an obvious DNA sequence similarity to other genes but are not expressed; that is, they are "silent." Pseudogenes seem to be common in eukaryotes, but their functions, if any, are unknown. Possible explanations of pseudogenes range from views that pseudogenes are remnants of evolutionary mistakes (ancient duplications that were simply inactivated by mutation rather than deleted) to views that pseudogenes are genes caught in the act of evolving (temporarily out of commission but perhaps one day to be reactivated to carry out a novel function), to views that pseudogenes, while not expressed as poly-peptides, play an essential role in the expression of their normal counterparts.

The numbers beneath the β-like map in Figure 9.14 show the positions of 16 potentially methylated cytosines whose methylation can be detected with appropriate restriction enzymes. Interesting patterns emerge when the extent of base modification is compared with the expression of the genes. In cell types such as sperm or fetal brain that do not express hemoglobin, virtually all the cytosines in Figure 9.14 are modified, with the sole exception of site 7, which is methylated in sperm and brain but not in some other nonexpressing cells. Relative to fetal brain or adult liver cells, fetal liver cells have less methylation, particularly at sites 2, 4, 6, and 9. This undermethylation correlates well with expression of the γ chains in fetal liver but not in fetal brain or adult liver. Similarly, hemoglobin-producing bone marrow cells from adults are undermethylated relative to cells from bone marrow that do not express hemoglobin, particularly at sites 12 and

15, which are near the δ and β genes. Thus, there is some connection between relative lack of methylation and gene expression, but methylation cannot be the whole story. The complication comes from the placenta and from certain tissue culture cells that have uniformly low levels of methylation at all sites yet fail to express hemoglobin. A present, it appears that undermethylation is just one of several prerequisites for gene expression.

DNA ALTERATIONS DURING DEVELOPMENT

DNA methylation is one way that the organism can physically change its DNA during the course of development, but it is reversible; the cell can as easily remove methyl groups from cytosines as put them on. Are there physical changes to DNA that are irreversible, that would forever limit the genetic potentiality of any cell carrying them? If so, then the chords-on-a-piano analogy mentioned earlier would be quite inappropriate for the origin of specialized cell types, because the cells in an organism would not be genetically equivalent. Although a few cases of permanent, irreversible DNA changes in certain specialized cells are known (one important example in humans will be discussed in Chapter 12), such DNA changes seem to be the exception. Most cells in an organism do seem to retain their full complement of genes.

If DNA does not change irreversibly during development, then the nucleus of any somatic cell should be able to support normal embryonic development. That is to say, the somatic cell nucleus should be able to substitute for the zygote nucleus of a newly fertilized egg and guide the egg successfully through development. Indeed, in plants, such nuclear replacement is unnecessary; when properly cultured, adult plant cells from differentiated tissues of carrot, tobacco, and other plants are able to divide and give rise to completely normal seedlings and fertile adults. In plants, therefore, irreversible DNA changes do not seem to accompany normal development—at least in the adult tissues examined.

In animals the experimental procedures are more difficult because nuclear replacement (called **nuclear transplantation**) is necessary. The animal egg is a marvelously complex cell with regional differences in its cytoplasm. After fertilization, when the chromosomes replicate and the cells divide, the cytoplasm of the daughter cells is not identical because each has a different region of cytoplasm from the original egg. The different cytoplasm evidently triggers the initial differentiation of the cells, and later in development, interactions between differentiated cells lead to still further differentiation. The embryo eventually comes to have lung cells, liver cells, brain cells, kidney cells, blood cells, and many more types of cells—each type differentiated, distinct from the others. The protein hormone insulin, for example, is produced in specialized cells in the pancreas but not elsewhere.

In vertebrates, the first nuclear transplantation success was achieved in the frog, *Rana pipiens,* and later in the African clawed toad, *Xenopus laevis.* The experimental protocol, outlined in Figure 9.15, begins with the destruction (by ultraviolet light) of the nucleus in an unfertilized egg; at the same time, the

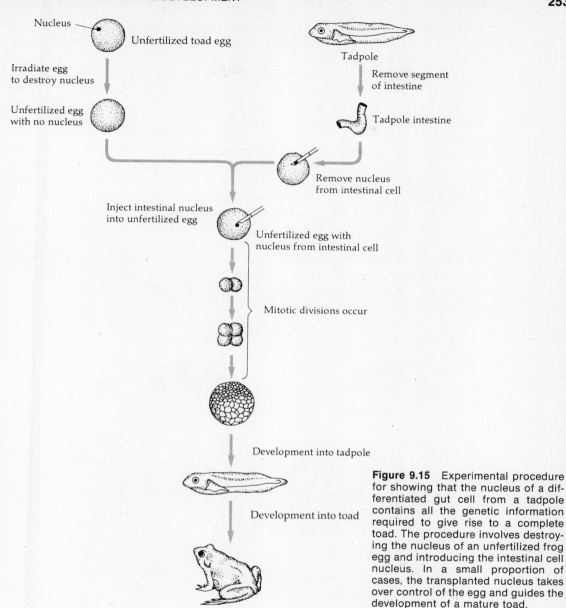

Nucleus — Unfertilized toad egg

Irradiate egg
to destroy nucleus

Unfertilized egg
with no nucleus

Tadpole

Remove segment
of intestine

Tadpole intestine

Remove nucleus
from intestinal cell

Inject intestinal nucleus
into unfertilized egg

Unfertilized egg with
nucleus from intestinal cell

Mitotic divisions occur

Development into tadpole

Development into toad

Figure 9.15 Experimental procedure for showing that the nucleus of a differentiated gut cell from a tadpole contains all the genetic information required to give rise to a complete toad. The procedure involves destroying the nucleus of an unfertilized frog egg and introducing the intestinal cell nucleus. In a small proportion of cases, the transplanted nucleus takes over control of the egg and guides the development of a mature toad.

nucleus from a tadpole intestinal cell is removed by microsurgery. The intestinal cell nucleus is then carefully injected into the cytoplasm of the nucleus-deficient egg. In most cases the procedure does not work; the egg with the transplanted nucleus does not develop. The reason for the failures seems to be that the surgery often mechanically damages the nucleus or the egg or both. However, in about

1 percent of cases, the intestinal nucleus does successfully take over development and gives rise to a normal, fertile animal. Differentiated tadpole intestinal nuclei therefore seem to have undergone no irreversible DNA changes.

Similar nuclear transplantation experiments are much more difficult in mammals because mammalian eggs are much smaller than those of amphibians. Success has nevertheless been achieved in the mouse by means of the procedure outlined in Figure 9.16. The first step is the isolation of blastocysts from the donor parent (gray mouse—see Figure 5.11 for close-up views of the blastocyst). After blastocyst isolation and removal of the membranous envelope, cells from the inner cell mass—the part of the blastocyst destined to give rise to the embryo itself—are isolated and dissociated into single cells. In the meantime, newly fertilized eggs are obtained from the recipient mouse (black) prior to the fusion of the **male pronucleus** (the haploid nucleus contributed by the sperm) and the **female pronucleus** (the haploid nucleus contributed by the egg itself). Then a nucleus from a single cell from the inner cell mass is carefully removed and transplanted into the zygote; with the same tiny needle, the male and female pronuclei originally in the zygote are both removed. The transplanted cells are cultured in vitro, where about 35 percent of them divide and produce new blastocysts. These blastocysts are then introduced into a foster mother (white

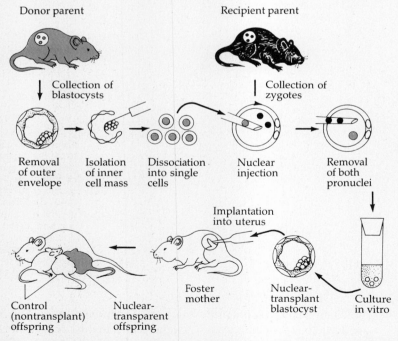

Figure 9.16 Nuclear transplantation in the mouse. A nucleus from the inner cell mass of a gray strain of mice is introduced into the fertilized egg of a black strain, and the male and female pronuclei in the egg are removed. The nuclear-transplant egg is then cultured to the blastocyst stage and introduced into the uterus of a white foster mother, where, if successful, it gives rise to a gray mouse.

mouse), where further development proceeds. In the first experiment of this kind, 3 normal baby mice were obtained from 16 implanted blastocysts; these 3 mice grew into fertile adults, and their own coat color as well as the coat-color genes transmitted to their offspring proved that they had indeed arisen from the nucleus of the donor mouse. Thus, mammalian blastocyst cells from the inner cell mass are able to support embryonic development. It is interesting, though, that nuclei from blastocyst cells not associated with the inner cell mass cannot support development.

Early embryonic cells not only have nuclei capable of supporting development, but they can also be disassociated and nevertheless reaggregate properly. That is to say, cells from two or more genetically different preblastocyst embryos can be disassociated and mixed together and introduced into the uterus of a foster mother. This mixture of cells will reorganize and form a normal blastocyst that gives rise to a normal mouse, but the mouse is a mosaic because cells from genetically distinct embryos have contributed to its makeup. Such mixed mice are known as **chimeric** or **allophenic** mice. Figure 9.17 is a photograph of a chimeric mouse derived from a mixture of embryonic cells from a black strain and a white strain.

Two principal issues are raised by the success of mammalian nuclear transplants. The first is a social issue. A successful nuclear transplant represents a clone of the donor organism, and the technical procedures would not be much more difficult in humans than they are in mice. Widespread cloning in humans raises the specter of Aldous Huxley's *Brave New World,* where children with

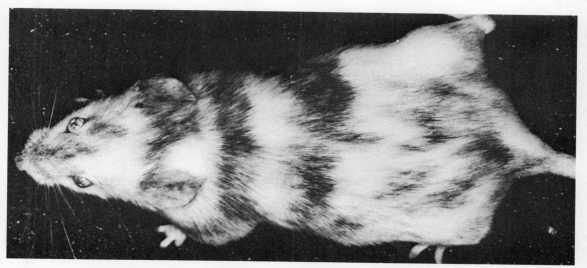

Figure 9.17 This mouse is composed of two genetically different types of cells: one type gave rise to dark-coat pigmentation and the other type to light-coat pigmentation. The mouse was produced artificially by combining cells from very young embryos of two different strains of mice (one with dark-coat pigmentation and the other with light) and implanting the mixed group of cells into the uterus of a female mouse where it then developed normally, giving rise to the individual shown here.

prescribed qualities befitting their future assignments to suitable castes in the social and economic hierarchy are made to order in uniform batches of thousands by genetic technicians and mass-produced in test tubes in baby hatcheries. It provokes visions of a world in which evil men like Adolf Hitler could perpetuate and multiply their genotypes by the thousands.

However, research in human cloning is not now being carried out, nor may it ever be. The research is directed toward other mammals, and the ability to clone farm animals would have tremendous benefits. The genetic segregation that occurs during meiosis is both a help and a hindrance to an animal breeder. In the early stages of breed improvement, segregation is helpful because it creates many different combinations of genes, and the best animals can then be chosen for breeding. Once a truly prize animal is encountered, however, segregation becomes a nuisance. The natural offspring of a prize animal will not, in general, be as good as the parent: first, because one parent contributes only half the genes to each offspring and, second, because the beneficial combination of genes in the parent is broken up and reshuffled by recombination and segregation. Both these problems could be avoided by cloning the prize animal, to the great benefit of breeders and the world meat supply. The animal technology might then be applied to humans, with the only obstacles being conventions of decency and law. Nevertheless, there will always be mavericks willing to defy convention or break the law, and some may be clever enough to succeed. There is no foolproof defense against such violations, only the hope that decent men and women involved in any large-scale human cloning project would quietly sabotage the procedures or blow the whistle.

The second issue raised by mammalian nuclear transplantation relates back to regulation. Since only embryonic cells have been studied, it could be that some adult somatic cells have undergone such irreversible DNA changes that their nuclei would be unable to support development. Experience with amphibians and plants suggests the opposite, however. In spite of their physical and functional differences, most cells in the body seem to have an identical set of genes. Differentiation thus results from differences in gene regulation in the various types of cells, not from irreversible changes in DNA. Despite their genetic similarity, a great variety of cell types can arise in the embryo by the expression of only a restricted group of genes in each type—in much the way that a great variety of different chords can be played on a piano by depressing different groups of keys, to invoke the analogy used earlier. How the coordinated regulation of large groups of genes can occur is one of the great unsolved mysteries of genetics.

REGULATION OF TRANSCRIPTION

A principal avenue of gene regulation in eukaryotes is **transcriptional regulation**, which refers to those processes that prevent or promote transcription. Indeed, transcriptional regulation is of prime importance in all organisms, and one of the first discovered and best understood examples of transcriptional regulation involves the genes for utilizing milk sugar (**lactose**) in *E. coli*. The essentials of

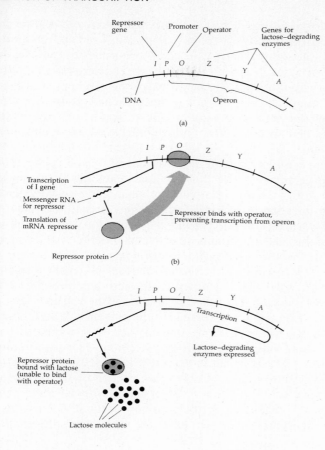

Figure 9.18 (a) Segment of the long, circular DNA molecule of the bacterium *E. coli* showing the genes involved in lactose metabolism. *I* codes for repressor protein; *P* (promoter) and *O* (operator) are controlling regions of DNA; and *Z, Y,* and *A* code for enzymes involved in lactose breakdown. (b) Control of the genes in the absence of lactose. The repressor protein binds with the *O* region on the DNA and prevents transcription. (c) Control of the genes in the presence of lactose. Lactose molecules bind with the repressor protein, preventing it from binding with *O*; transcription and expression of the lactose-degrading enzymes occurs. When all the lactose has been utilized, the system reverts to the state depicted in part (b).

transcriptional regulation of the lactose enzymes are outlined in Figure 9.18. Part (a) illustrates the organization of the relevant genes on the bacterial DNA. The *I* gene codes for a **repressor** protein (the function of which will be discussed shortly), *P* stands for promoter (the DNA region to which RNA polymerase binds), *O* represents the **operator** region of DNA, and *Z, Y,* and *A* are the genes coding for proteins required in lactose utilization.

The repressor gene is always "on" and is transcribed at a very low level. (Such continuously active genes are said to be **constitutive**.) As shown in part (b), the *I* gene produces an mRNA that is translated to produce the repressor protein, which, in the absence of lactose, recognizes the operator region, binds to the DNA, and so prevents passage of the RNA polymerase from the promoter. Consequently, in the absence of lactose, the repressor protein prevents expression of the lactose genes by binding with the operator.

The repressor protein can also bind with lactose, however, and, as shown in part (c), when lactose is present in the cell's environment, the lactose binds with the repressor and so prevents the repressor's binding with *O*. In this situation, transcription and translation of the lactose-utilizing enzymes will occur. At some later time, when all the lactose in the environment is used up, the repressor will again become free of lactose, and the situation will revert to the one shown in part (*b*), where the lactose genes are unexpressed. Transcriptional control of the lactose genes thus provides an example in which the production of lactose-utilizing enzymes is ultimately controlled by the presence of lactose. In the absence of the sugar (*b*) the enzymes are not produced; in the presence of lactose (*c*) they are produced; and so the lactose-utilizing enzymes are produced only when needed by the cell.

Lactose regulation is an example of **negative control**, which means that the lactose genes are preprogrammed to be "on" unless something (in this case the repressor) turns them "off." There are also many examples of **positive control**, in which the genes in question are turned "off" unless something turns them "on." An example of positive control involves the genes for utilizing the sugar **arabinose** in *E. coli*. In this case the arabinose **activator** protein (analogous to the lactose repressor) must bind with the DNA at a region near the promoter in order for the promoter to be able to bind with RNA polymerase and so allow transcription of the arabinose-utilizing genes. However, the activator protein by itself cannot bind to the DNA, so in the absence of arabinose the relevant genes are not transcribed and hence remain unexpressed. In the presence of arabinose, the activator protein combines with the sugar, and this activator-arabinose combination is the molecule that binds to the DNA and so promotes transcription and expression of the genes. The end result is the same as in the case of lactose— i.e., the arabinose enzymes are produced only when needed by the cell—but the details of positive control are just the reverse of those with negative control. Both positive and negative regulation of transcription occur in eukaryotes, but positive regulation in eukaryotes is thought to be the more common situation.

FEEDBACK INHIBITION

There is one other level of regulation that is important enough to be mentioned here, and this is a type of regulation occurring at the level of the completed, functional enzyme molecule. As mentioned earlier, many molecules produced in cells are synthesized in an assembly-line fashion by a number of enzymes acting in sequence, each enzyme working on the product of the preceding step in the pathway. In many such cases the molecule produced as the final product inactivates or inhibits the enzyme that carries out the first step in the pathway; thus the end product regulates its own production.

This kind of regulation is known as **feedback** or **end-product inhibition**. For example, human cells synthesize adenine and guanine ribonucleotides by a long sequence of reaction steps, each step requiring its own particular enzyme. The enzyme required in the first step of the pathway is sensitive to the level of adenine and guanine ribonucleotides in the cell. When the ribonucleotides are present in sufficient quantity, the first enzyme in the pathway is inactivated, and further production of the ribonucleotides is thereby prevented. When the level of ribonucleotides in the cell is low, on the other hand, the enzyme is not inactivated, which leads to synthesis of the ribonucleotides. (See Figure 9.19.)

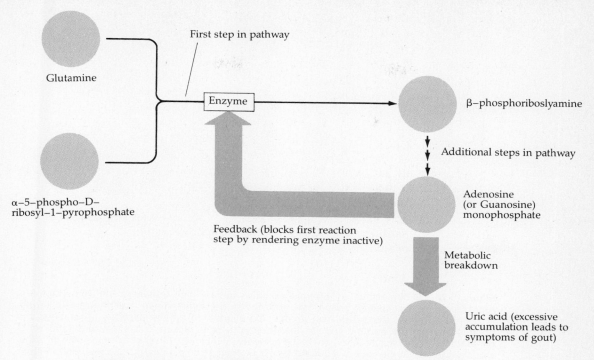

Glutamine

First step in pathway

Enzyme

β–phosphoriboslyamine

Additional steps in pathway

α–5–phospho–D–
ribosyl–1–pyrophosphate

Adenosine
(or Guanosine)
monophosphate

Feedback (blocks first reaction
step by rendering enzyme inactive)

Metabolic
breakdown

Uric acid (excessive
accumulation leads to
symptoms of gout)

Figure 9.19 Diagram of feedback inhibition in the biochemical pathway that produces adenosine and guanosine monophosphate. The horizontal arrow represents the first step in the metabolic pathway; this step is catalyzed by an enzyme that is sensitive to the level of adenosine and guanosine monophosphate. When the latter ribonucleotides are present in the cell in sufficient quantity, they interact with the first enzyme (upward-pointing broad arrow) and render it inactive, thus preventing further production of the ribonucleotides. A certain inherited defect in the gene coding for the first enzyme renders the enzyme insensitive to feedback inhibition. Thus the ribonucleotides are overproduced, and excess uric acid resulting from their breakdown (downward-pointing broad arrow) leads to symptoms of gout. (The conversion of β-phosphoribosylamine to adenosine [or guanosine] monophosphate requires more than the three steps represented here.)

Certain families have an inherited form of gout that results from a mutation in the gene whose protein product is the first enzyme in the pathway for the synthesis of adenine and guanine ribonucleotides. This mutation renders the enzyme insensitive to feedback inhibition, and hence the ribonucleotides are overproduced. One of the breakdown products of the excess ribonucleotides is uric acid, which accumulates to abnormally high levels in the blood of persons carrying this mutation, and the high levels of uric acid in the blood often lead to the painful inflammation of the joints, especially those in the hands and feet, known as gout.

Feedback inhibition comes about because the first enzyme in the pathway has a built-in tendency to become physically stuck to the inhibitor (Figure 9.20). The three-dimensional structure of the enzyme contains a precisely shaped hole or pocket into which the inhibitor molecule will fit because the pocket is of exactly the right size, shape, and electrical charge. The presence of the inhibitor in this pocket causes an overall change in the three-dimensional structure of the

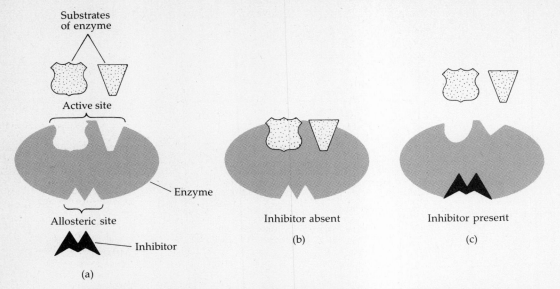

Figure 9.20   Schematic of an enzyme sensitive to feedback inhibition. (*a*) The enzyme is represented by the shaded shape, the inhibitor by the black shape, and the molecules the enzyme acts upon (the substrates) as stippled shapes. (*b*) In the absence of inhibitor, the conformation of the enzyme enables the enzyme to bind with its substrates and catalyze its characteristic chemical reaction. (*c*) In the presence of inhibitor, the enzyme binds with inhibitor and changes its shape so that it can no longer interact with its substrates. In relation to Figure 9.19, the inhibitor would represent adenosine or guanosine monophosphate, and the substrates would be glutamine and α-5-phospho-D-ribosyl-1-pyrophosphate. The enzyme would represent the first enzyme in the pathway in Figure 9.19, catalyzing the chemical reaction that leads to β-phosphoribosylamine.

enzyme, and this, in turn, prevents the enzyme from carrying out its reaction step in the pathway. When molecules of the inhibitor are not present, the enzyme regains its normal three-dimensional structure and can catalyze (accelerate) the reaction step.

Despite our present knowledge of various possible modes of regulation and our understanding of various special cases, we still know relatively little of how the tens of thousands of genes in a cell are *simultaneously* regulated. Yet they are regulated, and the upshot is that most normal individuals are more or less healthy most of the time. But genes can **mutate** (become altered) and the products of such mutated genes may be unable to function normally, causing the individuals who carry them to suffer discomfort or disease or perhaps even death. How mutations occur is the subject of the next chapter.

SUMMARY

1. **Proteins** are composed of chains of **amino acids** called **polypeptides**. The nucleotide sequence of most genes is used to specify the sequence of amino acids in a single type of polypeptide, and this amino acid sequence determines how the polypeptide will fold into its three-dimen-

sional configuration. However, several polypeptides from the same gene, or even polypeptides from different genes, may aggregate to form the finished protein.

2. **Inborn errors of metabolism** refer to mutations which lead to defective enzymes which disrupt vital chemical processes in the cell or organism. A classical example of an inborn error involves enzymes in the metabolism of the amino acid **phenylalanine**. One enzyme normally converts phenylalanine into another amino acid (tyrosine); when this enzyme is defective, harmful amounts of phenylalanine build up and result in the severe form of mental retardation called **phenylketonuria**.

3. Genetic information in DNA is used to specify the sequence of amino acids in a polypeptide by means of a two-step process in prokaryotes and a three-step process in eukaryotes. The first step is always **transcription**, in which the sequence of nucleotides in the sense strand of DNA is used as a template for the production of a complementary sequence of **ribonucleotides** in a molecule of **RNA (ribonucleic acid)** by means of base pairing. The final step is always **translation**, in which the RNA directs the sequence of amino acids to be incorporated into a polypeptide. Eukaryotes have a third step between transcription and translation, which is called **RNA processing**.

4. The structure of RNA is similar to that of single-stranded DNA except that (a) the sugar in RNA is **ribose**, and (b) the base **uracil** in RNA plays the role of thymine in hydrogen bonding with adenine.

5. The key enzyme in transcription is **RNA polymerase**. Transcription begins when the polymerase attaches to a DNA strand at a region called the **promoter**. The polymerase travels in the 3′ to 5′ direction along the sense strand of DNA, initiating a new RNA strand (called a **transcript**) and adding new ribonucleotides one by one to the 3′ end of the elongating transcript; this process ends when the polymerase encounters a termination signal farther along the DNA. The RNA transcript produced from DNA in prokaryotes corresponds to **messenger RNA (mRNA)**, the RNA molecule actually used in translation. In eukaryotes, the mRNA is produced in the nucleus from the primary transcript by means of RNA processing.

6. Translation begins when a **ribosome** particle attaches to the leader sequence of an mRNA and moves along the mRNA in the 3′ direction until it reaches the coding sequence. From there, the ribosome moves in exact steps of three bases at a time. Such groups of three bases are called **codons**, and each specifies a particular amino acid to be added to the polypeptide chain. The amino acids themselves are brought to the ribosome by molecules of **transfer RNA (tRNA)**. Each tRNA has at one end a three-base **anticodon** that can hydrogen bond with the complementary codon, and at the other end it has an attached amino acid that corresponds to the codon. The correspondence between codons and the amino acids they encode is called the **genetic code**. Codon 5′-AUG-3′ stands for methionine, and this is always the first (**amino terminal**) amino acid in a new polypeptide chain (although frequently the methionine is removed from the finished polypeptide later on). When the 5′-AUG-3′ codon and its tRNA are aligned, the ribosome brings in the tRNA corresponding to the second amino acid and attaches the methi-

onine to this amino acid by means of a **peptide bond**. The tRNA for methionine becomes detached from its amino acid in the process and falls free from the ribosome. The ribosome then shifts to the next codon, brings in the appropriate tRNA, attaches the growing peptide to this amino acid, and lets the previous tRNA fall free. This process continues codon by codon and amino acid by amino acid until one of the three **terminator codons** is encountered. These three are 5′-UAG-3′, 5′-UAA-3′, and 5′-UGA-3′, and each will cause the polypeptide to be detached from the last remaining tRNA and be released into the cytoplasm.

7. The genetic code is nearly **universal**. Virtually any translational apparatus, from any organism whatever, will translate the same coding sequence in the same way. The code also has synonyms—codons that code for the same amino acid. Most of the synonymous codons differ in the third base position: where one has a U, the other will have a C (i.e., both pyrimidines), or where one has an A, the other will have a G (both purines). The equivalence of pyrimidines (and purines) in the third codon position is thought to be due in part to the lack of rigidity at this position when the tRNA pairs with the codon. The translational apparatus of human mitochondria uses a slightly altered genetic code, different in several instances from the standard code.

8. In prokaryotes, the coding region of an mRNA corresponds to a continuous stretch of DNA. In eukaryotes, however, the coding region of a gene may be **split** into several smaller pieces between which noncoding regions are found. The smaller coding regions are called **exons** and the noncoding regions in between are called **introns** or **intervening sequences**. Some genes are split into a dozen or more exons, and the total length of the introns may be substantially greater than the total length of the exons. In the transcription process, both introns and exons are transcribed to produce a large RNA transcript. As part of RNA processing, the introns are spliced out of the RNA transcript to produce a molecule in which the formerly separated exons lie side by side. This splicing is evidently carried out by enzymes in the nucleus.

9. **Regulation** refers to the processes by which cells are able to express particular genes at the appropriate times. Since gene expression is a complex process involving transcription, RNA processing, translation, and so on, gene expression has many **control points**—places in the process at which control over gene expression can be exerted.

10. The regulation of **hemoglobin** synthesis is an example of gene regulation during development. Each hemoglobin molecule consists of two α (**alpha**)-like polypeptides and two β (**beta**)-like polypeptides. Humans have two α-like polypeptides, called ζ (**zeta**) and α; both chains are coded by duplicate genes, so there are two ζ genes (ζ1 and ζ2) and two α genes (α1 and α2) in addition to an unexpressed pseudogene for α (ψα1). (**Pseudogenes** are "silent" genes that are similar in nucleotide sequence to some other gene but are not expressed.) These genes are arranged on chromosome 16 in order of their expression during development, the ζ genes being upstream from the α genes; the ζ genes are expressed in early embryonic development, whereas the α genes are expressed primarily in the fetus and adult. Humans have four β-like polypeptides— ε (**epsilon**), γ (**gamma**, coded by two nearly identical genes called $^G\gamma$ and $^A\gamma$), δ (**delta**), and β (there is also a β pseudogene—ψβ1). The

expressed β-like genes are arranged on chromosome 11 in order of their expression, ε being expressed in embryos, γ in the fetus, and δ and β in the adult. However, δ expression is poor, so most adult hemoglobin consists of two α and two β chains ($\alpha_2\beta_2$).

11. DNA modification, particularly by the addition of methyl ($-CH_3$) groups to certain cytosines (**methylation**), is one control point of gene expression, and it has been implicated in hemoglobin regulation. In particular, active hemoglobin genes seem to be less methylated than inactive ones, although methylation by itself is only one aspect of hemoglobin regulation.

12. The importance of irreversible DNA changes during development is still unclear, but it can be tested by determining whether nuclei of differentiated cells can support embryonic development. In plants such as carrot and tobacco, cells from an adult plant can be induced by appropriate conditions to divide and undergo development into a normal plant; thus, irreversible DNA changes do not seem to be significant. In amphibians, nuclei of tadpole intestinal cells can by **nuclear transplantation** be introduced into egg cytoplasm, and a fraction of such transplanted nuclei will support normal development. In mice, transplanted nuclei from the **inner cell mass** of the **blastocyst** can sustain normal development into fully fertile adults; this experiment represents the first **cloning** of a mammal, with all its attendant moral and ethical issues should it be applied to humans. Previous experiments had shown that cells of disrupted mouse embryos that are combined and implanted into a female's uterus will reorganize and give rise to a morphologically normal offspring. Such offspring are called **allophenic** mice; they are **chimeras** (genetic mosaics) because they have two or more genetically distinct types of cells. On the whole, there is no convincing evidence that irreversible DNA changes play a major role in development.

13. Gene expression can also be regulated at the level of transcription, and two broad categories of transcriptional control can be distinguished. In **positive control**, a protein or other molecule must bind to a promoter region for transcription to be possible. In **negative control**, a protein or other molecule must *not* bind to a promoter region for transcription to occur. Negative control is exemplified in the regulation of the **lactose**-utilizing genes in *E. coli*. In this case, a **repressor** protein binds with a region of DNA near the promoter and so prevents transcription of the lactose-utilizing genes; in the presence of lactose, the repressor protein becomes bound with lactose and loses its ability to attach to DNA, thus allowing transcription and expression of the lactose-utilizing genes. Positive control is exemplified by the **arabinose**-utilizing genes in *E. coli*. In this case an **activator** protein binds with arabinose, and the activator-arabinose combination is able to bind to DNA near the promoter; this binding allows attachment of RNA polymerase and so promotes transcription and expression of the arabinose-utilizing genes. In both cases, the genes are expressed only when needed by the cell.

14. In many biochemical pathways concerned with the synthesis of molecules needed by the cell, the ultimate product of the pathway inhibits the first enzyme in the pathway. In this manner the end product of the biochemical pathway regulates its own production. This sort of regulation is known as **feedback inhibition**.

KEY WORDS

Amino acid	Inborn error of metabolism	Pseudogene
Amino end	Intervening sequence	Repressor
Anticodon	Intron	Ribonucleic acid
Carboxyl end	Kilobase	Ribose
Coding region	Messenger RNA	RNA polymerase
Codon	Metabolism	RNA processing
Colinear	Negative control	Terminator codon
Constitutive expression	Nuclear transplantation	Transcript
DNA methylation	Operator	Transcription
Exon	Peptide bond	Transfer RNA
Feedback inhibition	Phenylketonuria	Translation
Gene expression	Polypeptide	Uracil
Genetic code	Positive control	
Hemoglobin	Promoter	

PROBLEMS

9.1. What is meant by the term "metabolism" and what is an "inborn error of metabolism"?

9.2. Considering an analogy between translation and the writing of an ordinary English sentence, which codon corresponds to the capital letter at the beginning? Which codons correspond to the period at the end?

9.3. How many amino acids are coded by just one codon? How many by two codons? Three? Four? Five? Six? How many amino acids are there altogether?

9.4. Which end of a DNA strand, the 3' end or the 5' end, is transcribed first? Which end of an mRNA, the 3' end or the 5' end, is translated first?

9.5. If the DNA strand shown below were transcribed, would transcription occur from left to right or from right to left? Why?

5'-ATCCGTCAG-3'

9.6. If the DNA molecule shown below were transcribed from left to right, what would be the sequence of the corresponding transcript?

5'-ATTGCTGAATGAC-3'
3'-TAACGACTTACTG-5'

9.7. If the DNA strand shown below were transcribed, what would be the sequence of the RNA transcript? If translation of the transcript were to occur, what would be the corresponding amino acid sequence? (Note: Use the single-letter amino acid abbreviations and a word will appear.)

5'-AGCTATATGCGGGAGTTCGTCCGCTAAAATGTGGGG-3'

9.8. What amino acid sequence would correspond to the following mRNA sequence? What feature of the genetic code is illustrated by this example?

5'-UCUAGUUCCAGCUCAUCG-3'

9.9. The sense strand of a DNA molecule consists of alternating Ts and Cs. What would be the sequence of the corresponding RNA transcript? Of the corresponding polypeptide?

9.10. What amino acid sequence would correspond to the following mRNA sequence? (Note: Use the single-letter amino acid abbreviations to reveal a secret.)

5′-GCAUCAGAAUGUAGGGAGACA-3′

9.11. How many RNA transcripts could code for the amino acid sequence MWMW, including a proper termination codon at the end?

9.12. Assuming that the nucleotide sequence of an anticodon is the exact complement of its codon, what codons would correspond to the following anticodons and which amino acid would be carried at the other end of the tRNA? (Actually, many anticodons contain unusual ribonucleotides.)

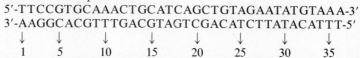

(*a*) 5′-AGU-3′	(*d*) 3′-GCU-5′
(*b*) 5′-GUU-3′	(*e*) 3′-CAU-5′
(*c*) 5′-UUC-3′	(*f*) 3′-GUC-5′

9.13. Imagine a mutant tRNA that misreads the 5′-AAU-3′ codon and inserts the amino acid T (threonine) in its place. With the mRNA shown here, what polypeptide sequence would normally be produced? What sequence would be produced in the presence of the mutant tRNA?

5′-CCGGAGAAUUCC-3′

9.14. The DNA sequence shown below is transcribed from left to right. If the sequence is in a coding region, what amino acid sequence would correspond to it? If nucleotides 7 through 26, inclusive, are actually an intervening sequence, what amino acid sequence would correspond?

```
5′-TTCCGTGCAAACTGCATCAGCTGTAGAATATGTAAA-3′
3′-AAGGCACGTTTGACGTAGTCGACATCTTATACATTT-5′
   ↓   ↓    ↓     ↓     ↓    ↓    ↓    ↓
   1   5    10    15    20   25   30   35
```

9.15. Imagine a mutation in the repressor gene involved in regulating the lactose-degrading enzymes in *E. coli,* and suppose that the effect of the mutation is to make the repressor nonfunctional (i.e., unable to attach to the operator) in cells grown at 42°C but to leave it fully functional in cells grown at 30°C. Would the lactose-degrading enzymes be produced in the presence of lactose at 30°? At 42°? Would they be produced in the absence of lactose at 30°? At 42°?

FURTHER READING

Anderson, S., *et al.* 1981. Sequence and organization of the human mitochondrial genome. Nature 290: 457–65. All 16,569 base pairs are presented, along with the mitochondrial genetic code.

Bank, A., J. G. Mears, and F. Ramirez. 1980. Disorders of human hemoglobin. Science 207: 486–93. Discussion of hemoglobin abnormalities with emphasis on gene regulation.

Brown, D. D. 1981. Gene expression in eukaryotes. Science 211: 667–74. Scholarly and readable review of gene expression.

Chambon, P. 1980 (May). Split genes. Scientific American 244: 60–71. On the sequence organization of genes in eukaryotic cells.

Crawford, M. d'A., D. A. Gibbs, and R. W. E. Watts. 1982. Advances in the Treatment of Inborn Errors of Metabolism. John Wiley and Sons, New York. How understanding inborn errors of metabolism contributes to treatment of disease.

Judson, H. F. 1979. The Eighth Day of Creation. Simon & Schuster, New York. A fascinating account of the rise of molecular biology.

Khoury, G., and P. Gruss. 1983. Enhancer elements. Cell 33: 313–14. DNA sequences sometimes found in intervening sequences that affect gene activity.

Maniatis, T., E. F. Fritsch, J. Lauer, and R. M. Lawn. 1980. The molecular genetics of human hemoglobin. Ann. Rev. Genet. 14: 145–78. The molecular arrangement and expression of the globin genes.

Mount, S. M. 1983. Sequences that signal where to splice. Nature 304: 309–10. How the cell recognizes an intron.

Perutz, M. F. 1978 (December). Hemoglobin structure and respiratory transport. Scientific American 239: 92–125. How the molecule functions as a "molecular lung."

Ptashne, M., A. D. Johnson, and C. O. Pabo. 1982 (November). A genetic switch in a bacterial virus. Scientific American 247: 128–40. How a lambda regulatory protein works.

Rich, A., and S. H. Kim. 1978 (January). The three-dimensional structure of transfer RNA. Scientific American 238: 52–62. A detailed look at this important molecule.

Scriver, C. R., and C. L. Clow. 1980. Phenylketonuria and other phenylalanine hydroxylation mutants in man. Ann. Rev. Genet. 14: 179–202. Detection and treatment of PKU and some related conditions.

Stent, G. S., and R. Calendar. 1978. 2d. ed. Molecular Genetics. W. H. Freeman, San Francisco. An advanced textbook covering virtually all aspects of molecular genetics in a readable format.

chapter **10**

Mutation

Writers of science fiction often portray mutant individuals as twisted, misshapen monsters who lust for destruction and crave human flesh. Poetic license as expressed in scripts of wee-hour television movies has little to do with what mutations are actually like, of course; real mutations are not nearly so colorful. A **mutation**, as seen by a geneticist, is any heritable change in the genetic material. Mutation is one of the three fundamental genetic processes, the other two being **gene transmission**, which refers to the manner in which genes are passed from generation to generation, and **gene expression**, which refers to the processes by which genes exert their effects on cells and organisms.

Mutation refers to a spontaneous, random change in the genetic material. Thus the statement that an individual is a "mutant" is only a kind of abbreviation of the correct expression—that the individual in question carries one or more mutant genes. This distinction may seem to be splitting hairs, but actually it is quite important. It is a distinction between a mutant gene itself and the effects it may have on its carriers, a distinction between an individual's genotype and phenotype. The newborn who suffers from prolonged lack of oxygen during delivery and thereby sustains permanent brain damage may have symptoms very similar to the child homozygous for a mutation that causes mental retardation. In the terminology of Chapter 3 we would say that the oxygen-deprived child is a **phenocopy**. This distinction between mutational and nonmutational causes of disorders is scant consolation to the affected child, of course, but to the parents, who may want to know the chance of having another retarded child, the distinction could make all the difference in the world.

Another feature of mutation worth emphasis is that mutations are heritable; they are faithfully replicated and capable of being passed from one cell to its descendants. If the cell's descendants include sperm or eggs, then the mutation can be transmitted from generation to generation. Many mutations occur in the **somatic cells** of the body, cells whose descendants do not include sperm or eggs. These mutations are known as **somatic mutations**, and they are not transmitted from parent to offspring. Quite a number of people have a brown sector in an otherwise blue eye, for example, or even one blue eye and one brown. Mosaic eye color often arises from somatic mutation, and an effect so strikingly visible in the eyes may lead one to believe that mosaics due to mutations are common. However, the eyes are not ordinarily mosaic, nor are the germ cells. On the other hand, mutations that occur in the germ cells are of great importance in genetics because these mutations may be transmitted from parent to offspring.

Traits acquired during the lifetime of an individual are not mutations, and they are not inherited. Toil as we do to master knowledge, our children are born ignorant. Grunt and sweat as we do to develop muscles, our children are born weak. Skills in art, science, athletics, and all other acquired abilities will rest with their possessors in the grave. This is not to deny that certain *propensities* may be inherited. There may well be inherited abilities toward, say, music. But the practice and perfection of these abilities is still a lonely, individual task. There is a good side to this story, too. The tragic accidents that afflict people during their lives are not transmitted either. Disease, accidental loss of eyes or limbs, surgical mutilations—all these and other acquired disabilities are burdensome to those afflicted but their children are born free.

TYPES OF MUTATIONS

Mutations can occur in any gene. They can therefore affect any part of the body at any time in life, from embryonic development to old age. Mutations have specific effects related to the gene involved. For example, a mutation at the testicular-feminization locus causes the many separate symptoms of the testicular-feminization syndrome, but such a mutation never leads to sickle cell anemia. Similarly, a mutation of the Tay-Sachs locus causes Tay-Sachs disease but not cystic fibrosis. On the other hand, mutations in different genes that affect the same organs or the same biochemical pathways can lead to similar symptoms. Dozens of different genes, when mutated, can cause deafness, for example, and even more mutations can result in mental retardation.

Mutations can be classified in many ways. Sometimes it is useful to classify mutations according to their effects. One such classification is based on **age of onset**—the age at which the phenotype associated with the mutation typically appears. At one extreme of age of onset are mutations that are expressed early in life, such as Tay-Sachs disease and cystic fibrosis, in which the phenotype is present at birth or can even appear during embryonic development. At the other extreme are mutations expressed later in life, such as the mutation leading to premature baldness. A well-known serious disorder caused by a mutation expressed relatively late in life is **Huntington disease**. This rare dominant mutation causes an incurable and relentless degeneration of the nervous system, usually beginning some time in middle age and progressing rapidly, so that in a few years the victim, once healthy and vigorous, becomes bedridden and helpless. Famed folk singer Woody Guthrie, who died in 1967, was afflicted with this tragic disorder.

Mutations can also be classified according to the severity of their phenotypic effects. The most severe mutations are, of course, **lethal**, which means that they are incompatible with life; Tay-Sachs disease is a lethal disorder. Equally severe from an evolutionary point of view (but not from a social point of view) are **male-sterile** or **female-sterile** mutations, which render the affected individual unable to reproduce; in some instances of mutations that cause sterility, the affected individual may be completely normal except for the inability to produce functional gametes. At the other end of the spectrum of severity are mutations that seem to have no marked effect on the ability of individuals to survive and reproduce. Included in this category are mutations like those listed in Table 3-1 in Chapter 3—common baldness, chin fissure, ear pits, Darwin tubercle, type of ear cerumen, and so on.

To classify mutations by their effects is only one way of looking at them. Mutations can also be subdivided by whether they are dominant or recessive. Recall that a dominant mutation expresses itself when heterozygous (that is, in single dose); a recessive mutation is expressed only when homozygous (that is, in double dose). Examples of dominant mutations are those that cause achondroplastic dwarfism, certain forms of polydactyly, and Huntington disease; examples of recessive mutations are those causing cystic fibrosis, Tay-Sachs disease, albinism, phenylketonuria, and hemophilia.

Some mutations are **conditional mutations**, which means that they are expressed only in certain environments. In *Drosophila,* for example, it is easy to find **temperature-sensitive lethals**—mutations that cause death at one temperature (called the **restrictive temperature**—typically 25°C for *Drosophila*) but are compatible with life at a different temperature (called the **permissive temperature**—typically 18°C for *Drosophila*). The usual interpretation of temperature-sensitive mutations is that they alter the amino acid sequence of a key protein or enzyme in such a way that the protein's three-dimensional configuration is sensitive to temperature. At the low, permissive temperature, the protein can assume an approximately normal configuration and carry out its function in the cell, but at the high, restrictive temperature, the protein unfolds (**denatures**) and is unable to carry out its function. In humans, the G6PD deficiency discussed in Chapter 5 is a sort of conditional mutation because the associated anemia occurs in response to such environmental triggers as eating raw fava beans or inhaling naphthalene fumes. Conditional mutations can occur in virtually any gene that codes for a protein; such mutations are used in many genetic studies, ranging from viruses that infect bacteria to human cells in culture.

Although classifying mutations as dominant or recessive is useful for certain purposes, it is admittedly rather crude. Dominance and recessiveness are not so much actual properties of genes as they are reflections of the way the phenotype is examined. Consider, for example, the sickle cell mutation β^S and its normal allele β. We can inquire which of the three genotypes—β^S/β^S, β^S/β, and β/β— causes severe anemia and premature death. Only an individual with β^S/β^S is so affected, and by this criterion β^S is recessive, exhibiting its effect only when homozygous. We may also ask whether the red blood cells can be made to undergo sickling with very low oxygen tension. The cells from both β^S/β and β^S/β^S people will sickle, so by this criterion β^S is dominant, its effect showing up when heterozygous. We may also ask what is the proportion of abnormal hemoglobin in the red cells. β/β has essentially 0 percent, β^S/β^S has 100 percent, and β/β^S has some amount in between. By this criterion neither β nor β^S is dominant. This example shows that dominance and recessiveness have meaning only with reference to a particular way of examining the phenotype. As ordinarily used, the term ''recessive mutation'' means a mutation for which heterozygotes (or carriers) do not have the disorder characteristic of the mutation. Recessive genes are seldom recessive in all respects, however, and in most cases a more careful examination of phenotypes will reveal subtle differences between carriers and noncarriers.

THE GENETIC BASIS OF MUTATION

The most natural classification of mutations is based on the sort of changes that occur in the chromosomes or DNA. Heritable changes in chromosomes include translocations, inversions, deficiencies, and duplications, the genetic consequences of which were examined in Chapter 7. Heritable changes in chromosomes also include alterations in chromosome number, such as trisomy, monosomy, and triploidy, which were discussed in Chapter 6. Many of these changes are associated with gross upsets in relative gene dosage and are often lethal in the embryo and not transmitted from generation to generation. All these chromosomal changes are heritable from cell to cell by means of mitosis, however, and this qualifies them as mutations.

Chromosomal changes that are visible in the light microscope are necessarily rather large; otherwise, they could not be seen. Consequently, most visible chromosomal aberrations involve a deficiency or duplication of many genes simultaneously or a change in the relative positions of large blocks of genes. But mutations occur at a much smaller level, too, when changes occur in the nucleotide sequence of DNA. These changes are not visible in the microscope, but they may have profound phenotypic effects in the individuals who carry them. Such subtle changes in DNA are far more common than visible chromosomal abnormalities among mutations that occur spontaneously.

Mutations at the level of DNA include the substitution of one base for another or the addition or deletion of bases in the molecule. If an addition or deletion is large enough, the alteration may be visible through the microscope; thus, the dividing line between mutations called chromosomal changes and changes at the level of DNA is sometimes fuzzy.

One relatively common kind of mutation is a simple **base substitution**, which occurs when one base pair at some position in a DNA duplex is replaced with a different base pair. If a base substitution occurs in a region of the DNA molecule that codes for a polypeptide, then the mRNA transcript produced from the gene will carry an incorrect base at the corresponding position; when this mRNA is translated into a polypeptide chain, the incorrect base may cause an incorrect amino acid to be inserted in the polypeptide chain. Hence, a base substitution in DNA can lead to an **amino acid substitution** in the corresponding polypeptide.

The effect an amino acid substitution might have on the functioning of a polypeptide depends on where in the chain the substitution occurs and on which amino acid is replaced by which other one. Sometimes a substitution may occur in a part of the molecule that can tolerate the change without undergoing a marked change in overall three-dimensional structure; the function of the molecule will then be quite normal. At other times the substitution replaces an amino acid, such as leucine, with a chemically similar one, such as isoleucine, and the net effect on the polypeptide may again be small. At still other times a base substitution in DNA may be "silent" and lead to no amino acid substitution at all; this can happen because the genetic code has synonyms, so several different codons can denote the same amino acid in translation (see Table 9-1). Codons 5'-UUA-3', 5'-UUG-3', 5'-CUU-3', 5'-CUC-3', 5'-CUA-3', and 5'-CUG-3' all code for leucine, for example. Mutations that cause slight or no changes in the functional capacity of the polypeptide have no obvious phenotypic effects; individuals who carry such mutations are normal or very nearly normal.

The base substitutions of concern to most people are those that cause dramatic changes in the functional ability of the affected protein and therefore cause inherited disorders. One well-known example is the mutation responsible for sickle cell anemia. Recall that the major adult hemoglobin is a molecule with four polypeptide chains: two α chains and two β chains. The sickle cell mutation is in the first coding region of the β gene. This gene codes for a polypeptide chain consisting of 146 amino acids, and the sequence (from the amino end of the molecule) is val-his-leu-thr-pro-glu-glu-lys- . . . . (The amino acids are abbreviated using the three-letter format in Table 9.1, and the ellipsis represents the other 138 amino acids.) The sickle cell mutation leads to an amino acid substitution at exactly one position in the chain—position number 6—at which a valine (val) replaces a glutamic acid (glu). The β^S chain therefore has the sequence val-his-leu-thr-pro-val-glu-lys- . . . .

This seemingly trivial change profoundly alters the molecule. Molecules of

$\alpha_2\beta_2{}^S$ tend to crystallize or become stacked like long rods under conditions of little oxygen; this causes the red blood cells to assume their characteristic sickle, or half-moon, shapes. Sickled cells tend to become rigid, clump together, and lose their ability to squeeze through the tiny capillaries (which are narrower than the diameter of normal red blood cells). This, in turn, often leads to clogging of these microscopic vessels, resulting in loss of blood supply to the tissues nourished by them. Thus, one strategically placed wrong amino acid in one polypeptide can cause numerous and varied abnormalities throughout the entire body (see Figure 10.1).

MISSENSE, NONSENSE, SILENT, AND FRAMESHIFT MUTATIONS

Various types of mutations can conveniently be discussed in relation to the β-globin gene. Recall from Chapter 9 that the human β-globin gene is split into three coding regions interrupted by two intervening sequences. Figure 10.2(a)

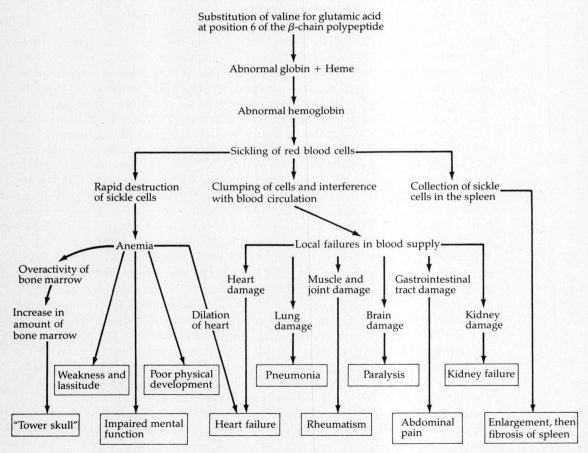

Figure 10.1 Manifold pleiotropic effects arising from the sickle cell hemoglobin mutation.

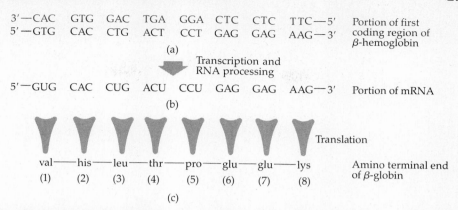

Figure 10.2 (*a*) DNA sequence coding for the first eight amino acids of human β-globin. Small gaps separate adjacent codons for ease of reading. The sense strand is at the top. (*b*) Corresponding mRNA sequence. (*c*) Amino terminal end of β-globin polypeptide. [The first amino acid is actually methionine, but it is later cleaved off the molecule and so its codon is not shown in (*a*) and (*b*).]

illustrates the nucleotide sequence of the beginning of the first coding region. The sense strand is at the top, and transcription occurs from left to right (from the 3′ end of the sense strand to the 5′ end). The spaces that separate adjacent triplets of bases do not exist in the actual molecule, of course; they are present in Figure 10.2 only for convenience of discussion. Figure 10.2(*b*) shows the corresponding portion of the β-globin mRNA produced after transcription and RNA processing. Although it is not shown in the figure, the next upstream codon from GUG is the AUG initiation codon, which codes for methionine. Figure 10.2(*c*) shows the first eight amino acids at the amino terminal end of normal β-globin; the initial methionine with which translation begins is not represented in the finished polypeptide because it is enzymatically removed from the chain.

As noted in the preceding section, a **base substitution** mutation is one in which a base pair in a DNA molecule is replaced with a different base pair. A base substitution that leads to an amino acid substitution in the corresponding polypeptide is called a **missense** mutation. An example of a missense mutation is illustrated in Figure 10.3. Part (*a*) again shows a portion of the first coding region of the β-globin gene, and part (*b*) shows the result of a base substitution in which the normal ^T_A pair indicated is replaced with an ^A_T pair. Transcription and RNA processing of the mutant DNA lead to the mRNA shown in Figure 10.3(*c*); note that the sixth codon, which normally reads GAG, now reads GUG. Translation of this mRNA leads to a substitution of valine for glutamic acid at the sixth position in the polypeptide (black arrow), as can be verified by examination of the genetic code in Table 9.1 in Chapter 9. The mutation illustrated in Figure 10.3 is indeed the actual molecular change responsible for β^S hemoglobin.

Not all base substitutions are missense mutations. Figure 10.4 illustrates a type of mutation called a **nonsense** mutation. A nonsense mutation is one that creates a chain-terminating codon (i.e., UAA, UAG, or UGA) in a coding region. The name *nonsense mutation* may seem odd, but it comes from the fact that

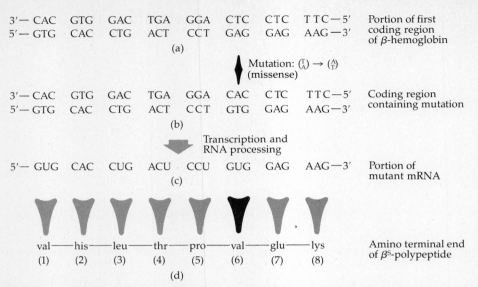

Figure 10.3 Molecular basis of the β^s-hemoglobin mutation. (*a*) Normal DNA sequence. (*b*) Mutated sequence with $_T^A$ base pair replacing $_A^T$ base pair at position 17. (*c*) Corresponding mRNA. (*d*) Corresponding polypeptide with valine replacing normal glutamic acid at position 6.

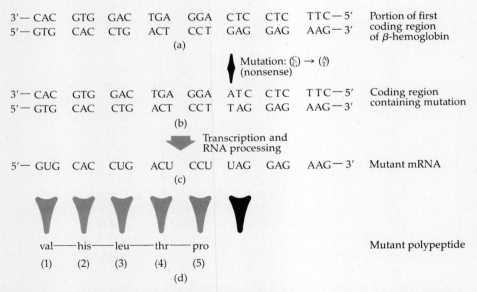

Figure 10.4 A nonsense mutation creates a chain-terminating codon, here illustrated with a hypothetical example involving the β-globin gene. (*a*) Normal sequence. (*b*) Mutant sequence with $_T^A$ replacing $_G^C$ at position 16. (*c*) Corresponding mRNA; note UAG terminator codon. (*d*) Mutant polypeptide consists of just five amino acids because translation terminates at UAG.

chain-terminating codons are at times referred to as *nonsense codons*. In the hypothetical example in Figure 10.4, the base substitution ^C_G to ^A_T leads to a UAG terminating codon at the sixth position in the mRNA. Thus, translation of this mRNA leads to a truncated polypeptide because translation terminates at the UAG codon (black arrow).

Because the genetic code (see Table 9.1) contains many synonymous codons, some base substitutions leave the amino acid sequence of the corresponding polypeptide unaltered. Such mutations are said to be **silent**; an example is illustrated in Figure 10.5. In this case, a ^C_G-to-^T_A base substitution at the third position in codon 6 of the β-globin gene leads to an mRNA that has GAA as its sixth codon. However, GAA is a synonymous codon for glutamic acid, so the amino acid sequence of the resulting polypeptide will be completely normal.

As a brief inspection of the genetic code will reveal, most silent substitutions in coding regions would be expected to involve base substitutions in the third position of a codon. The situation may be very different for intervening sequences, however. Since a particular length and base sequence of an intervening sequence do not seem to be essential for proper gene function, many mutations in intervening sequences—even inversions, deletions, or insertions—may be silent.

A final example of a type of mutation that has a relatively simple molecular basis is known as a **frameshift** mutation. A frameshift mutation alters the reading frame of an mRNA during translation and thereby greatly alters the amino acid sequence of the corresponding polypeptide. Frameshift mutations are not caused

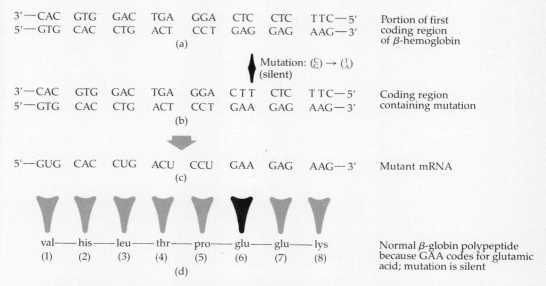

Figure 10.5 Silent mutations change one codon into a synonymous codon and thus lead to no change in amino acid sequence. (*a*) Normal DNA sequence. (*b*) Mutant sequence with ^T_A instead of ^C_G at position 18. (*c*) mRNA; note GAA codon replacing normal GAG. (*d*) Polypeptide has normal amino acid sequence because both GAA and GAG code for glutamic acid.

by simple base substitutions but by deletions or additions of a small number of nucleotides in a coding region; since the genetic code consists of triplets of nucleotides, any deletion or addition of a number of nucleotides other than an exact multiple of three will cause a shift in reading frame.

An example of a frameshift mutation associated with a single-nucleotide deletion is illustrated in Figure 10.6. Part (*a*) is again a portion of the first coding region of the human β-globin gene, and the left side of the figure illustrates the normal course of transcription and translation of this sequence. As in previous figures involving this gene, small gaps are introduced in part (*c*) to show the normal triplet reading frame. The right side of the figure shows the consequence of a deletion of the indicated $\frac{A}{T}$ nucleotide pair. When this deletion-bearing sequence is examined in terms of its triplet reading frame, it can be seen that all triplets beyond the second one are different from those in the normal sequence. This lack of correspondence occurs, of course, because the single-nucleotide deletion causes the triplet reading frame beyond the deletion to be shifted one nucleotide to the left. The extensive lack of correspondence due to the frameshift mutation is also evident in the mRNA, and the resulting amino acid sequence in the polypeptide is hardly recognizable as a β-globin sequence. Such out-of-frame translation will continue along the mutant mRNA until a termination codon is encountered.

RATES OF MUTATION

Missense, nonsense, and frameshift mutations are the simplest changes that occur in DNA. Other, more complex mutations include deletions, duplications, inversions, and, as will be discussed shortly, insertions. Certain questions about mutations inevitably arise: Where do mutations come from? How do mutations happen? How frequently do mutations occur? Relative to these issues it is important to distinguish between newly arising mutations and preexisting mutations. Most mutations in an individual are not the result of new mutations in that person; most such mutations are inherited from the previous generation, and they may have originated many generations in the past. There is a vast reservoir of genetic damage in normal, apparently healthy individuals, but it is concealed because the mutations are recessive and heterozygous and therefore hidden by their corresponding normal alleles.

In every generation, each of these mutations segregates in sperm and eggs, and each runs the risk of becoming homozygous if it meets a gamete that carries a similar mutation at the same locus. In most instances the recessive mutations do not become homozygous, however. Rather they are passed from heterozygous parent to heterozygous child, from generation to generation, seldom giving warning of their presence. In addition to these recessive mutations, there are the many dominant mutations that have incomplete penetrance and are therefore not always expressed. These, too, can be transmitted from generation to generation. And on top of all this are the offspring of parents who themselves have inherited disorders. Color blindness, hemophilia, sickle cell anemia, Huntington disease, and many other inherited disorders to not always prevent reproduction by af-

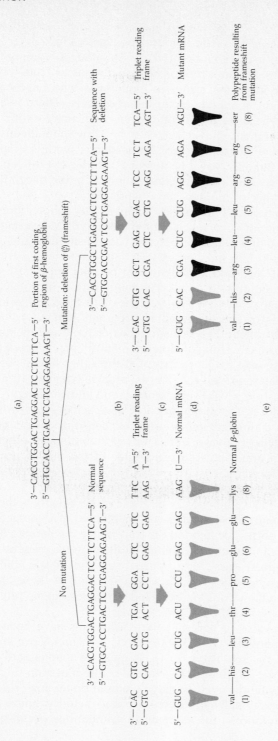

Figure 10.6 A frameshift mutation results from a small insertion or deletion that alters the downstream translational reading frame. Here a hypothetical frameshift mutation in the β-globin gene is illustrated. *Left side*: the normal situation; *right side*: the mutant situation. (*a*) Normal DNA sequence. (*b*) Deletion of A̲ base pair at position 8. (*c*) DNA sequences separated into codons by gaps for ease of reading. (*d*) Corresponding mRNAs. (*e*) Translation; note that all codons (and thus amino acids) downstream from and including position 3 are altered as a result of the frameshift mutation (black arrows).

fected individuals. Many, and indeed most, mutations present in individuals in any generation are inherited from the previous generation.

The ultimate source of mutation is the spontaneous, unpredictable change of a normal gene into a mutant form. These mutations are said to be **spontaneous** because they occur in normal genes in individuals who have had no previous contact with radiation or mutation-causing chemicals. There is a small amount of radiation everywhere on earth, of course, and many exotic chemicals are present in the environment. Nevertheless, only a few spontaneous mutations can firmly be attributed to these sources. Spontaneous mutations are random and unpredictable. The famous "royal hemophilia" among the descendants of Queen Victoria probably arose as a spontaneous mutation in the queen herself or in the germ cells of her father or mother (see Chapter 5).

Although spontaneous mutations happen only rarely, they are frequent enough that their rate of occurrence can be measured. Measuring **mutation rates** in such experimental organisms as bacteria, yeast, fruit flies, and mice is straightforward. The procedure in the case of yeast and bacteria is merely to introduce a known number of cells into a medium that permits only those cells that carry a particular new mutation to grow; the mutation rate is then calculated from the proportion of cells that survives. In fruit flies and mice, mutation rates can be measured by means of special breeding programs that allow certain kinds of mutations to be identified. The measurement of mutation in humans is not so straightforward, however. Human beings can hardly be expected to arrange their marriages for the benefit of geneticists! On the other hand, people do tend to keep detailed marriage and medical records, especially in industrialized countries, and hospital records are a vital source of genetic information.

In some instances the mutation rates of specific genes in humans can be measured directly. Dominant mutations that cause specific abnormalities not mimicked by nongenetic abnormalities or by mutations in other genes can be studied with this in mind, but the penetrance of the mutations must be nearly complete; that is, for the measurement of the rate to be accurate, virtually all individuals who carry the mutation must express the trait.

One disorder that fulfills nearly all these requirements is the type of dwarfism known as **achondroplasia**, which was mentioned in Chapter 3. In a study of 94,075 Danish newborns, 8 exhibited the trait. Since the parents of these children were normal, one can assume no dominant alleles were present in them, so the affected children must carry new mutations. The 94,075 children represent 188,150 sets of chromosomes (i.e., $2 \times 94,075$). In 8 of these chromosome sets a new mutation to achondroplasia occurred, so the estimated rate is $\frac{8}{188,150}$, which is about 0.00004 or 4×10^{-5} per generation.

The mutation rates of several genes in humans have been estimated, and the rate averages between 1 per 100,000 (i.e., 10^{-5}) and 1 per million (i.e., 10^{-6}) genes per generation (see Table 10.1). It must be emphasized that this range represents an average, a "ballpark" estimate. The mutation rate actually varies from gene to gene, and the average may be biased on the high side because genes that mutate at exceptionally low rates are not included among those studied. Despite these reservations, the estimate that mutations occur spontaneously in

Table 10.1   ESTIMATES OF SPONTANEOUS MUTATION RATES IN HUMANS

Trait	Estimated number of spontaneous mutations per 100,000 gametes per generation
Epiloia	0.8
Aniridia	0.5
Microphthalmus	0.5
Wardenberg syndrome	0.4
Facioscapular muscular dystrophy	0.5
Pelger anomaly	0.9
Myotonia dystrophica	1.6
Myotonia congenita	0.4
Huntington disease	0.2
Retinoblastoma	0.4
Neurofibromatosis	13–25
Hemophilia	2.7
X-linked muscular dystrophy	5.5

Note: Estimates of mutation rates in humans can be obtained by a variety of methods. In many cases the estimates are quite imprecise, yet most estimates are between 10^{-5} and 10^{-6} mutations per gene per generation. Except for hemophilia and X-linked muscular dystrophy, all the traits listed are due to autosomal dominant mutations.

10^{-5} to 10^{-6} genes per generation seems to be fairly reliable. It also agrees well with information derived from laboratory animals.

A rate of 10^{-5} to 10^{-6} mutations per gene per generation can be interpreted in two ways. One way is to focus on a specific locus and inquire how many normal alleles at this locus undergo mutation in any one generation. Imagine a collection of between 100,000 and 1 million sperm produced by a man homozygous for the normal allele at, say, the phenylketonuria locus. Then a mutation rate of 10^{-5} to 10^{-6} means that, on the average, one sperm will actually carry an allele of this locus that has undergone a spontaneous mutation. The same reasoning would apply to a collection of 100,000 to 1 million eggs from homozygous normal women. This interpretation suffers from the fact that the estimated mutation rate is only an average. The calculation is in error to the extent that the phenylketonuria locus or whatever locus is under consideration may have an atypical mutation rate that deviates from the average for all loci.

A second interpretation of the meaning of the human mutation rate is obtained by focusing on all the genes in a single cell at once. If a human sperm or egg contains, say, very approximately, 50,000 genes, then what percentage of these gametes will carry a gene that has undergone a spontaneous mutation? The answer is somewhere between $50,000 \times 10^{-5} = 0.50 = 50$ percent and $50,000 \times 10^{-6} = 0.05 = 5$ percent. In other words, roughly between 5 and 50 percent of all sperm or eggs will carry at least one new mutation. This may seem like a large number, but it is actually only a small fraction of the mutations already present in the sperm and eggs of normal individuals.

INSERTION MUTATIONS AND TRANSPOSABLE ELEMENTS

For many years it was thought that most spontaneous mutations involved relatively simple changes at the nucleotide level—nucleotide substitutions or perhaps an occasional small insertion or deletion or inversion. Recombinant DNA techniques have changed this picture completely. It is now clear that many spontaneous mutations involve massive changes at the nucleotide level, changes such as the insertion of thousands of nucleotides right in the middle of a gene. In many cases the nucleotide sequence of the large insertion can be recognized as one of a **family** (or group) of similar or identical sequences present in 20 or 30 or more copies and **dispersed** (scattered) throughout the chromosomes.

In *Drosophila melanogaster,* for example, one such family of sequences is called **copia**; the copia sequence is about 5 kb long and it is found at about 30 widely scattered chromosomal locations (Figure 10.7). What is unique about copia and other analogous **dispersed repeated gene families** (as they are sometimes called) is that the sequences are able to change location within the chromosomes; that is, they are able to move from one place in a DNA molecule to a different place in the same molecule or to a different molecule altogether. Because of their mobility, such sequences are often called **transposable elements**. Transposable elements seem able to become inserted at virtually any site in a DNA molecule; when one transposes into a site in an existing gene, the insertion often disrupts the gene's ability to function. Transposition thus becomes a source of spontaneous mutation, and recent evidence suggests that it is an important source.

Figure 10.8 illustrates the molecular structure of the copia transposable element, and it exhibits several features that are characteristic of virtually all transposable elements. Copia is about 5 kb (5000 base pairs) long, and within the

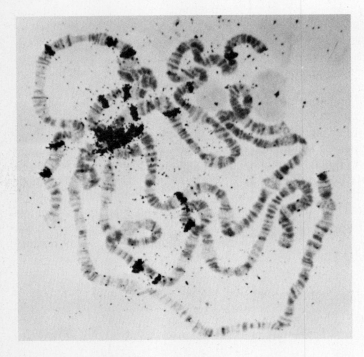

Figure 10.7 Autoradiograph showing many sites of hybridization of copia to *Drosophila* salivary-gland chromosomes.

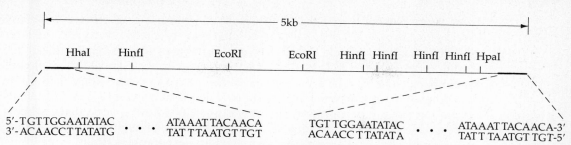

Figure 10.8 Restriction map of 5 kb (5000 base pair) element copia and partial nucleotide sequence of the 276 base pair direct-repeat sequence at its ends.

central region (thin black line) are at least three coding regions that are transcribed into RNA. The ends of the element (thick black lines) are remarkable because the nucleotide sequence of 276 base pairs at the left is virtually identical to the 276 base pairs at the right. Such **terminal repeats** are characteristic of transposable elements. Although the terminal repeats in copia are **direct repeats** (i.e., repeated in the same orientation at both ends), the terminal repeats of some other elements are **inverted repeats** (i.e., repeated in opposite orientation at the ends). (However, careful examination of Figure 10.8 will reveal that the five base pairs at the very tips of copia form an inverted repeat.)

The origin and functions of transposable elements are unknown, but they seem to be related to certain types of viruses (retroviruses), to be discussed in Chapter 11. Transposable elements seem to account for a substantial fraction of a class of eukaryotic DNA called moderately repetitive DNA. **Moderately repetitive DNA** refers to about 30 percent of mammalian DNA, which consists of relatively short (about 600 base pairs) sequences, each repeated tens or hundreds of times and scattered throughout the chromosomes. Mammalian DNA contains about 1500 different sequences of moderately repetitive DNA, and much of this now seems to be related to transposable elements. In humans, the largest family of transposable elements is the **Alu family** (so called because the transposable elements in this family all share a common restriction site for the restriction enzyme AluI). The Alu transposable element is 300 base pairs long, and normal human cells carry some 300,000 scattered copies of this sequence, which in the aggregate accounts for about 3 percent of human DNA. Whatever transposable elements may do—and some commentators refer to transposable elements as **selfish DNA** or **parasitic DNA** to suggest that they may function only toward their own increased replication and transposition—it is clear that transposable elements constitute a major class of eukaryotic DNA and contribute significantly to spontaneous mutation.

RADIATION AS A CAUSE OF MUTATION

Although many mutations occur spontaneously, mutations can also be induced by certain kinds of radiation and chemicals. Agents that can cause mutations, called **mutagens**, are of great concern. Even a small increase in the mutation rate from these sources is amplified into an enormous number of unnecessary birth defects because of the large number of genes that can be mutated and the large size of the human population. It has been estimated, for example, that if the mutation rate were increased by 20 percent for one generation and then allowed to slip back to its normal level, the consequences of that temporary increase in the United States alone would be about 400,000 new, severe genetic defects spread over many generations, with perhaps 40,000 new genetic defects being expressed in the first generation. This is a staggering biological and social cost, and it should be emphasized that corresponding genetic benefits from new mutations are so rare as to be virtually nonexistent.

X-rays and other forms of ionizing radiation are potent mutagens. These penetrating rays pierce through cells and along their paths cause electrons to be ejected from their usual locations in atoms and molecules; the affected electrons dart off and collide with other molecules, causing more damage, and many of the ions (charged atoms or molecules) produced by molecules that lose electrons become extremely reactive. In the nucleus these ions may combine with bases in the DNA and cause mistakes in base pairing during replication, leading ultimately to base substitutions; they may also sever the sugar-phosphate backbones of DNA and thereby cause physical breaks in the chromosomes (see Figure 10.9, for example).

Nowadays the dangers of x-rays are so widely recognized that it is hard to believe they were at one time considered to be safe and were even used as playthings. According to Hilgartner, Bell, and O'Connor,

> By 1898, personal X-rays had become a popular status symbol in New York. The *New York Times* reported that "there is quite as much difference in the appearance of the hand of a washerwoman and the hand of a fine lady in an X-ray picture as in reality. . . ." The hit of the exhibition season was Dr. W. J. Morton's full-length portrait of "*the X-ray lady*," a "fashionable woman who had evidently a scientific desire to see her bones." The portrait was said to be a "fascinating and coquettish" picture, the lady having agreed to be photographed without her stays and corset, the better to satisfy the "longing to have a portrait of well-developed ribs." Dr. Morton said women were not afraid of X-rays: "After being assured that there is *no danger* they take the rays without fear."
>
> The titillating possibility of using X-rays to see through clothing or to invade the privacy of locked rooms was a familiar theme in popular discussions of X-rays and in cartoons and jokes. Newspapers carried advertisements for "*X-ray proof underclothing*" for those seeking to protect themselves from X-ray inspection. . . .
>
> The luminous properties of radium soon produced a full-fledged radium craze. A famous woman dancer performed *radium dances* using veils dipped in fluorescent salts containing radium. . . . *Radium roulette* was popular at New York casinos, featuring a "roulette wheel . . . washed with a radium solution, such that it glowed brightly in the darkness . . . an unseen hand cast the ball on the turning wheel and sparks marked its course as it bounded from pocket to glimmery pocket." A patent was issued for a process for making women's gowns

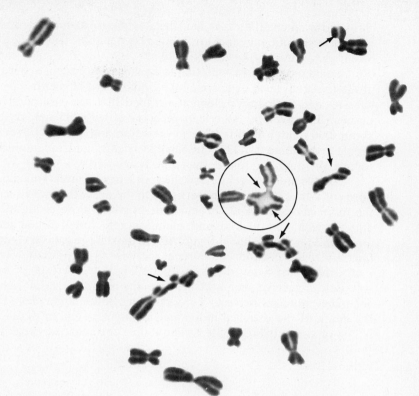

Figure 10.9 Human chromosomes showing breaks (arrows) induced by 270 r (roentgen) of x-rays. The circled structure arose from interchange and restitution of broken chromatids in two chromosomes; the positions of the breaks before restitution are denoted by arrows within the circle. A dose of only 20 r of x-rays is sufficient to produce one visible chromosome break per cell in human chromosomes in tissue culture.

luminous with radium, and Broadway producer Florenz Ziegfeld snapped up the rights for his stage extravaganzas. . . .

Even while this unrestrained use of X-rays and radium was growing, evidence was accumulating that the new forces might not be so benign after all. Hailed as tools for fighting cancer, they could also cause cancer [and mutations]. Doctors using X-rays were the first to learn this bitter lesson.*

A unit frequently used for measuring the amount of damage done by radiation to human tissue is the **rem**. Other units are the **rad** and the **roentgen** (r). For x-rays and other forms of ionizing radiation, these units are very nearly the same. An appreciation of the quantity that one rem of radiation represents may be gleaned from the fact that an average diagnostic x-ray exposes the patient to about 1 rem. (The amount varies from 0.1 to 10 rem, depending on the type of

*Hilgartner, S., R. C. Bell, and R. O'Connor. 1982. Nukespeak. Sierra Club Books, San Francisco, pp. 2–4, 8.

x-ray being taken.) The U.S. Nuclear Regulatory Commission sets the permissible level of occupational exposure at 1.5 rem per three-month period, not to exceed 5 rem per year.

Some exposure to radiation is unavoidable because there are certain background sources such as cosmic rays, radioactive elements in the earth and on its surface, and radioactive elements in food that are incorporated into the body's tissues. The average American receives about 0.05 rem/yr from cosmic rays, about 0.06 rem/yr from terrestrial radiation, and about 0.02 rem/yr from ingested radioactive elements. The inescapable background radiation is thus about 0.13 rem/yr, or roughly 6 rem between birth and age 45.

X-rays are mutagenic, and the frequency of mutations induced by moderate doses of x-rays is strictly proportional to the dose; the more x-rays a cell receives, the more mutations and chromosome breaks will be produced. The proportionality between x-ray dose and mutation rate for X-linked recessive lethals in *Drosophila* is illustrated in Figure 10.10; the **dose-response relationship** is a straight line, and there is about a 3 percent increase in the mutation rate for every 1000 r of x-rays. The x-ray doses in Figure 10.10 are very large because adult *Drosophila* can survive such massive amounts of x-rays. Adult humans, by contrast, are much more sensitive to x-ray damage; doses larger than 450 r usually cause the death of the exposed individual. In human cells in tissue culture, exposure to 20 r of x-rays induces at least one visible chromosome break per cell. However,

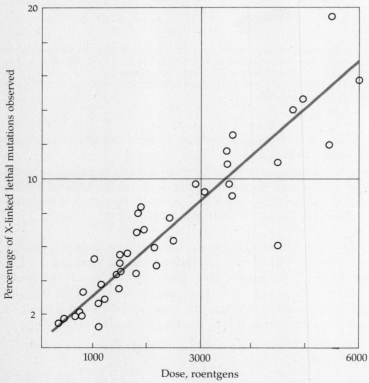

Figure 10.10 Summary of studies on the induction of lethal mutations in the X chromosome of *Drosophila* by x-rays. Note that most of the points fall close to the straight line. Thus, the rate of induction of new lethal mutations is proportional to the dose of x-rays over the range of doses shown here.

the amount of mutational damage produced by x-rays depends on whether the radiation is **acute** (delivered all at once) or **chronic** (delivered as a series of smaller doses). Acute doses are generally three or four times more effective in producing mutation; the difference apparently is that chronic irradiation gives the cells more opportunity to repair certain kinds of genetic damage in the DNA.

Radiation that hits the testes and ovaries—the **gonadal dose**—is obviously of greatest importance to future generations. Radiation also affects somatic cells, however. People cannot usually survive acute doses over 450 r, and 100 r will produce **radiation sickness**—a condition marked by nausea, vomiting, headache, cramps, diarrhea, loss of hair and teeth, decrease in the number of blood cells, and prolonged hemorrhage. Dividing cells, such as those in the bone marrow, are generally more sensitive to radiation than are nondividing cells.

People accidentally exposed to high doses of radiation generally experience for some time after exposure a much lowered resistance to infectious disease. This is evidently because the the rapidly dividing white blood cells, which play a central role in immunity, are very sensitive to x-rays. Certain cancers are also treated with ionizing radiation, often the gamma rays from radioactive cobalt; the reason is that cells in a tumor are rapidly dividing and therefore more sensitive to radiation than the surrounding cells, which are not dividing. The embryo, loaded with rapidly dividing cells, is also very sensitive to x-rays; this is why women who are pregnant (or who might be) must not casually allow themselves to be exposed to ionizing radiation.

A convenient measure of the mutagenic effects of radiation is the **doubling dose**—the dose that induces a frequency of mutation exactly equal to the spontaneous frequency so that the overall frequency is doubled. This is a useful measure despite the fact that the relative proportions of specific kinds of mutations caused by x-rays are somewhat different from those arising spontaneously; x-rays generally produce more deletions and more lethal mutations, for example. Based on family studies of Japanese individuals exposed to radioactive fallout from the atom bombs dropped on Hiroshima and Nagasaki in August 1945, the doubling dose in humans has been estimated as 156 rem.

How much radiation do we receive over and above natural background radiation? A significant source of radiation is the radioactive debris trapped in the upper atmosphere from previous above-ground atom bomb tests. This debris leaks slowly back to earth, and it exposes everyone to about 0.008 rem/yr. This exposure may not seem large, but it is about 6 percent of the background level, and the cumulative dose from birth to age 45 due to this source alone is about 0.4 rem. On the other hand, this level of exposure is about the same as that incurred by watching color television, owing to radioactive emissions from the set, or by leaving sea level to spend one month a year in Denver, owing to the greater exposure to cosmic rays because of high altitude.

For most people, the most important single source of ionizing radiation other than background is medical and dental x-rays. On the average, the exposure from these sources is about 0.073 rem/yr, but there is wide variation from individual to individual. People who are chronically ill and in frequent need of diagnostic x-rays receive doses many times larger than the average, whereas others may never have had an x-ray in their lives. As always, the gonadal dose is the most important for future generations, and this risk is greatest in x-rays of the abdominal region. With proper shielding of the reproductive organs, the gonadal dose is negligible for dental x-rays or for x-rays of the skull, neck, upper chest, and limbs. On the average, a male in the United States will receive a testicular dose of about 1.3 rem from diagnostic x-rays from birth to age 45; the ovarian dose to an average female would be less, about 0.3 rem, since the ovaries are inside the body cavity and therefore more protected. Although the gonadal dose from dental and medical x-rays is moderate, it should be emphasized that unnecessary x-rays are always to be avoided.

Another mutagenic form of radiation is **ultraviolet light**—a short-wavelength component of ordinary sunlight that causes the skin to tan or, more painfully, burn. Ultraviolet rays are not very penetrating, so the damage they do directly is usually confined to cells in the skin. Ultraviolet light affects DNA in at least two ways. The bases of DNA tend to absorb these wavelengths and become energized, which increases the incidence of mispairing. Ultraviolet light can also cause chemical crosslinks to form between adjacent thymines in a DNA strand. These linked thymines constitute a **thymine dimer**; the formation of a thymine dimer is illustrated in Figure 10.11(*a*). Thymine dimers are ordinarily repaired by a process called **excision repair**, which is illustrated in Figure 10.11(*b*). Essentially

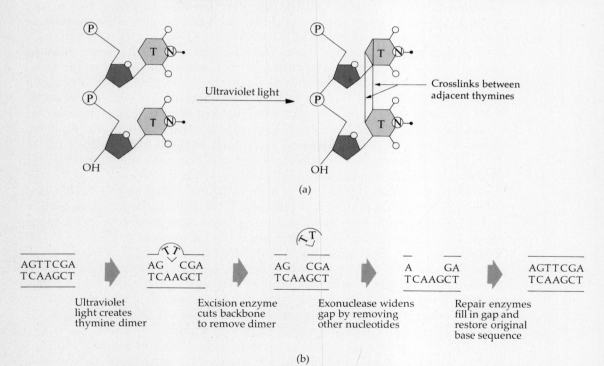

Figure 10.11   (*a*) Ultraviolet light can cause chemical crosslinks (covalent bonds) between adjacent thymines in a strand of DNA, as shown here. Recall that the rings of the bases in DNA are rather flat and are stacked on top of one another; thus, the atoms shown crosslinked are in close physical proximity. (*b*) Mechanism of repair of crosslinked thymines involving excision, widening of the gap, and resynthesis.

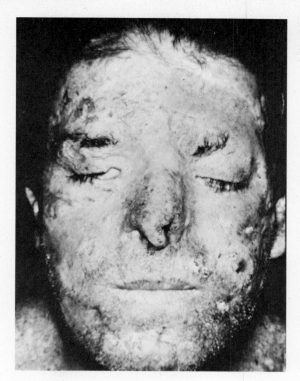

Figure 10.12 This person has xeroderma pigmentosum and thus has multiple skin tumors.

a cut-and-patch process, repair of a thymine dimer begins when a particular **excision enzyme** recognizes the dimer and cleaves the DNA backbone to remove the dimer. Other enzymes then widen the gap by removing additional nucleotides, and repair enzymes take over and restore the original base sequence by resynthesizing the excised piece, using the intact strand as a template.

Ultraviolet light is usually of no concern genetically because it does not penetrate to the germ cells in the testes or ovaries. Ultraviolet light does damage the genetic material in the living cell layers in the skin, although this damage is often repaired by the appropriate enzymes. However, some individuals have a rare disorder known as **xeroderma pigmentosum**, which is inherited as an autosomal recessive. Patients with xeroderma pigmentosum are extremely sensitive to ultraviolet light, and in childhood or early teens most of them develop multiple skin cancers (Figure 10.12), which usually lead to death before age 20. Although there appear to be several forms of the disorder, it is now known that at least some xeroderma pigmentosum patients have defective DNA repair systems so they are unable to repair the damage induced by ultraviolet light. Xeroderma pigmentosum provides one of the many links between mutation and cancer. Other links will be discussed in a later section.

CHEMICALS AS MUTAGENS

Many chemicals are also mutagenic, and some have rather specific effects. An agent known as nitrous acid [Figure 10.13(a)] clips an —NH$_2$ group from a base

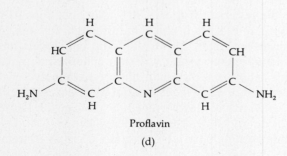

Figure 10.13 Some chemical mutagens. (a) Nitrous acid causes deamination and converts cytosine to uracil. (b) 5-bromouracil is incorporated into DNA in place of thymine but often pairs with guanine. (c) Mustard gas is an alkylating agent and so a potent mutagen. (d) Proflavin inserts itself between the bases of a DNA molecule, causes distortions of the molecule during replication, and so leads to single-base deletions or insertions (frameshift mutations).

such as cytosine or adenine and substitutes an oxygen atom. This converts cytosine into uracil, for example, so a $\frac{G}{C}$ base pair momentarily becomes a $\frac{G}{U}$ pair. At the next replication one daughter molecule will have a $\frac{G}{C}$ base pair whereas the other will have an $\frac{A}{U}$ base pair (because A pairs with U). In the next round of replication, T will pair with A and A will again pair with U, so that the descendants of the original "U" strand will have undergone the equivalent of a $\frac{G}{C} \to \frac{A}{T}$ base substitution. Nitrous acid changes adenine into a base that pairs with C, causing $\frac{A}{T} \to \frac{G}{C}$ substitutions.

Certain other substances chemically similar to bases and therefore known as **base analogues** may be incorporated into DNA in place of one of the usual bases. Base analogues mispair more frequently than do normal bases, so they increase the mutation rate. An example of a base analogue is 5-bromouracil [Figure 10.13(b)]. This compound is incorporated into DNA in place of thymine, but it often pairs with guanine, causing $\frac{A}{T} \to \frac{G}{C}$ substitutions. The substance 2-aminopurine is incorporated into DNA in place of either A or G, and it can pair with either T or C. If you drink coffee, you may be interested to know that caffeine is also a base analogue that causes mutations in bacteria and yeast, but it does not seem to be mutagenic in mammals.

Many highly reactive chemicals are mutagenic, as you might expect. These chemicals apparently react with bases to create abnormal forms that cause errors in replication, although some of their effects evidently can be repaired in the cell. Among this class of mutagens are the **alkylating agents**, of which mustard gas [Figure 10.13(c)], used systematically by each side against the other in World War I, is a prime example. Still another chemical mutagen is a class of compounds called the **acridines** [Figure 10.13(d)]. Acridines interact with DNA in such a way as to cause deletions or additions of one or more bases (usually just one). Acridines therefore cause frameshift mutations, apparently by squeezing themselves between the bases in a DNA molecule and causing distortions of the molecule and mistakes in replication.

THE AMES TEST

The testing of chemicals for mutagenic activity has been greatly aided by the development of a procedure called the **Ames test**, after its inventor. The organisms used in the Ames test are special strains of the bacterium *Salmonella typhimurium*. These strains are highly sensitive to mutagens because they are defective in the excision-repair system for correcting DNA damage, and they also have cell membrane defects that make them highly permeable to chemicals; some tester strains also carry a factor that enhances mutagen sensitivity.

The Ames test relies on certain mutations in genes for the synthesis of the amino acid histidine. Some tester strains have a known base-substitution mutation, so mutagens that cause base substitutions will revert these mutations and thus permit growth on media lacking histidine. Other tester strains have a known frameshift mutation, so frameshift mutagens will revert these and permit growth on media lacking histidine. In addition, the chemical to be tested is first mixed with an extract of liver, typically rat liver, because many chemicals are not themselves mutagenic but are converted into mutagenic forms by enzymes in the liver. (The liver is the primary organ in mammals for metabolizing foreign chemicals.)

A typical Ames test involves mixing the suspected mutagen with the liver extract and with about 10^9 cells of one of the tester strains. This mixture is then spread (**plated**) onto a Petri dish (the **plate**) containing a histidine-less medium, and the plates are incubated for two days. Each colony that appears on the agar surface is a clone consisting of the descendants of a single cell whose histidine mutation had been reverted (see Figure 10.14). The number of colonies appearing on the experimental plates with mutagenized cells is compared with the number appearing on control plates with nonmutagenized cells; a significant increase in the reversion rate is taken as evidence of the mutagenic activity of the suspected chemical. In practice, suspected mutagens are tested in a series of concentrations, and the dose-response relationships are determined. These dose-response curves are almost always linear (see Figure 10.15 for some examples).

The Ames test is simple, rapid, inexpensive, and exquisitely sensitive. Some chemicals can be detected to be mutagenic in amounts as small as 10^{-9}g, and a condensate of as little as $\frac{1}{100}$ of a cigarette can be shown to be mutagenic. Moreover, the test is quantitative. Chemicals need not be classified simplistically as "mutagenic" or "nonmutagenic." They can be classified according to their potency as mutagens because more than a millionfold range in potency can be detected in the *Salmonella* test.

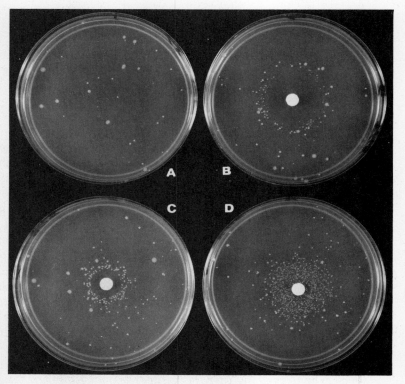

Figure 10.14 Ames test. Petri dish (*a*) contains no mutagens: the few bacterial colonies result from spontaneous mutation. The white paper disk in the center of the other dishes contains a chemical shown to be mutagenic by the increased number of bacterial colonies. The mutagens are AF-2(*b*), aflatoxin (*c*), and 2-aminofluorente (*d*).

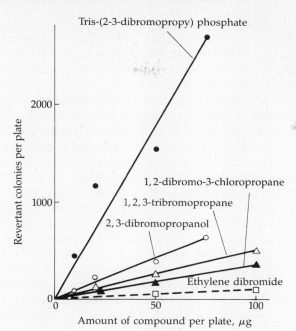

Figure 10.15 Straight-line dose-response relationship obtained with various chemical mutagens in the Ames text. The broken line indicates that the concentration of the chemical was 10 times that indicated on the horizontal axis.

MUTAGENS AND CARCINOGENS IN THE ENVIRONMENT

The Ames test is only one of a group of short-term tests now favored for the rapid screening of chemicals. Other tests measure the potency of chemicals to induce mutations in *Drosophila,* to induce mutations in cultured animal cells, or to induce chromosome breaks in laboratory animals such as mice or rats.

The need for screening potential mutagens and **carcinogens** (agents that cause cancer) is acute. More than 50,000 untested synthetic chemicals are already in use, and about 1000 new ones are produced commercially each year. In the past, only a small fraction of these chemicals was adequately tested prior to use, partly because thorough testing of the carcinogenic potential of a single substance, using rodents, is expensive in terms of both capital (up to $500,000) and time (about three years). Such tests are also not feasible on an appropriate scale. For example, a widespread carcinogen that caused cancer in just 1 percent of all people in the United States would result in more than 2 million new cases of cancer. Yet to detect such a potency in test animals would require an experiment involving 10,000 rats or mice; this scale is 100 times the number of animals usually involved in such studies. Partly as a consequence of such difficulties, and partly out of ignorance, carelessness, callousness, and greed, our environment has been contaminated with many substances that are now known to be carcinogenic or mutagenic. Among a group of 168 Canadians, for example, almost all were found to have in their tissues significant quantities of chlorinated hydrocarbons such as DDT, DDE, PCB, BHC, oxychlordane, heptachlor epoxide, and dieldrin. All are carcinogens. More ominously, all these chemicals are found in significant quantities in human milk.

An example of the damage of inadequate testing involves a chemical known as tris-BP. This substance was used as a flame retardant in children's polyester pajamas between 1972 and 1977. Then it was discovered that tris-BP is a potent mutagen in *Salmonella* and *Drosophila*; it interacts with human DNA and damages mammalian chromosomes; it was found to be a carcinogen in experimental tests in rats and mice; and it was found to be capable of causing sterility in laboratory animals. Moreover, the substance was shown to be absorbed through the skin, and its breakdown products could be detected in the urine of children who were wearing the treated sleepwear. Unfortunately, before all this was known, more than *50 million* children were exposed to the chemical through contact with their nightclothes. The price to be paid for the tris-BP mistake is not known. Cancer has a latent period of 20 to 30 years before it is expressed, so any increase in the cancer rate will not be detectable until the 1990s or later. Any mutagenic consequences will take generations to be revealed.

The development of rapid and inexpensive procedures such as the *Salmonella* system for detecting mutagens has eased the testing crisis somewhat. Of particular good fortune and importance is that tests for mutagenicity have proven to be extremely useful for identifying carcinogens as well. In tests involving hundreds of chemicals in *Salmonella,* for example, over 90 percent of all known carcinogens were also mutagenic. Similarly, a very high proportion of mutagens was also carcinogenic. In addition, there is a good correlation between a chemical's potency as a mutagen and its potency as a carcinogen. It therefore seems likely that some weak mutagens are in fact carcinogens but were improperly classified as noncarcinogens because of inadequate numbers of test animals. In short, most carcinogens are mutagenic, and most mutagens are carcinogenic. This discovery implies that many carcinogens produce their effects by direct interactions with DNA, and it establishes yet another link between mutation and cancer.

From the standpoint of environmental health, the high mutagen-carcinogen correspondence means that tests as simple as the *Salmonella* test can be used to identify virtually all mutagens and most carcinogens prior to their commercial use. Because the test is so rapid and inexpensive, suspects among the 50,000 synthetic chemicals already in use can readily be examined. In one case, for example, an antibacterial food additive called AF-2 used widely in fish and soybean products in Japan was found to be extraordinarily mutagenic in *Salmonella*. Subsequent tests showed it to be carcinogenic in mice, rats, and hamsters, although previous tests for carcinogenicity had had negative results. Use of AF-2 as a food additive has, of course, been prohibited, but this does not undo the damage it caused during the nine years in which it was used in large amounts. As with tris-BP, it is too early to know the effects of the error.

These words are being written in St. Louis, and as I write a controversy is swirling around Times Beach, Missouri, a small community a few miles to the southwest where the Center for Disease Control has confirmed that deadly dioxin, a highly toxic by-product of chemical manufacturing, was illegally and apparently inadvertently dumped on the roads. Later flooding spread the contaminant throughout the community, and levels up to 5 times the recommended maximum exposure level are currently found in homes and up to 300 times this level in the town's roadways and ditches.

Legal toxic waste dumps are hazardous in themselves because their contents can be disturbed by such occurrences as weathering, floods, tornados, or earthquakes, and the toxic wastes can be spread over a wide area or percolate into the supply of ground water. Now it appears that illegal waste dumping has been much more common than originally thought, and Times Beach is merely one of the first of a series of such discoveries to be made. Times Beach is a small community, and the Environmental Protection Agency has decided to buy all its land and structures (at a cost of $33,000,000) so that the people who live there can move. What will happen to the vacated site and to its former residents is still unclear, but it is grimly fortunate that Times Beach is small. Buying Chicago or Los Angeles would have proven expensive.

When a mutagen, carcinogen, or **teratogen** (an agent causing birth defects) escapes notice and is released for use, it might be imagined that the agent would quickly be recognized by its effects on people. This is far from the truth. Unless the agent is exceptionally potent or causes an otherwise unusual disease, detection is extremely difficult. Two examples illustrate the point. Cigarette smoking became suspect as a carcinogen because it causes lung cancer, an otherwise relatively uncommon type of cancer. Even so, it took decades to accumulate convincing evidence because of the time delay involved in carcinogenesis. The teratogen thalidomide, which, incidentally, is not itself teratogenic but is converted into a teratogen by enzymes in the body, was identified because it causes an absence of arms or legs, an otherwise exceedingly rare birth defect (see Figure 5.13). Even so, it took five years to track down the agent responsible for the defect.

Detection is exceptionally difficult with mutagens because the mutations they cause are relatively nonspecific; that is, a general increase in the mutation rate among humans would create a diversity of genetic defects. Although a few of these might be unique, most would be abnormalities already well known. There would simply be slightly more of everything—more achondroplasia, more hemophilia, more cystic fibrosis, more schizophrenia, more mental retardation, more diabetes, and so on. No single birth defect might be striking enough or common enough to draw attention to itself and flag the fact that something had gone wrong. For this reason, a doubling or tripling of the mutation rate might go unrecognized for decades; a smaller increase might never be detected. But the price of even a minor increase in the mutation rate is measured in hundreds of thousands of children who are needlessly crippled, deformed, or mentally deficient. This conclusion emphasizes the importance of rigorous testing of chemicals before they are put into widespread use.

SUMMARY

1. A **mutation** is a heritable change in the genetic material. Mutations can involve any level of genetic organization, from gross changes in chromosome number, such as polyploidy or polysomy, to aberrations in

chromosome structure, such as inversions, translocations, and deletions, to fine-scale changes at the molecular level, such as **base substitutions** (the substitution of one nucleotide for another in the DNA). Mutations can occur in any cellular type; the most important categories are **somatic** mutations, which occur in somatic cells and are not transmitted to future generations, and **germinal** mutations, which occur in germ cells and can be transmitted.

2. Mutations can be classified in many ways, depending on the purpose of the classification. One classification is based on the time of life when a mutation is expressed; those with an early or late **age of onset** are expressed relatively early or late in life. **Huntington disease** is an example of a severe genetic disorder with a relatively late age of onset. The severity of phenotypic effects provides another means of classifying mutations; **lethal** mutations are incompatible with life, and **sterile** mutations render affected individuals unable to reproduce. In experimental genetics, an important class of mutations is **conditional** mutations, which are expressed under certain environmental conditions (the **restrictive** conditions) but are not expressed under other conditions (the **permissive** conditions). Mutations are also sometimes classified as **dominant** or **recessive** with respect to other alleles at the locus; although useful for some purposes, this classification depends on the level of sensitivity with which the phenotype is examined; a mutation that is recessive with respect to some criterion may have incomplete or complete dominance with respect to another criterion.

3. Mutations can conveniently be classified according to their genetic basis—whether they involve gross changes in chromosome number or structure or more subtle changes at the DNA level. An important category of molecular-level mutation is **base substitutions**, which occur when one nucleotide in the DNA is replaced with a different nucleotide. The effect of a base substitution depends on where in the gene the substitution occurs and which nucleotide is substituted for which other one. Relative to intervening sequences, the effects of base substitutions are at present unknown. Relative to coding regions, base substitutions may leave the corresponding amino acid sequence unaltered (**silent** substitutions) because one codon may be changed into a synonymous codon. Other substitutions lead to nonsynonymous codons, and the corresponding polypeptide undergoes an **amino acid substitution** (the replacement of one amino acid with another). Such amino-acid changing mutations are called **missense** mutations.

4. Not all base substitutions are silent or missense mutations. Some substitutions create new chain-terminating codons within coding regions and thereby cause premature termination of polypeptide synthesis during translation. Such substitutions are called **nonsense** mutations. **Frameshift** mutations involve the addition or deletion of a number of nucleotides, but not an exact multiple of three, in a coding region of DNA. Such mutations alter the translational reading frame of the mRNA and thus change the entire amino acid sequence downstream from the site of the mutation. Additions or deletions of an exact multiple of three nucleotides do not alter the reading frame and are not frameshift mutations.

5. **Spontaneous mutations** occur in the absence of known **mutagens** (agents that cause mutations). On the whole, the spontaneous mutation rate in humans ranges between 10^{-5} and 10^{-6} mutations per locus per generation; that is, between 1 in 100,000 and 1 in 1,000,000 genes at a locus will undergo spontaneous mutation each generation. These rates are averages, however, and there is wide variation from locus to locus.

6. Although spontaneous mutation often involves simple nucleotide substitutions, recent findings suggest that many spontaneous mutations actually result from insertion of long (up to several kilobase) sequences of DNA in or near the locus in question. The inserted sequences are often members of **families** of similar or identical sequences that are found scattered at many chromosomal sites (**dispersed repeated multigene families**). These sequences are able to move from one chromosomal site to another, and they are called **transposable elements**. Examples of transposable elements include the **copia** element of *Drosophila* and the highly repeated (about 300,000 copy) **Alu** family in humans.

7. **Induced** mutations occur in response to known mutagens. One potent mutagen is **ionizing radiation**, of which x-rays are a prime example. X-rays cause breaks in DNA molecules and in chromosomes. For low doses of x-rays, the **dose-response relationship** is a straight line; that is, the number of mutations is strictly proportional to the dose. In addition, the amount of mutation caused by a given dose of x-rays depends on whether the dose is **acute** (delivered all at once) or **chronic** (split into a series of smaller doses). Only a small proportion of spontaneous mutations can be attributed to background radiation. In humans, the **doubling dose** for x-rays (the amount required to double the mutation rate) is thought to be about 156 rem; the level of background radiation we receive in a 45-year reproductive lifetime averages about 6 rem.

8. **Ultraviolet light** is an example of a nonionizing radiation that is mutagenic. Most damage is done to the skin because the light is not very penetrating, but the energy is absorbed by DNA and causes chemical crosslinks between adjacent thymines (**thymine dimers**). These are usually repaired by means of a cut-and-patch process known as **excision repair**. However, certain patients with the recessive disorder **xeroderma pigmentosum** have a defective excision-repair system and are especially prone to ultraviolet-light-induced skin cancers.

9. Certain chemicals are potent mutagens. Examples are **nitrous acid**, which removes —NH_2 groups from cytosine or adenine and therefore causes mispairing of bases; **base analogues**, which are incorporated into DNA in place of their normal counterparts but mispair more frequently; **alkylating agents**, which chemically react with DNA and so cause mutations; and the **acridines**, which become inserted between adjacent bases in DNA and so cause single-base additions or deletions, thus leading to frameshift mutations.

10. Detection of mutagens has been greatly aided by the development of the **Ames test**, which uses certain highly mutagen-sensitive strains of *Salmonella typhimurium* that contain known types of mutagens. Suspected mutagens are usually mixed with an extract of liver enzymes, because many chemicals are not themselves mutagenic but are converted into mutagenic forms by liver enzymes. The Ames test

is highly sensitive (as little as $\frac{1}{100}$ of a cigarette can be shown to be mutagenic), and over a million-fold range of mutagenic potency can be assessed quantitatively.

11. Screening for mutagens, **carcinogens** (agents that cause cancer), and **teratogens** (agents that cause birth defects) is particularly important because more than 50,000 untested synthetic chemicals are already in use and more than 1000 new ones are produced commercially each year. Potential carcinogens are especially difficult to screen because the tests are expensive and lengthy and must often be based on a limited number of animals. The Ames *Salmonella* test has proven to be useful as a preliminary test of carcinogens because in the Ames test most carcinogens are mutagenic and most mutagens are carcinogenic. Examples of widely used chemicals that have proven to be mutagenic and carcinogenic are tris-BP, used for six years in the United States as a flame retardant in children's polyester pajamas, and AF-2, used for nine years as a food additive in fish and soybean products in Japan. Prescreening of potential mutagens is essential because a mutagen in widespread use might not produce sufficiently striking effects that it could even be discovered, yet hundreds of thousands or millions of new, harmful mutations would result.

KEY WORDS

Achondroplasia	Dose-response rela-	rad
Acridine	tionship	rem
Alkylating agent	Doubling dose	Silent mutation
Alu family	Excision repair	Somatic mutation
Ames test	Frameshift mutation	Spontaneous mutation
Base analogue	Huntington disease	Terminal repeat
Base substitution	Missense mutation	Thymine dimer
Carcinogen	Mutagen	Transposable element
Conditional mutation	Mutation	Ultraviolet light
Dispersed repeated	Nonsense mutation	Xeroderma pigmentosum
multigene family	Phenocopy	

PROBLEMS

10.1. Why would a dental x-ray technician have patients wear a lead-lined apron when taking x-rays of the teeth?

10.2. What is the Ames test used for and what organism is employed?

10.3. Interpret the meaning of a mutation rate of 5×10^{-5} mutants per gene per generation in terms of the number of mutations at the locus in question per million gametes per generation.

10.4. How many x-ray-induced chromosome breaks are required to produce a single chromatid break? An inversion? A translocation? A deletion of a chromosome tip?

10.5. If the following DNA sequence in an organism were irradiated with ultraviolet light, what would be the most likely site and type of damage?

5'-AGATTCAGCCGGA-3'
3'-TCTAAGTCGGCCT-5'

10.6. If the spontaneous mutation rate of a gene is 4×10^{-6} mutations per gene per generation and the doubling dose is 156 rads, what mutation rate would occur with 156 rads of radiation? Assuming that the rate of induced mutation is strictly proportional to radiation dose, what mutation rate would be expected with 15.6 rads?

10.7. What is the *minimum* number of simple nucleotide substitutions that would be necessary to result in the following amino acid substitutions?

    (*a*) A in place of V   (*d*) C in place of M

    (*b*) D in place of E   (*e*) R in place of K

    (*c*) W in place of M   (*f*) I in place of W

10.8. The DNA strand shown below is the antisense strand of a coding region, but the sense strand is not shown for the sake of simplicity. What amino acid sequence would correspond to this DNA strand?

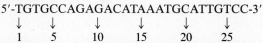

$$5'\text{-TGTGCCAGAGACATAAATGCATTGTCC-}3'$$

$$\downarrow \quad\quad \downarrow \quad\quad \downarrow \quad\quad \downarrow \quad\quad \downarrow \quad\quad \downarrow$$

$$1 \quad\quad 5 \quad\quad 10 \quad\quad 15 \quad\quad 20 \quad\quad 25$$

10.9. If the A at position 7 in the DNA strand in Problem 8 becomes replaced with C, what amino acid sequence would result? What kind of mutational change does this illustrate?

10.10. If the T at position 22 in the DNA strand in Problem 8 becomes replaced with A, what amino acid sequence would result? What kind of mutational change does this illustrate?

10.11. If the G at position 24 in the DNA strand in Problem 8 becomes replaced with C, what amino acid sequence would result?

10.12. If the DNA strand in Problem 8 undergoes an insertion of an A between nucleotides 3 and 4, what amino acid sequence would result? What kind of mutation does this illustrate?

10.13. If the DNA strand in Problem 8 undergoes an insertion of an A between nucleotides 3 and 4 and simultaneously undergoes a deletion of the T at position 14, what amino acid sequence would result?

10.14. If the DNA strand in Problem 8 has the A at position 7 replaced with U, what amino acid sequence would result? What kind of mutation does this illustrate?

10.15. A certain tRNA has the anticodon $5'\text{-AUC-}3'$ as a result of a mutation in its normal anticodon $5'\text{-AGC-}3'$, although the mutant tRNA still picks up the same amino acid as its unmutated counterpart. What codon would the mutant tRNA recognize, and what amino acid would it insert into the polypeptide?

10.16. The principal effect of nitrous acid is to cause deamination of C, forming a base that pairs like T. What possible mutant DNA sequences could be formed by nitrous acid treatment and replication of the sequence shown below?

$$5'\text{-AT CT TAGAT-}3'$$

$$3'\text{-TAGAAT CTA-}5'$$

FURTHER READING

Ames, B. N. 1979. Identifying environmental chemicals causing mutations and cancer. Science 204: 587–93. An excellent review of all aspects of the problem.

Cohen, S. N., and J. A. Shapiro. 1980 (February). Transposable genetic elements. Scientific American 242: 40–49. Discusses these "jumping genes" in prokaryotes and eukaryotes.

Cairns, J. 1981. The origin of human cancers. Nature 289: 353–57. Expresses the view that most human cancers may be the result of genetic transpositions.

Crow, J. F., and C. Denniston. 1981. The mutation component of genetic damage. Science 212: 888–93. Discusses the assessment of genetic damage expected from an increased mutation rate.

Haseltine, W. A. 1983. Ultraviolet light repair and mutagenesis revisited. Cell 33: 13–17. Complications in DNA repair.

Lewin, R. 1983. Promiscuous DNA leaps all barriers. Science 219: 478–79. Jumping genes may be the rule, not the exception.

McCann, J., and B. N. Ames. 1976. Detection of carcinogens as mutagens in the *Salmonella*/microsome test: Assay of 300 chemicals: Discussion. Proc. Natl. Acad. Sci. U.S.A. 73: 950–54. Discusses public-health aspects of mutagen screening.

McElheny, V. K., and S. Abrahamson (eds.). 1979. Banbury Report-1. Assessing Chemical Mutagens: The Risk to Humans. Cold Spring Harbor Laboratory. Cold Spring Harbor, New York. Proceedings of a conference on the effects of chemical mutagens in humans.

Nagao, M., T. Sugimura, and T. Matsushima. 1978. Environmental mutagens and carcinogens. Ann. Rev. Genet. 12: 117–59. Technical aspects of testing are reviewed.

Schull, W. J., M. Otaki, and J. V. Neel. 1981. Genetic effects of the atomic bombs: A reappraisal. Science 213: 1220–27. Thorough summary of the genetic effects of the atomic bombs dropped on Hiroshima and Nagasaki in August 1945.

Spradling, A. C., and G. M. Rubin. 1981. Drosophila genome organization: Conserved and dynamic aspects. Ann. Rev. Genet. 15: 219–64. Reviews types of DNA including transposable elements and their possible roles in development and evolution.

Sun, M. 1983. Missouri's costly dioxin lesson. Science 219: 367–69. Times Beach and other horrors of waste disposal.

Weatherall, D. J., and J. B. Clegg. 1982. Thalassemia revisited. Cell 29: 7–9. Molecular basis of the disorder.

Weisburger, J. H., and G. M. Williams. 1981. Carcinogen testing: Current problems and new approaches. Science 214: 401–407. How mutagen and carcinogen testing can be improved.

chapter *11*

Viruses and Cancer

Human beings play host to an enormous variety of other living things. Like other organisms, our bodies are attractive homes for all manner of parasitic plants, animals, and microbes. Various fungi, tiny mites, and bacteria eke out their existence in habitats between our toes, in our hair or teeth, under our nails, or in small folds of our skin. Many other parasites find homes in our insides, particularly in the gut. Included here are roundworms and tapeworms, for example, and many single-celled organisms such as the species of amoeba that causes amoebic dysentery and the organism responsible for "traveler's diarrhea." Many types of bacteria that live in the gut are completely harmless most of the time, and some may be beneficial in digestion. But some bacteria invade other parts of the body—tuberculosis bacillus and plague bacillus, for example—and they are not nearly so innocuous.

Our bodies are also prey to even smaller forms of life—viruses. These tiny things (one hesitates to call them "organisms" because they cannot reproduce outside living cells) are able to subvert the biochemical machinery of cells to accomplish their own ends and cause diseases such as the common cold, measles, mumps, chicken pox, smallpox, influenza (flu), polio, and many others. This chapter is about viruses, including certain viruses that can induce genetic change in cells that lead to their becoming cancerous. This chapter is also about the ability of certain viruses to transport genes from one cell to another in an elegant (but potentially hazardous to the host) form of genetic surgery.

VIRUSES

Viruses are submicroscopic parasites of cells, and hundreds of different kinds are known. Generally speaking, viruses are **host specific**; that is, a particular kind of virus is able to attack the cells of only one or a few related species of micro-organisms, animals, or plants. Indeed, some viruses are able to infect only cells of certain tissues that carry on the cell surface specific molecules to which the virus can attach. Viruses may be relatively innocuous when they infect one type of cell but extremely harmful when they infect another. One example of a virus with such a dual personality is **herpesvirus**, actually a family of viruses. Some types of herpesvirus live in the subsurface layer of skin in humans or in the protective and insulating sheath of cells surrounding certain nerves. Herpesvirus is usually harmless, and 5 to 10 percent of all normal individuals are carriers of it. The carriers occasionally suffer a herpes outbreak, a characteristic type of runny cold sore around the lips or nose, but the outbreak usually subsides in a few days and the sore heals.

What keeps the virus in check, or at least subdued, is the body's immune system. The general function of the immune system is to seek out, recognize, and destroy cells, tissues, virus particles, or large molecules that are "foreign" in the sense that they are not part of the body itself. In particular, herpesvirus cannot usually invade the interior of the body but is confined to cells where it does little real harm, such as cells in the skin. Most runny cold sores are produced by a type of herpesvirus called type I.

More serious is herpesvirus type II, which is associated with **general herpes**—a venereal disease with symptoms such as swelling, itching, and throbbing in the genital area followed by local inflammation and an eruption of painful blisters. Although type I herpes usually lives above the waist, about 20 percent of genital herpes cases are due to type I virus.

Genital herpes is at epidemic proportions in the United States. Between 5 and 10 million Americans are infected, which is more than the combined total of all curable types of venereal disease. Unlike such venereal diseases as gonorrhea and syphilis, sexual contact is not necessary for the spread of genital herpes. Complications of genital herpes include an increased risk of urogenital cancer in long-term infections. Particularly serious is life-threatening **neonatal herpes**— herpes infection of the newborn through contact with the mother's genital tract. Complications of neonatal herpes include herpes of the eye, which causes blindness, and herpes of the brain membranes, which causes death. Genital herpes is a public health problem of immense proportions because there is as yet no cure.

CHARACTERISTICS OF VIRUSES

Although viruses are small and can usually be examined only through the electron microscope, they do have an internal structure. The structure consists basically of the viral genetic material (DNA or, in many cases, RNA) surrounded by a protective protein coat called a **capsid**. The capsid of most viruses that infect animal cells is surrounded by still another coat, called the **envelope**, composed of lipid, protein, and carbohydrate, which is acquired as the viruses exit from the host cell by budding through special regions of the host cell membrane. These features of viral structure are illustrated for herpesvirus in Figure 11.1.

The genetic material of all viruses is nucleic acid, but the nucleic acid may be DNA or RNA, and it may be double stranded (two complementary strands held together by base pairing) or single stranded. The nucleic acid molecules of viruses are hundreds or thousands of times shorter than those in cells, so viruses have relatively few genes. Polyoma virus, one of the smallest viruses, has six genes; influenza virus has eight genes. The virus **lambda** (symbolized λ), which infects cells of *Escherichia coli,* is relatively large and complex; it has sufficient genetic material to have roughly 50 genes. (λ contains about 50 kb of DNA, and its nucleotide sequence has been completely determined.)

Viruses vary enormously in size and shape. Influenza virus, to take an example, is roughly spherical; it has a diameter of about 1000 Å (the volume of

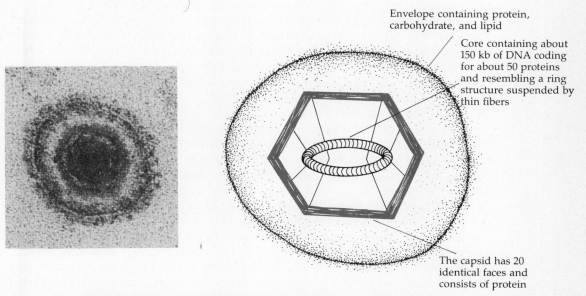

Envelope containing protein, carbohydrate, and lipid

Core containing about 150 kb of DNA coding for about 50 proteins and resembling a ring structure suspended by thin fibers

The capsid has 20 identical faces and consists of protein

Figure 11.1 Electron micrograph of herpesvirus and interpretative sketch of its structure. The diameter of the virus is 1300 to 1600 Å.

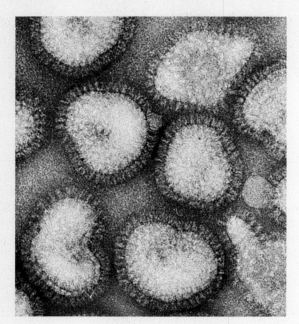

Figure 11.2 Electron micrograph of an influenza virus magnified about 200,000 times. The "halo" that surrounds each virus particle is formed by molecules projecting out from the viral surface.

an average cell is about 1 million times larger). These viruses have a curious appearance as seen through the electron microscope (Figure 11.2). Protruding out from their surface layer, composed mainly of lipid, are some 600 or more "spikes" made of a complex of carbohydrate and proteins. Some of these spikes are evidently the structures that recognize cell surfaces and permit infection.

Among the largest viruses are the pox viruses, such as smallpox, which are ovoid or brick-shaped and are almost as large as a small bacterial cell. Among the smallest viruses at the other end of the size scale are polyoma viruses, which are **oncogenic** (cancer-causing) in animals, including humans. Polyoma viruses are very small indeed, having a diameter of about 50 Å. The virus λ is almost tadpole-shaped (Figure 11.3).

Viruses are consummate parasites, totally unable to reproduce outside living cells. A major reason for their total reliance on cells is that viruses lack ribosomes, transfer RNAs, and much of the other machinery required for protein synthesis; viruses also lack mitochondria, chloroplasts, and other cellular machinery required for capturing or converting energy. The most important thing a virus *does* contain is its genetic material, which in some viruses is DNA and in others is RNA. The genetic material of the virus carries information for the manufacture of viral proteins. Some viral proteins function in replication of the viral nucleic acid, others form the protective protein coat of the virus, and still others are key elements in establishing mastery over the internal workings of a cell. When a virus enters a living cell, it can steal control of the cell from the cell's own nucleus and subvert the cellular machinery into such tasks as making viral proteins and replicating the viral nucleic acid. The products of this synthesis are then assembled into new offspring viruses.

Figure 11.3 Electron micrograph of λ, a virus that infects *E. coli*. The diameter of the head is about 50 Å.

LAMBDA: A VIRAL PARASITE OF *ESCHERICHIA COLI*

Because the life cycle of the virus λ, which infects *E. coli,* is similar to those of many other viruses, λ has been the object of much study. Indeed, among the **bacteriophage** (viruses that infect bacteria, also called simply **phage**), λ has been among the most popular objects of investigation. At one time knowledge of λ was confined to a relatively small group of professional "lambdologists," but this situation has changed because λ has become a popular vector for the cloning of prokaryotic and eukaryotic DNA fragments. (See Chapter 8 for a general discussion of molecular cloning.) One reason for λ's popularity is that λ phage can carry much larger fragments of DNA than can, for example, plasmids.

λ has two alternative life cycles; in one of them, called the **lytic** cycle, illustrated in Figure 11.4, the viral DNA takes over the biochemical machinery of the cell and causes the cell to produce progeny viruses instead of cellular constituents. The first step in the infection of a bacterium by λ is the adherence of the tip of the tail of the phage onto the cell surface [Figure 11.4(*a*)]. Then, rather like a hypodermic syringe releasing its contents, λ injects its DNA into the cell (*b*). (It should be noted here that most animal viruses do not infect cells by injecting their DNA in this manner; animal viruses generally enter cells directly by passing through the cell membrane, often aided by the tendency of some cells to engulf viruses and other small particles.) λ DNA is a linear molecule inside the phage head, but once inside the cell it becomes circularized (i.e., formed into a ring) by an enzyme that joins its ends together. This circular molecule is the principal intracellular form of λ DNA.

In the lytic cycle of λ, the viral DNA takes over the biochemical machinery

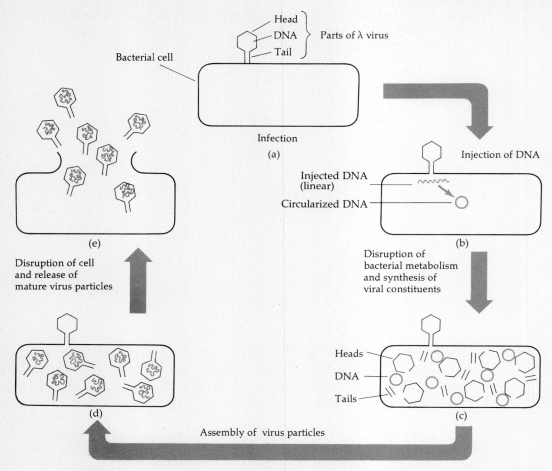

Figure 11.4 Lytic cycle of bacteriophage λ, showing how the virus absorbs to the cell (*a*), injects the DNA (*b*), and establishes control, leading subsequently to the production of viral constituents (*c*), which are then assembled into mature virus particles (*d*) and released as the cell is split open (*e*).

of the bacterial cell; it induces this machinery to make viral proteins, the DNA of the virus is replicated, and the DNA and protein components are assembled into whole progeny viruses [see Figure 11.4(*c*) and (*d*)]. The bacterial cell then breaks open (*e*), which is the "lysis" of the lytic cycle, and the approximately 100 progeny viruses inside are liberated. (The exit of λ virus from infected cells is not typical of the exit of animal viruses from their host cells; animal viruses often push outward through the cell membrane one by one and the host cell is not burst.)

Bacteriophage λ is one of a class of phage called **temperate phage** that have an alternative life cycle. (Bacteriophage that do not have this alternative life cycle are said to be **virulent**.) The alternative to the lytic cycle is called **lysogeny**, and it involves a process in which the DNA of the virus becomes incorporated into the DNA of the bacterial cell without disrupting the metabolism of the host. This physical integration of λ DNA is illustrated in Figure 11.5. After the λ DNA has become circularized, a special region, denoted *att*, is brought into close proximity with a corresponding region on the bacterial DNA called the λ **attachment site** [see Figure 11.5(*a*)]. As shown in the figure, this attachment site is between the genes for galactose utilization (*gal*) and biotin synthesis (*bio*). The juxtaposition of the two molecules involves base pairing of some sort because both the *att* site in λ and the λ attachment site in the bacterial DNA have an identical sequence of 15 nucleotides. Pairing of the two DNA molecules is followed by a site-specific recombination [breakage and reunion—see Figure 11.5(*b*)], which results in the physical insertion of the λ DNA into that of the host (*c*). In its inserted state, λ DNA is said to constitute a **prophage**.

Once inserted, the prophage is carried passively along by the bacterial cell. Every time the bacterial DNA replicates, the λ DNA is replicated, and every time the bacterial cell divides, each daughter cell receives a copy of the viral

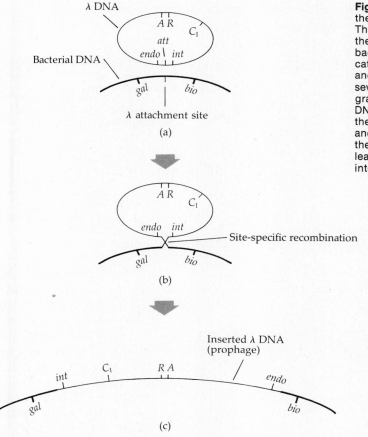

Figure 11.5 Integration of λ DNA into the DNA of *E. coli* during lysogeny. The gray line represents a segment of the long, circular bacterial DNA; two bacterial genes (*gal* and *bio*) are indicated. The black line represents λ DNA and shows the relative positions of several viral genes. During the integration process, the circularized λ DNA aligns with the bacterial DNA at the λ attachment site (*a*). A breakage and reunion (recombination) between the two DNA molecules occur (*b*), leading to the insertion of the λ DNA into the bacterial DNA (*c*).

DNA as part of its genetic endowment. The physiology of the bacterium is not disrupted by the presence of the viral DNA linked to its own DNA, and the phage-bacterial association may persist through many bacterial cell divisions.

However, there is always a small chance that prophage will excise from the bacterial DNA and proceed to enter the lytic cycle. The likelihood that this will happen depends on a variety of conditions, including the nutritional state of the host, and it may also be greatly enhanced by treating the bacterial cells with ultraviolet light. The excision process is roughly the reverse of the integration process (Figure 11.6). The regions of the virus and host DNA that were in proximity during integration are again brought together, and the viral DNA loops into a circle, producing a molecule that looks like a figure 8 except that the loop on one side, the bacterial DNA, is much larger than the loop on the other side [see Figure 11.6(b)]. Site-specific recombination again occurs at the position of the cross in the figure 8, and the viral DNA becomes free, as if it had just been injected into the cell and become circularized [Figure 11.6(c)]. A lytic cycle now takes place and less than an hour later the bacterium lyses and 100 or more λ virus particles are liberated.

Occasionally the process in Figure 11.6 by which prophage λ is excised from the DNA of *E. coli* is slightly inexact. In these cases, the viral DNA picks up one or a few bacterial genes in its loop as it excises, leaving behind in the

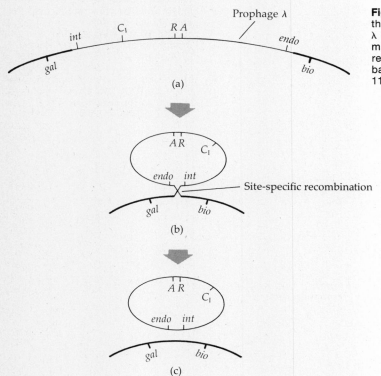

Figure 11.6 Excision of λ DNA from the bacterial host DNA. (*a*) Prophage λ attached to bacterial DNA. (*b*) Formation of loop containing λ DNA and recombination. (*c*) λ DNA freed from bacterial DNA. (Compare with Figure 11.5.)

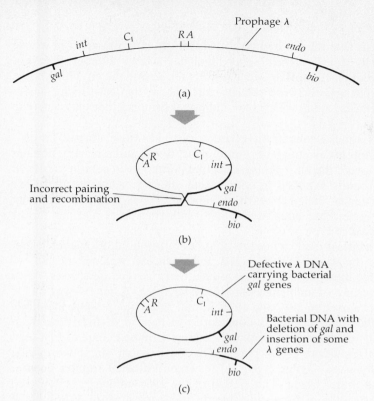

Figure 11.7 Defective excision of λ leading to a virus capable of transduction. (*a*) Prophage λ attached to bacterial DNA. (*b*) Defective loop formation and recombination. (*c*) λ DNA freed form bacterial DNA. Note that the viral DNA carries certain bacterial sequences (in this case *gal*) and is deficient for certain viral sequences (in this case a region including *endo*). (Compare with Figure 11.6.)

bacterial DNA a few viral genes. The process of defective excision is illustrated in Figure 11.7. Instead of the attachment sites of λ and the bacterium coming into proximity to form the figure 8, the pairing sites are displaced; bacterial genes become associated with the λ loop and viral genes with the bacterial loop [Figure 11.7(*b*)]. Recombination in the paired region then produces a loop of λ DNA with certain viral genes replaced with genes from the host; conversely, the host chromosome has some of its genes replaced with genes from the virus [see Figure 11.7(*c*)]. Although Figure 11.7 illustrates how λ can pick up the *gal* genes, displacement of the loop to the other side would allow it to pick up the *bio* genes.

When a defective phage such as the one illustrated in Figure 11.7 infects a new bacterial host, it carries along the *gal* genes it now contains. These *gal* genes can replace those on the host chromosome by means of recombination events flanking the *gal* region, so λ can mediate the transfer of *gal* (or *bio*) genes from one bacterial cell to another. Such bacteriophage-mediated genetic exchange is known as **transduction**, and it is a naturally occurring form of "recombinant DNA" that prokaryotes evolved a very long time ago. Although λ phage can transduce only the *gal* or *bio* genes (or genes nearby), other bacteriophages are able to transduce any genes on the bacterial chromosome. Indeed, some phage can transduce up to 2 percent of the entire bacterial chromosome in a single phage "headful" of DNA.

INFLUENZA VIRUS

An example of a virus that infects human cells is **influenza virus**—a virus whose envelope is essentially a lipid-protein sphere that has a large number of carbohydrate-protein projections or **spikes** (see Figure 11.2). Inside the envelope is the protective protein coat and the viral genetic material, in this case eight single-stranded RNA molecules, along with a few additional proteins that play a role in establishing infection.

Influenza virus attacks cells of the respiratory tract—nose, throat, lungs—and causes such symptoms as coughing, runny nose, fever, and a general aching, commonly referred to as "the flu." The virus is transmitted from person to person in the tiny water droplets expelled in coughs and sneezes. Actually, there are several types of influenza, called types A, B, and C. Types B and C generally produce a milder infection than does A, and major influenza epidemics are almost always due to varieties of type A virus.

With modern medical care, influenza alone is rarely fatal. However, especially in the young, the weak, and the aged, influenza can weaken the patient and make him or her more susceptible to other, more severe infections. As a result, the complications of influenza can often be fatal. A major worldwide influenza epidemic occurred in 1917 and 1918, killing 20 million people, including 548,000 in the United States. [This is considered to be one of the three most devastating epidemics ever to inflict mankind, ranking with the Plague of Justinian (541–593) and the Black Death (1348–1352).] The most recent major worldwide epidemic was in 1968—the "Hong Kong flu"—which killed 51 million people, 20,000 in the United States alone. The spread of the 1968 epidemic is depicted in Figure 11.8. The strain of virus responsible for the disease arose in Southeast

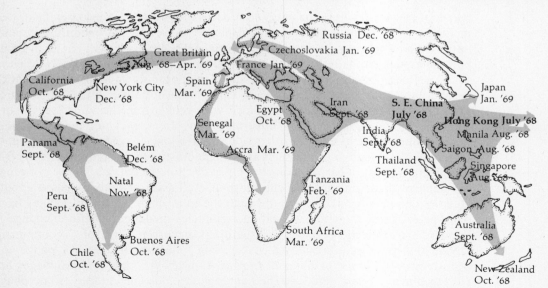

Figure 11.8 Map showing the spread of Hong Kong flu in 1968 and 1969.

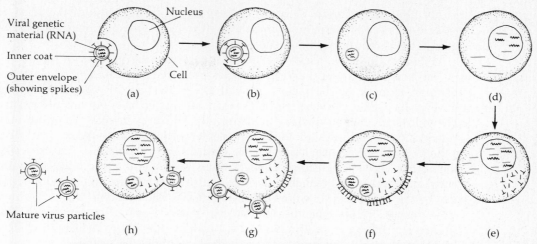

Figure 11.9 Life cycle of influenza virus showing how the virus enters the cell [(a) and (b)], shedding its envelope in the process (c). Viral proteins (gray lines) and RNA (wavy lines) make their way to the nucleus (d), where control over the cell is established and viral RNA is replicated. Viral coat proteins, produced in the cytoplasm, migrate to the nucleus (e), where a protein-RNA complex is formed, which then migrates back to the cytoplasm (f). In the meantime, other viral proteins, which will form part of the outer envelope of the virus, are synthesized in the cytoplasm and become attached to the cell membrane (e), (f). Then the protein-RNA complex buds off the infected cell at regions where the viral "spikes" have previously been attached (g), (h).

China in July 1968. From there it spread, reaching California in October 1968 and New York in December.

Influenza virus infects susceptible cells by adhering to their surface and literally passing through the cell membrane [Figure 11.9(a), (b), and (c)]. What enters the cell is the viral RNA and associated proteins; the envelope of the virus is left behind at the cell membrane. The RNA and proteins of the virus make their way to the cell nucleus [Figure 11.9(d)], where they establish control over the cell and cause it to manufacture viral RNA and viral proteins [Figure 11.9(e)].

Once influenza virus establishes an infection in a cell, the viral RNA is replicated in the nucleus. (The viral material entering the cell includes specific proteins that function in replicating the RNA.) Each of the eight RNAs is transcribed into a single mRNA coding for one viral protein (influenza thus has only eight genes and eight proteins), and the viral proteins are produced on ribosomes in the cytoplasm. Proteins destined to become part of the viral envelope migrate to the cell membrane, while other viral proteins migrate to the nucleus [see Figure 11.9(f)]. There, in the nucleus, an RNA-protein complex is formed; this will eventually become the interior of the mature virus particle. The RNA-protein complex leaves the nucleus and makes its way to the cell membrane, where the carbohydrate-protein spikes of mature virus particles have already become attached. The RNA-protein complex is included in a small bud that pinches off from the rest of the membrane [see Figure 11.9(g)]. This bud forms the mature virus particle (h), and its lipid-protein envelope is composed partly of constituents that are cellular, not viral, in origin. It may be that other parts of the virus are also cellular in origin. In any event, the first progeny viruses from infected cells are released three to four hours after infection.

From the point of view of the influenza virus, the process of infection is very inefficient. Many of the virus particles produced are defective because they lack one or more of the pieces of viral RNA or some of the viral proteins. Between 100 and 1000 virus particles must be applied to a culture of sensitive cells to obtain one productive infection. In a healthy human being or other organism, the virus faces an additional problem: the **immune system**. The spikes on the viral envelope provide the virus with a sort of "signature"; these spikes have certain three-dimensional configurations of atoms that the body's immune system recognizes as belonging to a foreign invader. The configurations recognized by the immune system are known as **antigenic determinants**, and the immune system responds to foreign antigenic determinants by producing specific proteins that can recognize the determinants and become attached to them (see Chapter 12). These proteins are known as **antibodies**, and they aid in the destruction of invading viruses. A virus actually carries many different antigenic determinants, with each one individually recognized by the immune system. A particle, such as a virus, that carries antigenic determinants and stimulates the immune system is known as an **antigen**.

Viruses rendered inactive—unable to infect or to reproduce in cells—may still carry viral antigenic determinants that stimulate the immune response, and individuals exposed to such inactive viruses produce antibodies against them. Subsequent invasions by active viruses are less likely to lead to full-blown infections because antibodies to combat the viruses will already be present. (Influenza vaccinations are based on this principle.) Similarly, an infection caused by the active virus itself stimulates the immune response so that the virus cannot easily reinvade the same person once he or she has recovered. Such people are said to be **immune**. An influenza virus that sweeps through a population in an epidemic will leave in its wake a great number of immune individuals. Then the same strain of virus can no longer produce an epidemic in that population because the proportion of people who are immune is too high; not enough nonimmune people come into contact for rapid transmission of the virus from person to person.

Acquired immunity to influenza is the principal reason influenza is usually not a significant problem in public health for a few years following a major epidemic. However, the antigens present on the envelope of the virus are encoded by viral genes, and these genes mutate like any other genes. Eventually there arises a mutated form of the virus that has sufficiently altered antigenic determinants to escape destruction by the immunity in the population. This new strain of virus is free to sweep through the population in another epidemic.

An **epidemic** is a localized outbreak of disease, and influenza epidemics are usually due to mutated viruses that have minor alterations in their antigenic determinants. At intervals of 10 to 30 years there is an influenza **pandemic**—a worldwide outbreak of disease. The antigenic differences in viruses associated with pandemics are much greater than in those associated with epidemics. Such major antigenic changes are thought to arise when a human virus and an animal virus infect the same animal, because the progeny viruses can carry many possible combinations of RNA molecules of human and animal origin. Some of these progeny viruses are able to infect humans, but their antigens may be very different because they are coded for by RNA molecules of nonhuman viruses. For example, the Hong Kong flu virus carries an RNA molecule thought to be derived from a virus that normally infects birds or horses.

CANCER

Viruses are responsible for many human diseases, and one of the most widespread and insidious of these is **cancer**, an unrestrained growth of cells. Certain cancers are almost certainly due to the action of viruses, but it is important to bear in mind that cancer is not a single disease. There are at least 100 varieties of cancer, and some, such as lung cancer, in which cigarette smoking is a major risk factor, are associated with environmental agents. In addition, there also seems to be genetic variation among people in their susceptibility to cancer-causing agents.

Cells in the body of an adult are not, for the most part, actively dividing. They are metabolically active, which is to say alive, but they are not always in the process of dividing. Only a few cell types, such as white blood cells and cells in the bone marrow, are continuously producing daughter cells. In most tissues, only a small percent of cells are engaged in mitosis, just enough to replace cells that may die or become defective. Conditions can occur, however, that stimulate normally quiescent cells to divide. At the site of a wound, for example, the cells are somehow stimulated to undergo rapid division, and this is partly what allows the wound to heal.

Cancer may arise from a single cell that has undergone a genetic change, making it insensitive to the mechanisms that ordinarily control the rate of cell division. This cell begins to divide rapidly and since the change in the cancer cell is hereditary, the daughter cells also divide, unaffected by the mechanisms that control cell division in normal cells. The growing mass of cells pushes normal cells aside and steals nutrients from the surrounding tissue. Eventually this process results in a whole group of malignant (cancerous) cells—a tumor. (Not all, or even most, tumors are malignant. Sometimes a group of cells proliferates and forms a tumor, but then cell divisions cease. These tumors are called *benign* in contrast to *malignant* tumors.) If a malignant tumor is detected sufficiently early, it can sometimes be removed surgically with little risk of recurrence. Cancer can also be treated with radiation or with drugs that preferentially kill dividing cells, and recent methods involve an attempt to turn the full force of the immune system against the malignant cells.

The membranes of cancer cells are not normal. They develop characteristic antigens. These can stimulate the production of antibodies against themselves, and the immune system, by attacking the cancer cells, plays an important role in the body's defense against cancer. Current evidence suggests that cancerous cells arise frequently, perhaps even daily, in normal people and that these cells are usually destroyed or kept in check by the immune system. This point of view is buttressed by the fact that people afflicted with certain hereditary defects in the immune system are especially prone to malignancies. The development of a malignancy requires two events in any case: the alteration of a normal cell into a cancerous one and a failure of the immune system to attack the tumor effectively.

The causes of various types of cancer are generally unknown. Since there are many kinds of cancers that arise in many types of tissue, there may be as many different causes. Most cancers are said to arise "spontaneously," which is a graceful way to say their cause is unknown. But there is a suspected link between mutations and cancer.

GENETICS OF CANCER

Certain rare cancers are inherited. Among these are xeroderma pigmentosum, inherited as an autosomal-recessive condition associated with malignancies of the skin (see Chapter 10), and **retinoblastoma**, an autosomal-dominant condition marked by tumors in the retinas of the eyes. Inherited cancers are individually quite rare, however. Most common cancers are not known to be inherited, and the risk of cancer in close relatives of cancer victims is only slightly higher than the risk in nonrelatives. Nevertheless, the more than 50 inherited forms of cancer do establish a connection between mutations and cancer.

Another link between cancer and mutated genes comes from the relationship between carcinogens and mutagens discussed in Chapter 10 in connection with the Ames *Salmonella* test—the fact that most cancer-causing agents are mutagenic and most mutation-causing agents are carcinogenic. For some years this link could not be established clearly because many carcinogens seemed not to be mutagenic in cultured cells. Then it was discovered that these carcinogenic chemicals are not themselves carcinogens but are modified by chemical processes in the body, particularly by enzymes in the liver, and are converted into their active, carcinogenic forms. These active modified forms of the carcinogens are also the forms that are mutagens. Enzymatic activation is the reason that chemicals to be examined in the Ames test are first mixed with an extract containing liver enzymes.

A final link between genes and cancer is found in the relationship between certain cancers and chromosomal abnormalities. For example, the blood cells of patients with a particular type of leukemia (*chronic myelogenous leukemia*) frequently have a reciprocal translocation involving an interchange of parts between the long arms of chromosomes 9 and 22. (The chromosome 22 in this translocation is often called the **Philadelphia chromosome** after the city of its discovery.) The breakpoint of this translocation occurs in chromosome 9 at the site of a normal chromosomal gene that is sometimes also found in cancer-causing viruses. When the cellular gene escapes its normal controls, or when it is introduced into cells by a virus, the abnormal gene function can result in cancer. Such cancer-causing genes are known as **oncogenes**, and more than a dozen different chromosomal oncogenes and their virus-borne counterparts are known. In the case of chronic myelogenous leukemia, translocation of the gene seems to allow the oncogene to escape its normal controls and ultimately to lead to the leukemia. Similarly, a different type of leukemia (*acute myeloblastic leukemia*) is often associated with a specific translocation involving chromosomes 8 and 22 and, again, a chromosomal oncogene is known to be located at the site of breakage in chromosome 8.

POLYOMA: A DNA TUMOR VIRUS

Two broad groups of cancer-causing viruses can be distinguished by the nature of their genetic material—DNA or RNA. Among the DNA tumor viruses is **polyoma**, which is among the smallest of known viruses (see Figure 11.10). Polyoma consists of an outer protein coat that surrounds a double-stranded, circular molecule of DNA. The viral DNA is small (5292 base pairs) and codes for just six proteins. The normal host of polyoma is the mouse, and it is widespread in both wild and laboratory mice. The virus does not seem to cause tumors in its wild host species, but it does cause them in newborn mice of some laboratory strains and in newborn rats, rabbits, and hamsters. Although polyoma is not known to infect humans, a closely related virus (**SV40**) does.

Polyoma normally enters host cells by being ingested by the cells (Figure 11.11). Once inside the cell, one of three outcomes involving the virus can occur. First, the infection may be abortive. In **abortive infection**, the process of infection stops in its early stages for reasons not entirely understood; the polyoma virus is digested by enzymes and the cell goes on as before [Figure 11.11(*a*)]. Abortive infections are the most common kind with polyoma, and this is also true of influenza and many other viruses.

A second possible sequence of events constitutes polyoma's **lytic cycle**. Between 1 and 3 percent of the engulfed virus particles will enter the lytic cycle upon infection, and almost all the remainder will undergo abortive infection. In

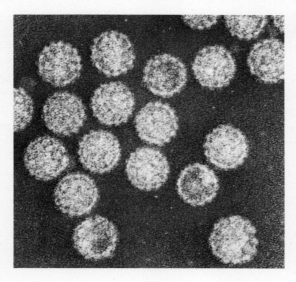

Figure 11.10 Polyoma virus magnified 270,000 times.

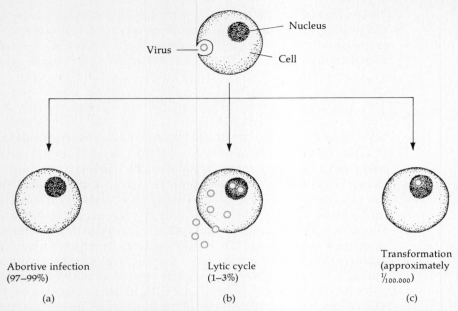

Figure 11.11 The possible outcomes of infection by polyoma. (*a*) In abortive infection, which occurs in 97 to 99 percent of infections, the virus is destroyed by the cell. (*b*) In the lytic cycle (1 to 3 percent of infections), the virus takes over the cell, leading to the production and liberation of numerous new virus particles. (*c*) In transformation, which occurs in about 1 in 100,000 infections, the infected cell does not produce new viruses, yet it retains the virus and becomes transformed in a manner similar to many types of cancer cells.

the lytic cycle, the viral DNA takes over cellular metabolism. Some 10 to 12 hours after infection, three virus-coded "early" proteins begin to appear. DNA replication commences 12 to 15 hours after infection, and shortly thereafter three virus-coded "late" proteins are produced. Infected cells undergo pronounced changes in their metabolism, including enhanced activity of the cellular DNA polymerase responsible for viral DNA replication, increased synthesis of ribosomal RNA, increased rate of protein synthesis, increased rate of transport of six-carbon sugars across the cell membrane, and altered cell surface properties. The first progeny viruses appear 20 to 25 hours after infection. Virus production continues for approximately 36 to 48 hours, after which the infected cell dies.

The third possible outcome of infection of a cell by polyoma is called **transformation** [Figure 11.11(*c*)]. In this case no new virus particles are produced by the infected cell, but the cells do undergo changes (a "transformation") similar to those of cancer cells. When such cells are injected into a live animal, there follows, in a matter of weeks, the development of a tumor at the site of inoculation. Small pieces of such tumors can be transplanted into other animals and soon give rise to full-blown tumors. However, animals that have developed an immune response to polyoma virus by being exposed to the virus are found to be resistant to the transplantation of polyoma-induced tumors because the transplanted tumor pieces are destroyed by the resistant animal's immune system.

The antigens responsible for this immune response, called **transplantation antigens (T antigens)**, are the products of the three "early" polyoma genes. Transformation of cells by polyoma is rare. Roughly 1 in 100,000 virus particles will lead to transformation of the host cell.

In the process of transformation, the viral DNA becomes incorporated into the DNA of the host cell. The details of the integration process are not known, but it may be analogous to the manner in which eukaryotic transposable elements change position in the chromosomes (see Chapter 10). In any case, transformed cells can have from one to several copies of viral DNA integrated at sites scattered throughout the chromosomes.

Transformed cells do not produce the "late" viral proteins, nor do progeny viruses appear. However, the "early" proteins (the T antigens) are produced in transformed cells. One of these T antigens is primarily responsible for inducing transformation. How transformation is brought about is not known, but the T antigen responsible for it has an enzymatic activity of a **protein kinase**; that is, it can transfer the terminal phosphate of an ATP onto an amino acid in a different polypeptide and therefore affect the function of other proteins in the cell. Such control of enzyme activity by means of phosphorylation seems to be an important process of cellular regulation, and the T antigens of polyoma interfere with this process. In any case, transformation of cells is accompanied by the appearance of T antigens on the cell surface. The immune system now gets a chance to defend itself against the cancerous cells by attacking cells with the virus-specified T antigens. If this attack fails, the transformed cells will continue to proliferate and cause a tumor.

RETROVIRUSES, ENDOGENOUS VIRUSES, AND SELFISH DNA

Here we consider tumor viruses with genetic material that consists of a single-stranded RNA molecule. Such viruses are usually small, consisting of their RNA surrounded by a protein capsid in an outer lipid-containing envelope composed partly of cellular constituents. They infect sensitive cells by passing through the membrane, much like polyoma, and inside the cell the protein coat is removed to release the RNA. One of the enzymes brought into the cell by the virus is a **reverse transcriptase**, which produces a DNA strand using the viral RNA strand as a template. A complementary DNA strand is then produced by other enzymes, forming a duplex DNA molecule, which then becomes covalently closed to create a circular DNA containing the viral genetic information. This circular DNA is called a **provirus**, and it proceeds to the nucleus, where it is incorporated into the host DNA. Because of their capacity to produce DNA from RNA, RNA tumor viruses are often called **retroviruses**.

Retroviruses have a very different life cycle from DNA viruses such as polyoma, not only because of their RNA genetic material and reverse transcriptase but also because retroviruses do not have a lytic cycle. All successful retrovirus infections involve integration of the provirus into the host DNA. Moreover, provirus integration does not necessarily upset normal functions of thé host cell. The provirus DNA is transcribed, the RNA is processed, and virus-specified proteins are produced in the cytoplasm. Some of these proteins form the capsid around viral RNA transcript. Other proteins associated with the viral envelope migrate to the cell membrane and are inserted in it. Then, as in the case of influenza virus, the core particles bud through the cell membrane at the regions tagged with viral envelope protein, thus becoming mature progeny viruses. As reproduction of the retrovirus is going on, the host cell can be behaving normally and even undergoing successive cell divisions. Unlike the case with polyoma, cells with integrated provirus do not necessarily die. Thus, a single infected cell can produce thousands of progeny retroviruses.

Many kinds of retroviruses are known; they infect such animals as chickens, mice, rats, hamsters, cats, pigs, deer, monkeys, and baboons. Indeed, many vertebrate species contain multiple copies of provirus DNA as a constituent of the normal genome. Viral genetic information can therefore be transmitted from parent to offspring through the germ line, but it can also be transmitted from individual to individual by means of infection with virus particles.

Such normally occurring integrated proviruses are known as **endogenous viruses**. Although typically present in 5 to 50 copies per haploid chromosome set, endogenous viruses are usually not expressed. Expression of the viral genetic information seems to be controlled by the host cells, and there can be a prolonged latent period. Sometimes infectious virus particles are produced in the absence of disease; at other times production is associated with disease (cancer). Endogenous viruses seem to persist in species for long periods of time, as closely related species tend to have closely related endogenous viruses. On occasion, cross-species transmission seems to occur. For example, one endogenous virus in domestic cats has been found to be more similar to a virus in primates than to viruses in other cats, as if a primate-to-cat transmission had occurred sometime in the distant past. Endogenous viruses may be an example of what has been called **selfish DNA**—DNA that maintains itself in a species by virtue of its own intrinsic characteristics, including the ability to integrate and transpose. Transposable elements (Chapter 10) are also often considered examples of selfish DNA.

AVIAN SARCOMA VIRUS AND CANCER

From the standpoint of human health, retroviruses are important because they can cause malignant transformations. One well-studied example of a cancer-causing RNA virus is the chicken virus called **avian sarcoma virus**. (A **sarcoma** is a tumor of connective tissue.) A genetic map of the RNA of avian sarcoma virus is shown in Figure 11.12. Structurally it resembles a eukaryotic mRNA. Highlighted at the top of the figure is a 20-nucleotide direct repeat occurring just downstream from the 5′ end and just upstream from the stretch of consecutive A's (poly-A) at the 3′ end. Such repeated sequences are reminiscent of transposable elements (see Chapter 10), with which retroviruses share several other characteristics, such as presence in the genome in multiple copies.

Avian sarcoma virus has four genes coding for four proteins. These are shown in Figure 11.12 along with the approximate number of nucleotides in each gene. (The entire RNA molecule consists of about 9700 nucleotides.) The gene

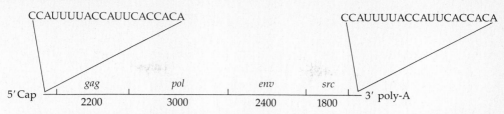

Figure 11.12 Genetic map of avian sarcoma virus. Note 5′ cap, 3′ poly-A tail, and 20-bp direct repeat near the ends. Genes *gag* and *pol* are translated in different reading frames. Genes *pol* and *env* actually overlap and are translated in different reading frames.

designated *gag* codes for the capsid protein, *pol* codes for reverse transcriptase, and *env* codes for the virus-specified protein of the outer envelope. The gene of interest in transformation is the one designated *src*.

The *src* gene codes for a protein kinase. (As noted earlier, protein kinases are implicated in several viral and nonviral types of cancer.) The mechanism by which *src* induces transformation is not understood in detail, but the overall outlines are illustrated in Figure 11.13. Initially, an *src*-containing transcript is produced from the integrated provirus. After RNA processing in the nucleus, the *src* mRNA is released into the cytoplasm, where it is translated. The *src* protein kinase, which phosphorylates other proteins, then initiates an unknown series of cellular events (indicated by the question mark), which ultimately has manifold effects on cellular metabolism. These changes include alterations in a cellular pump concerned with the transport of glucose across the cell membrane and alterations in certain aspects of cellular architecture involving tubules and filaments. Such changes may well lead to the initiation of cell division and release from normal inhibitory processes, which is to say malignancy.

The *src* gene is an example of an oncogene that has a normal chromosomal counterpart, similar to the chromosomal oncogenes discussed earlier in connection with certain human leukemias. However, the normal chicken *src* gene is relatively inactive and the protein kinase product is produced in very small amounts. Introduction of avian sarcoma virus increases the amount of phosphorylation by 50 to 100 times because of the contribution from the viral *src* gene, and this increase is sufficient to initiate the key events. The nucleotide sequence of the viral *src* gene is almost identical with that of its chromosomal counterpart.

Such close similarities are common among oncogenes. For example, the product of a human chromosomal oncogene (*ras*) differs by only three amino acids from its viral homologue found in Harvey murine sarcoma virus. Moreover, certain spontaneous bladder cancers have been found to be associated with a single nucleotide substitution in the chromosomal *ras* gene, leading to a single amino acid substitution in the *ras* protein. This finding further strengthens the correspondence between spontaneous mutations and cancer. In any case, the similarity between chromosomal oncogenes and their viral homologues, along with other similarities between retroviruses and transposable elements, suggests that transforming genes like *src* and *ras* were once normal cellular genes that have become mobilized (transposable). Of course, once mobilized on a retrovirus,

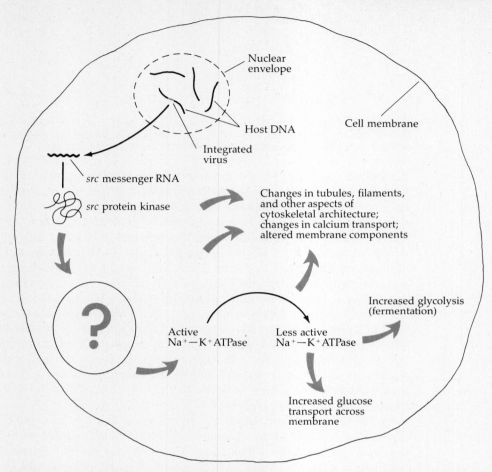

Figure 11.13 The *src* gene of avian sarcoma virus codes for a protein kinase. By an as yet unknown mechanism (question mark), this enzyme produces manifold changes in cellular metabolism, including changes in glucose transport and increased glycolysis associated with many (but by no means all) tumors.

the gene can be reintroduced into the chromosomes by provirus integration, and there is evidence that some pseudogenes have been formed by this sort of gene hopping.

SUMMARY

1. **Viruses** are submicroscopic parasites of cells that are unable to reproduce outside cells. Most viruses are **host specific**, which means they are able to infect just one or a few related species of bacteria, animals, or plants. Although some viruses seem to be relatively harmless, some can cause serious disease. Examples of disease-causing viruses are **rubella** (German measles, birth defects in embryos), the **herpesvirus** family (cold

sores, genital herpes venereal disease associated with birth defects and very possibly urogenital cancer), **influenza**, mumps, chicken pox, smallpox, encephalitis, the common cold, and certain types of cancer.

2. Viruses are extraordinarily diverse. Their genetic material may be DNA or RNA, single stranded or double stranded, linear or circular, and with one molecule or several. The genetic material is packaged inside a protective protein coat called a **capsid** and in animal viruses, the capsid is surrounded by an **envelope** composed of lipid, protein, and carbohydrate, which is acquired as the virus exits through the host cell membrane. Viruses vary enormously in size. Some, like the pox viruses, are almost as large as a small bacterial cell and have enough nucleic acid to code for hundreds of proteins; at the other end of the size spectrum are viruses like **polyoma** (a mouse tumor virus), which has only six genes.

3. The virus **lambda** (λ), which infects cells of *Escherichia coli,* has provided several fundamental concepts related to viral function. (λ has also become a popular vector for the cloning of eukaryotic DNA fragments.) This bacterial virus, or **bacteriophage**, has two alternative life cycles: the **lytic cycle**, in which the virus takes over control of the host metabolism and produces about 100 progeny viruses that are released when the cell undergoes **lysis** (bursting open); and **lysogeny**, in which the λ DNA is integrated into the host DNA by a site-specific recombination event between circularized λ DNA and the bacterial DNA. The lytic cycle–lysogeny options make λ a **temperate** phage, in contrast to a **virulent** phage, which has only a lytic cycle.

4. Sometimes the integrated λ DNA (called **prophage**) undergoes defective excision from the host DNA and incorporates a piece of host DNA in place of some of its own genes. A bacterial cell infected with such a virus can incorporate a gene or genes from the previous host by means of ordinary recombination. Such gene transfer mediated by a virus is called **transduction**.

5. **Influenza** virus infects the respiratory tracts of birds, many mammals, and humans. The virus consists of eight molecules of single-stranded RNA (each corresponding to a different gene) inside a protein capsid surrounded by a lipid-protein envelope festooned with virus-coded **spikes**. These spikes are **antigenic**—they elicit an immune response— and animals produce **antibodies** (immunity proteins) that recognize and attack the virus. Influenza outbreaks occur when the proportion of immune individuals is sufficiently low that the virus can spread from individual to individual. Local outbreaks (**epidemics**) are usually the result of minor changes in the spike antigens due to mutations in the corresponding genes. Worldwide epidemics (**pandemics**), such as the pandemic of Hong Kong flu in 1968, are usually due to major changes in the viral antigens brought about by recombination between viruses of human and nonhuman origin.

6. **Cancer** refers to a heterogeneous collection of diseases characterized by unrestrained cell division. Although many cancers are thought to be induced by environmental agents, there are nevertheless links between mutations and cancer. First, more than 50 inherited forms of cancer are known, all of them rare; examples are xeroderma pigmentosum (a simple Mendelian recessive) and **retinoblastoma** (retinal tumors associated

with a simple Mendelian dominant). Second, most carcinogens have been found to be mutagenic, and most mutagens have been found to be carcinogenic. Third, certain cancers are associated with chromosomal abnormalities, such as the **Philadelphia chromosome** (the chromosome 22 part of a 9–22 reciprocal translocation), which occurs in a certain type of leukemia. In some cases, the breakpoint of the chromosomal abnormality is known to correspond to the site of a chromosomal **oncogene** (a gene which, when functioning abnormally, can cause cancer). Counterparts of cellular oncogenes are also found in certain cancer-causing viruses.

7. **Polyoma** is an example of a DNA virus, which infects certain strains of mice, rats, rabbits, and hamsters. (A close relative, **SV40** virus, infects humans.) The DNA of polyoma consists of 5292 base pairs of circular, double-stranded DNA, and it codes for six proteins—three "early" proteins, produced early in infection, and three "late" proteins. Three outcomes of infection of cells with polyoma are possible. The infection may be **abortive** (failed), which is the most common outcome. In 1 to 3 percent of infections, polyoma undergoes its **lytic cycle**, in which progeny viruses are produced and the infected cell invariably dies. In a small proportion of cases polyoma induces the **transformation** of normal cells into cancer cells. In transformation, from one to several copies of polyoma DNA become incorporated into the host DNA. Only the early polyoma proteins (called the **T antigens**) are produced; one of these has the enzymatic activity of a **protein kinase** (it adds phosphate groups to other proteins) and is primarily responsible for the transformation. Transformed cells have T antigens on their surface, which provide the immune system with an opportunity to recognize and attack the tumor.

8. **Retroviruses** are single-stranded RNA viruses coding for a **reverse transcriptase** enzyme that produces a DNA strand using the viral RNA strand as a template shortly after infection. Following synthesis of the complementary DNA strand and circularization, the circular DNA (called the **provirus**) becomes integrated into the host chromosomes. The integrated provirus need not disrupt host cell functions, and the integrated provirus can be transcribed and lead to the formation of progeny viruses. Thus, an infected host cell can survive and undergo division and release thousands of progeny retroviruses. Retroviruses always become integrated in successfully infected cells; these viruses do not possess a conventional lytic cycle.

9. **Endogenous viruses** are integrated proviruses that occur as a normal constituent of the host chromosomes. Typically having 5 to 50 copies per haploid chromosome set, endogenous viruses are widespread in vertebrate species, including primates. Being a normal constituent of the chromosomes, endogenous viruses can be passed from generation to generation through the germ cells. Although provirus expression is controlled by the host and provirus genetic information is usually not expressed, endogenous viruses do carry the genetic information for producing mature retrovirus particles. Provirus expression may have a prolonged latent period, and cells that actively produce virus particles are sometimes cancerous and sometimes not. Cross-species transmission of endogenous viral DNA sometimes seems to occur, as illustrated

by one such virus in the domestic cat, which is very primatelike in its characteristics. Endogenous viruses may be an example of **selfish DNA**—DNA that maintains itself in a species by virtue of integration and transposition.

10. Many retroviruses can cause malignant transformations; one example is **avian sarcoma virus**, which causes **sarcomas** (connective tissue tumors) in birds. The RNA of avian sarcoma virus is about 9700 nucleotide pairs long, has a 20-nucleotide direct repeat at its ends, and carries four genes: *gag* (a capsid protein), *pol* (reverse transcriptase), *env* (an envelope protein), and *src* (a protein kinase responsible for malignant transformation). Transformation induced by avian sarcoma virus involves a series of events initiated by the *src* kinase. These events include changes in the cell membrane and cell structure, but the molecular details of the transformation are unknown. The *src* gene is almost identical to a normal chromosomal chicken gene that codes for the same protein kinase, which suggests that *src* may originally have been acquired from the host. Other retroviruses also carry genes that are similar if not identical to their normal host counterparts. These similarities, together with structural similarities between proviruses and transposable elements, suggest that cancer-causing viruses may represent transposable elements gone awry.

KEY WORDS

Abortive infection	Lysis	Provirus
Antibody	Lysogeny	Retinoblastoma
Avian sarcoma virus	Lytic cycle	Retrovirus
Capsid	Oncogene	Reverse transcriptase
Endogenous virus	Pandemic	Selfish DNA
Envelope	Phage (bacteriophage)	T antigen
Epidemic	Philadelphia chromosome	Temperate
Herpesvirus	Polyoma	Transduction
Influenza	Prophage	Transformation
Lambda	Protein kinase	Virulent

PROBLEMS

11.1. Why are cigarettes sometimes called "cancer sticks"?

11.2. What is the meaning of the term *transformation* in the context of cancer research?

11.3. A certain synthetic chemical is found to be mutagenic in mammalian cells. Lacking any other information, would you expect the chemical to be carcinogenic? Why or why not?

11.4. One of the three deadliest epidemics ever to hit mankind occurred toward the end of and following the First World War. What was the disease?

11.5. What is the difference between an epidemic and a pandemic?

11.6. Influenza epidemics are usually due to types of viruses that are antigenically quite different from the types of influenza virus previously in the same population. Why should this be so?

11.7. If there were only two possible alternatives of each of the eight RNA molecules of influenza virus, how many genetically different types of virus would be possible? How many would be possible if there were as few as 10 possible alternatives of each molecule?

11.8. In the case of polyoma, what is by far the most common outcome of infection?

11.9. What is a retrovirus? Is influenza virus a retrovirus?

11.10. Part of a retrovirus has the sequence 5'-AUCAUUUCAGU-3'. What DNA sequence would be produced by reverse transcriptase?

11.11. From where are retroviruses thought to have acquired their transformation-associated genes, such as *src* in the case of avian sarcoma virus?

11.12. Are any types of cancer inherited?

11.13. What is the Philadelphia chromosome?

11.14. What features of certain cancers allow them to be treated immunologically?

11.15. Cotransduction refers to the simultaneous transduction (i.e., gene transfer between bacterial cells by means of a virus) of two genes. How would you expect the frequency of cotransduction to depend on the distance between the genes on the bacterial DNA?

FURTHER READING

Bishop, J. M. 1981 (March). Oncogenes. Scientific American 246: 80–92. These cancer-causing genes were first found in viruses, but they are also normal constituents of cells.

———. 1983. Cancer genes come of age. Cell 32: 1018–20. Discusses genes known to be involved in cancer.

Cairns, J. 1980. Cancer: Science and Society. W. H. Freeman, San Francisco. An overall view of what is known about cancer.

Croce, C. M., and H. Koprowski. 1978 (February). The genetics of human cancer. Scientific American 238: 117–25. Reviews a very complex subject.

Epstein, S. S., and J. B. Swartz. 1981. Fallacies of lifestyle cancer theories. Nature 289: 127–30. Challenges the widely held view that lifestyle is a major cause of cancer and that people who develop cancer have brought it on themselves.

Finnegan, D. J. 1983. Retroviruses and transposable elements—which came first? Nature 302: 105–06. A chicken-or-egg problem.

Kitamura, N., *et al.* 1981. Primary structure, gene organization, and polypeptide expression of poliovirus RNA. Nature 291: 547–53. The 7433-nucleotide virus encodes one long polypeptide that is cleaved into 12 viral proteins.

Nash, H. A. 1981. Integration and excision of bacteriophage λ: The mechanism of conservative site-specific recombination. Ann. Rev. Genet. 15: 143–67. Details on the precision of integration and excision.

Nicolson, G. L. 1979 (March). Cancer metastasis. Scientific American 240: 66–76. This process is the real life-threatening danger in most cancers.

Perera, F., and C. Petito. 1982. Formaldehyde: A question of cancer policy? Science 216: 1285–91. A case study of scientific and political issues in evaluating and regulating a possible cancer-causing substance.

Reddy, V. B., *et al.* 1978. The genome of simian virus 40. Science 200: 494–502. Sequence of the 5226 nucleotides in the virus.

Roizman, B. 1979. The organization of the herpes simplex virus genomes. Ann. Rev. Genet. 13: 25–57. A detailed review of the genetics of herpesvirus.

Roizman, B. (ed.). 1982. The Herpesviruses. The Viruses, Vol. 1. Plenum, New York. Technical details of a diverse and important group of viruses.

Schimke, R. N. 1978. Genetics and Cancer in Man. Churchill Livingstone, Edinburgh. A slim but authoritative introduction to genetic aspects of cancer.

Simons, K., H. Garoff, and A. Helenius. 1982 (February). How an animal virus gets into and out of its host cell. Scientific American 246: 58–66. The process is described in illuminating detail.

Temin, H. M. 1983. We still don't understand cancer. Nature 302: 656. Complications in some simple theories.

Varmus, H. E. 1982. Form and function of retroviral proviruses. Science 216: 812–20. Details of several RNA tumor viruses are reviewed.

Webster, R. G., and W. J. Bean, Jr. 1978. Genetics of influenza virus. Ann. Rev. Genet. 12: 415–31. Details of influenza structure and genetics.

Weinberg, R. A. 1980. Origins and roles of endogenous retroviruses. Cell 22: 643–44. Where such retroviruses come from.

Weinstein, I. B. 1983. Protein kinase, phospholipid and control of growth. Nature 302: 750. How tumor promoters work.

Wiley, D. C. 1983. Neuraminidase of influenza virus reveals a flower-like head. Nature 303: 19–20. Structure of some of influenza's spikes.

chapter 12

Immunity and Blood Groups

We live in an ocean of hostile microorganisms such as the viruses associated with influenza and cancer, and immunity is a sort of life jacket that keeps us afloat. The immune response to invading pathogens, as found in humans and other mammals, is one of the sublime achievements of evolution. It is a system that recognizes and destroys viruses, bacteria, and transformed cancerous cells; at the same time it can identify our own normal tissues and refrain from attacking them. Unfortunately, the immune system is not perfect. Its failings are proclaimed by dysentery, influenza, and cancer; by allergies, rheumatism, and rheumatic fever; and by many degenerative disorders of aging. On the other hand, the immune system is so effective that it presents a major hindrance to the transplantation of tissues from one person to another. Even though a great deal is known about the immune system, the nature of immunity is complex and so sophisticated that it still defies complete description. The study of immunity is revealing new principles in biology while providing dramatic new procedures in the fight against disease.

THE IMMUNE RESPONSE

Immunity is rooted in a trillion or so white blood cells called **lymphocytes** and in related blood cells known as **macrophages**. Lymphocytes, which die and are replenished throughout life at the rate of about 10 million cells per minute, are produced by the division of certain **stem cells** in the mushy, reddish bone marrow where many blood components are manufactured. Once formed, these lymphocyte precursors undergo further differentiation in either of two distinct ways. About half the lymphocytes pass through or under the influence of the **thymus**— a small organ underlying the breastbone in children that grows until puberty and then gradually shrinks, virtually disappearing by adulthood. Lymphocytes that differentiate under the influence of the thymus are called **T cells** or **thymus-dependent cells**; these cells constitute one arm of the immune system.

The other half of the lymphocytes undergoes a different sort of processing, which takes place in bone marrow and perhaps in the spleen. The lymphocytes that undergo this alternative type of processing are known as **B cells** or **bone-marrow-derived cells**; B cells constitute a second arm of the immune system. Figure 12.1 is a photograph of a differentiated lymphocyte (a T or a B cell) from a normal mouse.

T cells circulate in the bloodstream, whereas B cells are concentrated in the lymphatic system, which consists of the spleen (the main blood filter), the lymph nodes (local lymph filters), and clumps of cells associated with the gut, respiratory tract, and urogenital tract (including cells in the tissues of the appendix, tonsils, and adenoids). (See Figure 12.2 for the anatomical relationships between some of the principal constituents of the immune system.) The immune system thus monitors all foreign invaders that come into the body through either normal body openings or wounds, and all invaders in the blood or lymphatic system. The T cells are known as agents of **cellular immunity**; they include **killer T cells**, which recognize and attack such alien intruders as bacteria and fungi, and they can even attack such abnormal cells as transformed cancer cells. Dividing rapidly and surrounding the foreign invaders, T cells produce a local inflammation. They can release poisons or toxins to kill the intruders, and they release chemical signals that summon legions of large scavenger white blood cells (macrophages) that devour and digest the invaders.

The other arm of the immune system, which relies on B cells, attacks invaders indirectly. Alien cells carry immunity-stimulating substances called **antigens** on their surface. These antigens trigger B cells to divide and secrete protein molecules called **antibodies**, which circulate in the blood and stick to the antigens of the foreign invaders. The antigen-antibody complex marks the invaders for

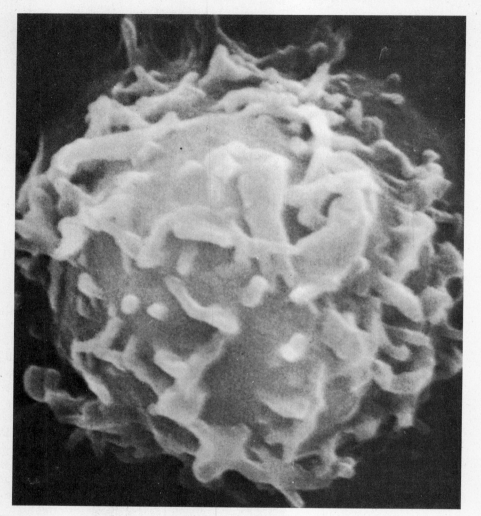

Figure 12.1 Scanning electron micrograph of a differentiated lymphocyte from a normal mouse.

destruction by blood proteins, macrophages, and other scavenger cells. Because antibodies circulate freely in the blood, B cells are known as the agents of **humoral immunity** after a now-outdated use of the word "humor" to mean "body fluid."

The crux of immunity is the recognition of foreign antigens by T cells and B cells and the stimulation of response. Normal recognition and response require **cell cooperation**—the interaction of cell types. Appropriate response of killer T cells requires interaction with a class of T cells called **helper T cells**. Appropriate response of B cells requires interaction not only with helper T cells but also with a macrophagelike cell sometimes called an **antigen-presenting macrophage** because it seems to attach to the antigen and "present" it to the lymphocyte.

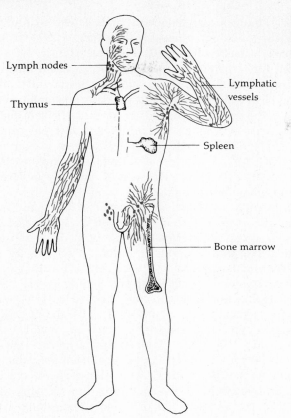

Figure 12.2 The relationship of some principal components of the immune system—lymph nodes and lymphatic vessels, thymus, spleen, and bone marrow.

Lymph nodes

Thymus

Lymphatic vessels

Spleen

Bone marrow

Details of cell cooperation aside, a multitude of substances are antigenic (that is, able to elicit an immune response). Indeed, almost anything as big as or bigger than a protein molecule is antigenic. Viruses, bacteria, and foreign cells are therefore antigenic, as are most protein molecules, large carbohydrates, and some nucleic acids introduced from outside the body. On the other hand, virtually no small molecules are able to elicit an immune response.

CLONAL SELECTION

What is it about an antigen that elicits an immune response? It is the detailed relief, the texture, of its surface. All large molecules, including those on the surface of viruses, bacteria, and cells of higher organisms, fold into precise three-dimensional configurations, providing a sort of landscape with identifiable features. In much the way that every human face can be recognized by its distinct surface features—the precise shape of the ears, the exact arrangement of skin creases around the eyes, the shape and placement of blemishes on the nose—a number of distinct surface features of an antigen can be recognized by their three-dimensional configuration of atoms. Each separate, recognizable feature of an antigen is known as an **antigenic determinant**; since as few as four amino acids in a protein can provide a distinct antigenic landmark on its surface, large molecules or cells may have tens or hundreds of different antigenic determinants. The only requirement for a material to be antigenic is that it have at least one surface landmark sufficiently distinct from substances normally present in the body to be recognized as foreign. Because large molecules and cells have many antigenic determinants, the vast majority of such substances from sources outside an individual have at least one antigenic determinant different from those normally present in the body; these identify the substances as foreign. Moreover, a typical antigen has many different antigenic determinants and so elicits the production of many different antibodies, each type of antibody in response to one type of antigenic determinant on the antigen.

Antigenic determinants, then, are the features of antigens that are recognized and attacked by the immune system. However, each B cell or T cell is able to recognize only one antigenic determinant. Every B cell has on its surface about 100,000 identical **receptor sites** whose surface relief exactly matches the specific antigenic determinant recognized by the cell; a T cell has about 10,000 receptor sites. The match between receptor site and antigenic determinant is complementary; they match like lock and key (Figure 12.3). When an antigen gains entrance into the body, therefore, not all T cells and B cells swing into action. The only lymphocytes that respond are the ones with receptors that happen to fit, or "recognize," an antigenic determinant on the antigen.

When the body is invaded by a foreign antigen, each antigenic determinant on the intruder stimulates the B cells and T cells that respond specifically to it. Any B cell whose surface receptors match one of the antigenic determinants becomes combined with the antigen, and this combining, along with functions provided by macrophages and helper T cells, stimulates the B cell to undergo mitosis. The cells produced by successive cell divisions constitute a **clone** because they all derive from the same parental cell and are thus genetically identical. The stimulation process is called **clonal selection** because the B cell that gives rise to

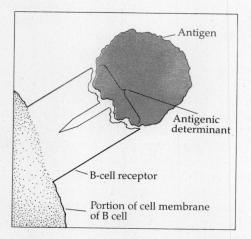

Figure 12.3 The complementary fit of an antigenic determinant and a receptor site on the surface of a B cell.

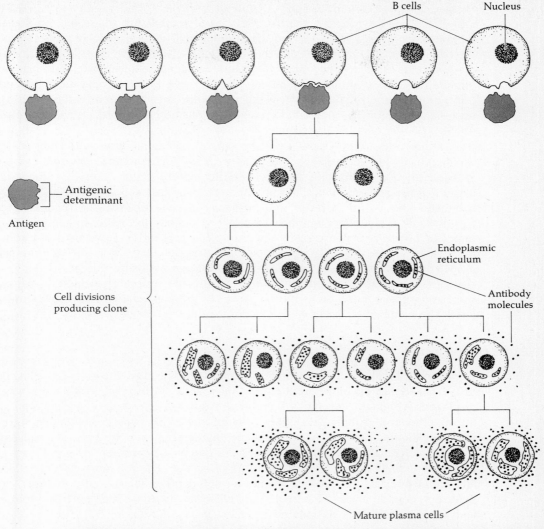

Figure 12.4 Clonal selection. An antigenic determinant on an antigen (shaded shape) combines with the complementary surface receptor of a B cell and stimulates the cell to undergo division. (Note that B cells whose receptors are not complementary to the antigenic determinant are not stimulated.) From the stimulated B cell, a clone of cells is produced. In this clone are cells whose endoplasmic reticulum is markedly distended; these cells, called *plasma cells*, produce and release large numbers of antibody molecules that attack the antigenic determinant on the antigen. Some cells in the clone do not immediately become plasma cells but cease dividing; these cells provide a more immediate response to a second assault of the same antigen.

the clone is selected from among billions of B cells by the antigenic determinant it fits.

The process of clonal selection is outlined in Figure 12.4. An antigenic determinant on an antigen (shaded shape) combines with the complementary surface receptor of a B cell and stimulates the cell to undergo division. (Note that B cells with receptors that are not complementary to the antigenic determinant are not stimulated.) From the stimulated B cell, a clone of cells is produced. Within this clone are cells that differentiate and begin to secrete antibodies—proteins able to recognize the same antigenic determinant that caused stimulation. These antibody-manufacturing cells are known as **plasma cells**, and their endoplasmic reticulum is covered with ribosomes and greatly distended by accumulated antibody molecules. An active plasma cell produces and secretes about 2000 identical antibody molecules every second.

Other cells of the clone do not become plasma cells, however. They cease dividing or divide very slowly. The cells in a clone that become quiescent can be induced to resume division and give rise to mature plasma cells by a second assault of the antigen. Thus, the immune response to a repeated infection or to a booster shot is much faster and produces more antibody. The readiness to mount such a secondary immune response is commonly called **immunity**. The initial or primary response of the immune system is rather slow, taking several days. This slowness is why people may fall prey to such fast-acting diseases as influenza if not previously exposed. However, one exposure to a particular antigen, such as a strain of influenza virus, stimulates antibody production, which wards off subsequent attacks by the same or very similar antigens. Once stimulated, the memory of the immune system may persist for decades.

VARIETIES OF ANTIBODY

A plasma cell makes only one kind of antibody, but because the body has plasma cells from thousands of different clones, thousands of different antibodies can be produced. Although each type of antibody is unique, antibodies can be classified into five general groups: IgG, IgM, IgA, IgD, and IgE, where the Ig stands for **immunoglobulin**. Table 12.1 outlines some characteristics of the antibody classes.

In molecular structure, antibodies are composed primarily of protein with a relatively small amount of carbohydrate attached. The fundamental unit of all antibodies consists of two long, or **heavy**, polypeptide chains and two short, or **light**, polypeptide chains. Each antibody class has its own, characteristic type of heavy chain, which is designated by a lower-case Greek letter (γ, μ, α, δ, ϵ) corresponding to an antibody **class** (G, M, A, D, E). Heavy chains range in size from about 500 amino acids (γ) to about 700 amino acids (ϵ). Although each

Table 12.1 SOME CHARACTERISTICS OF HUMAN IMMUNOGLOBULINS

Class	IgG	IgM	IgA	IgD	IgE
Heavy (H) chains	γ	μ	α	δ	ϵ
Light (L) chains	κ, λ	κ, λ	κ, λ	κ, λ	κ, λ
Number of subclasses	4	2	2	—	—
Serum concentration, mg/ml	8–16	0.5–1.9	1.4–4.2	<0.04	<0.007

Source: E. S. Golub, 1981, The Cellular Basis of the Immune Response, 2d ed., Sinauer Associates, Sunderland, Mass., p. 215.

antibody class has its own heavy chains, they all contain light chains of just one of two classes, called κ and λ, consisting of about 220 amino acids.

IgG is the workhorse of immunity and is the prevalent type of antibody in the blood serum. Each IgG molecule consists of two heavy chains and two light chains and thus has the structure $\gamma_2\kappa_2$ or $\gamma_2\lambda_2$ (never $\gamma_2\kappa\lambda$ because the light chains in an antibody are always the same). However, there are four nearly identical genes that can code for the γ chains in any one molecule. These four γ-chain genes lead to four **subclasses** of IgG; each subclass is determined by the particular γ chain present in the antibody. Although IgG is the most concentrated antibody class in blood serum, IgG molecules can also cross the placenta. Indeed, IgG is the only class of antibody able to cross the placenta.

Although IgG antibody has been the most intensively studied, a few words about the others are in order. IgM is a large and powerful antibody specialized to combat certain kinds of antigens. It seems to be involved in **rheumatoid arthritis**—a disorder of the immune system marked by swollen, stiff, and painful joints. IgA is the antibody found in the gut, respiratory tract, urogenital tract, and other mucous membranes. IgD and IgE appear in much smaller amounts in blood serum than do the other classes of antibody. The IgE antibody seems to be involved in allergic reactions. Although the functions of IgD are not well known, this class of antibody is thought to be involved in the recognition process.

A CLOSER LOOK AT IgG

As noted earlier, the fundamental unit of antibody structure consists of two heavy polypeptide chains and two light polypeptide chains. This structure is illustrated for the IgG molecule in Figure 12.5, where it can be observed that the two heavy and the two light chains are arranged in the shape of a Y—actually a double Y like this: \\Y/. The longest lines represent the heavy chains, which extend from the tines of the Y down into the stalk. The short lines in the tines of the Y represent the light chains. The carbohydrate part of the molecule, which differs for each class of antibody, is attached to the heavy chains in the stalk of the Y,

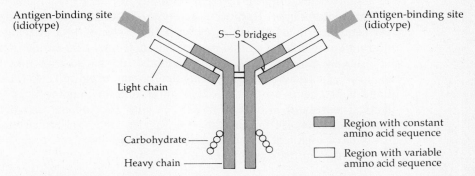

Figure 12.5 An IgG molecule showing the basic Y-shaped structure formed of two heavy and two light polypeptide chains held together by S—S bridges. Note the two antigen-binding sites; they are identical. The great diversity of types of antibodies arises from the variable amino acid sequences found in the parts of the heavy and light chains associated with the antigen-binding sites.

and the whole structure is held together by S—S bridges between the polypeptide chains (see Figure 12.5). The two heavy chains in any particular molecule are identical, as are the two light chains—a result of the fact that each plasma cell has only one set of genes actively making a particular antibody. The Y-shaped antibody molecule has two sites that bind with the corresponding antigenic determinant. These sites, located at the tips of the tines of the Y, are called **antigen-binding sites**.

Although Figure 12.5 illustrates the structure of IgG, a similar structure is found in all classes of antibody. Each class, of course, has its own heavy chain (see Table 12.1) and type of carbohydrate. IgG, IgD, and IgE consist of one such Y-shaped unit, IgA consists of a variable number, and IgM consists of five Y-shaped units joined at the bottoms of their stalks. In each case, the stalk portion of the Y is recognized by other components of the immune system, such as killer macrophages, thus allowing these components to recognize and attack antibody-coated antigens.

A large number of uniform samples of different IgG molecules has been isolated and their amino acid sequences determined. Such pure antibodies are currently obtained by means of the cell fusion procedures outlined in Chapter 4. An antibody-producing B cell is fused with a certain type of tumor cell, and the resulting hybrid cell will have the ability to divide in cell culture (like the original tumor cell) and at the same time will continue to produce antibody (like the original B cell). Such uniform antibody produced by a B cell–tumor cell hybrid is called **monoclonal antibody** because it derives from a single clone of B cells. Monoclonal antibodies are very useful in research because, being homogeneous, they react with only a single antigenic determinant on the corresponding antigen. With the heterogeneous antibody normally produced, many antigenic determinants on the antigen will be recognized by different antibodies in the mixture.

In any case, when different IgG antibodies are compared with respect to their amino acid sequence, the amino acid sequences of both the heavy and light chains in the upper half of the tines of the Y are different in every type of IgG antibody examined! These **variable sequences** (see Figure 12.5)—approximately 115 amino acids long—give each antibody its ability to recognize a specific antigenic determinant. Because every type of IgG molecule has a different amino acid sequence in its variable region, every type of IgG molecule will have a different surface relief in this part of the molecule. The antigen-binding sites of the antibody molecule have a surface landscape that just fits, like lock and key, the surface landscape of an antigenic determinant (Figure 12.6). The lower halves of the light chains are not variable, however; their amino acid sequences are identical (**constant**) in all IgG molecules with light chains that derive from the same gene. Likewise the **constant** (i.e., not variable) three-quarters of the heavy chains have identical amino acid sequences in all molecules with heavy chains that derive from the same gene.

GENE SPLICING AND THE ORIGIN OF ANTIBODY DIVERSITY

All normal individuals are able to produce at least 1 million (10^6) antibodies that differ in their antigen-binding specificity and hence in their amino acid sequence.

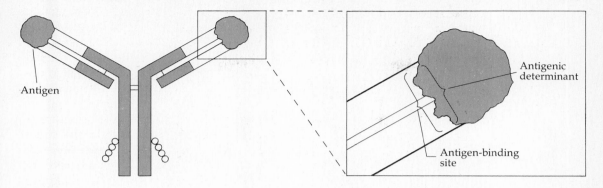

Antibody (IgG)

Figure 12.6 An IgG antibody molecule bound with antigen and showing the complementary fit of the antigen-binding site and the antigenic determinant. The B cell in Figure 12.4 is shown as being stimulated by this same antigen. The stimulated B cells give rise to plasma cells producing antibodies such as the one shown here, which are able to recognize the same antigen.

How many genes are involved in coding for this extensive repertoire of amino acid sequences? At one extreme, every possible light chain and every possible heavy chain would correspond to one conventional gene in the germ line. In this **germ-line theory** of antibody diversity, there would have to be at least 1000 (10^3) light-chain genes and 1000 (10^3) heavy-chain genes; random combinations of light chains and heavy chains could then produce the necessary 10^6 antibodies ($10^3 \times 10^3 = 10^6$). At the other extreme, there could be relatively few germ-line antibody genes, with antibody diversity being created anew during each individual's lifetime by somatic mutation of these genes or by some sort of recombinational process; this view, which emphasizes relatively few germ-line genes, is called the **somatic theory** of antibody diversity.

Application of recombinant DNA technology to the issue of antibody diversity has now revealed that neither the germ-line theory nor the somatic theory is completely correct, and neither is completely wrong. Relative to the constant regions of the light and heavy chains, there is virtually no variation. Human beings, for example, have 10 genes for the constant portion of the heavy chains (four coding for alternative forms of the γ chain of IgG, two for the μ chain of IgM, two for the α chain of IgA, and one each for the δ chain of IgD and the ε chain of IgE) and three genes for the constant portion of the light chains (one coding for κ and two for λ); some additional genes for the constant portions may yet be discovered, but the number is not likely to be large. The problem of the origin of antibody diversity pertains to the variable regions of the polypeptide chains, and a major source of the variation has turned out to be unprecedented. The variable region of antibody genes is pieced together from a large array of possibilities by means of **DNA splicing**.

The DNA splicing process involved in heavy chain formation is illustrated schematically in Figure 12.7. In most somatic cells, the DNA corresponding to various parts of the heavy chain is arrayed along the chromosome as shown at

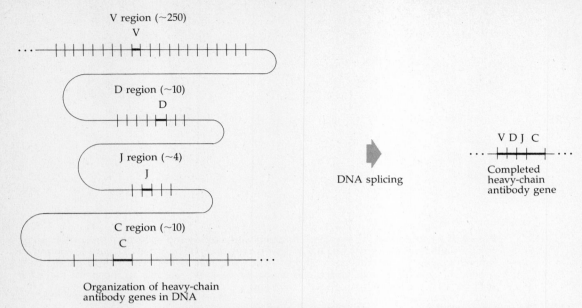

Organization of heavy-chain
antibody genes in DNA

Figure 12.7 Diversity among antibody heavy chains is brought about by DNA splicing of one representative from each of several regions corresponding to different parts of the gene. The variable region of a heavy chain involves one V region (about 250 possibilities), one D region (about 10 possibilities), and one J region (about 4 possibilities). The spliced variable region is also spliced onto one of the constant (C) regions, forming a completed heavy-chain gene. Light chains are formed similarly from their own V, J, and C regions (light chains lack a D region).

the left, but there are alternatives for various parts of the gene. The left end of the variable region is coded by one of the V regions, of which there are some 250 different possibilities; the central part of the variable region is coded by one of the D regions (about 10 possibilities); and the right end is coded by one of the J regions (about 4 possibilities). In the formation of a B cell, one of each of these regions (indicated by the heavy lines) becomes spliced together to create a complete variable region, and the whole thing is spliced onto one of the constant (C) regions (as noted above, having about 10 possibilities).

Such DNA splicing in antibody formation is called **combinatorial joining** (because there are so many combinations of ways in which the V, D, and J regions can be joined), and in the case of the variable region of a heavy chain there would be $250 \times 10 \times 4 = 10,000$ possibilities using the numbers in Figure 12.7. The variable region of the light chains is formed similarly by combinatorial joining, but the light chains do not carry a region corresponding to D (i.e., they are formed by V–J joining rather than V–D–J joining). Although they are referred to by the same symbols, the V and J genes corresponding to the heavy chains are completely different from the V and J genes corresponding to the light chains. The number of V regions for the light chains is about 250, and the number of J regions is 5, giving $250 \times 5 = 1250$ possibilities for the variable part, each of which can be connected to any of the three constant-region parts of the light chain. Altogether, combinations of heavy chains and light chains yield $10,000 \times$

$1250 = 1.25 \times 10^7$ possible antibody molecules, so combinatorial joining of a modest number of gene parts generates an enormous number of possible spliced molecules. Additional antibody diversity is generated by a high rate of somatic mutation in the V, D, and J regions of the heavy chains and the V and J regions of the light chains.

BREAKDOWNS OF IMMUNITY

Immunodeficiency diseases are disorders in which affected individuals lack T-cell functions or B-cell functions or both. Because affected individuals are susceptible to infectious diseases transmitted by bacteria and viruses, patients frequently require treatment with antibiotics. In severe cases the affected individual may be able to survive only in a germ-free artificial environment. The medical history of patients with immunodeficiency diseases highlights the division of labor in the immune system. Patients who lack T-cell function accept mismatched skin transplants (rejection is a T-cell function) and fall prey to viral infections, whereas patients who lack B-cell function accept mismatched blood transfusions and are prone to bacterial infections.

Some immunodeficiency diseases are caused by infectious agents. An example is **acquired immunodeficiency syndrome**, or **AIDS**, which is caused by a virus related to a family of leukemia viruses, leading to partial loss of T-cell function. AIDS was first identified because of its high incidence in male homosexuals with a history of sexual promiscuity, suggesting sexual contact as one of its modes of transmission, but later it was found in hemophiliacs who had received repeated blood transfusions, implying transmission of the agent through the blood. However, it should be emphasized that AIDS has not been reported to be acquired as a consequence of *donating* blood, so the marked decrease in blood donations that has occurred as a result of the AIDS publicity is unwarranted. Although AIDS is very rare in most populations, it does, for unknown reasons, have a much higher incidence among certain groups, most notably Haitian refugees.

All human beings are genetically unique, and this uniqueness includes various aspects of the immune response. No two persons have exactly the same immune response. Some people can respond to certain antigens while others cannot, or some may respond strongly while others respond weakly. Innocuous antigens like wool, egg albumin, cat fur, or house dust do not bother most people, yet they elicit speedy and sometimes disastrous responses from the immune systems of people who are allergic to them. It is common experience that certain people are more susceptible than others to allergies, colds, poison ivy, and many other agents, and such susceptibilities may run in families. However, the mode of inheritance of these susceptibilities involves many genes and environmental factors working simultaneously, and it is therefore very complex.

The immune system is a potent weapon against foreign tissues. It would be equally effective against one's own tissues were it not for **immunological tolerance**—the process whereby the immune system, especially T cells, comes to recognize and tolerate one's own antigens. Exactly how immunological tolerance comes about is not known. At some time during the embryological development of the immune system, those parts of the system that would be programmed to attack the specific antigenic determinants present in an individual's body are somehow destroyed or made inactive. What remain functional are only those lymphocytes that recognize "nonself," or foreign, antigenic determinants.

Sometimes immunological tolerance fails, and the immune system begins to attack one's own tissues—a condition known as **autoimmune disease**. Rheumatoid arthritis and several other diseases of aging are known or strongly suspected to involve a partial breakdown of immunological tolerance and are thus examples of autoimmune diseases. A particularly serious example in children is **rheumatic fever**. Rheumatic fever is caused by a bacterium, and certain antigens on the bacterium are similar to those on the heart. The bacterial antigens are different enough from heart antigens to stimulate antibody production; at the same time, they are similar enough to heart antigens that the antibodies produced against them will also attack the tissues of the heart. This is the cause of the heart damage that frequently occurs in children afflicted with rheumatic fever.

Why did humans and other mammals evolve an immune system that would tolerate "self" antigens and reject "nonself" antigens, that would reject even the antigenic determinants of exotic laboratory chemicals that no T cell or B cell would meet in millions of years? One plausible answer is found in the theory of **immune surveillance**, which suggests that the immune system evolved not only to fight foreign invaders but also to "police" the cells and tissues of the body and quell any "insurrections" (mutant clones) that might arise. According to this theory, newly mutated cells with new antigenic determinants are continually occurring in our bodies, being recognized as "nonself" by our immune system, and being destroyed. The theory of immune surveillance thus suggests that cancer represents a twofold failure: first, the origin of a transformed clone of cells, and second, failure of the immune system to destroy the clone or hold it in check. The theory is supported by the observation that individuals with immunodeficiency diseases are particularly prone to cancer. Indeed, some normal individuals with cancer *do* make antibodies against their cancer cells, but the cancer cells secrete molecules that neutralize the antibodies, presumably by filling up the antigen-binding sites.

TRANSPLANTS

The same powers that make the immune system a potent weapon against the onslaught of microorganisms make it a serious barrier to the success of organ transplants and tissue grafts. Human cells and tissues are antigens, of course. Their antigenic determinants are on the cell surfaces and are genetically determined by the alleles at more than a dozen loci known as **histocompatibility loci**. Histocompatibility antigens are produced according to the general rule that each allele at a locus produces a specific antigen. A homozygous locus produces tissue antigens corresponding to whichever allele is homozygous; a heterozygous locus produces two different tissue antigens corresponding to the two different alleles on the homologous chromosomes. Since each of the dozen-plus histocompatibility loci has many alleles in the human population, each of us has a unique combination of alleles; thus the combination of antigenic determinants on our cells is also unique. When any tissue is transplanted from a donor to an unrelated recipient, the recipient's lymphocytes will recognize the nonself antigens and set out to reject the tissue.

Two exceptions in which no immune barrier arises should be mentioned. Tissues from one part of an individual's body can be transplanted to another part without fear of rejection, as in the now commonplace skin grafts used in the treatment of victims of severe burns and in the hair transplants performed for cosmetic purposes. Identical twins, because they are genetically identical, have identical tissue antigens, and transplants between identical twins are not rejected.

A transplant will be rejected whenever the transplanted tissue has one or more antigenic determinants not present on tissues of the recipient, because the immune system is organized to attack nonself antigens. Finding completely compatible tissues among non-twins—tissues whose entire complement of antigenic determinants is also present in the recipient—is an almost impossible task; the enormous diversity of tissue types means that a perfect or nearly perfect match is unattainable. Luckily, not all histocompatibility loci are of equal importance. Some provide antigens that elicit a stronger immune reaction than others.

In the mouse, there is one major histocompatibility locus and a dozen or so minor ones. This most important histocompatibility complex is known as the **major histocompatibility complex (MHC)**, and it consists of a chromosomal region containing several genes controlling diverse functions in the immune system. In the mouse the MHC complex is designated **H-2**. Incompatibility at the H-2 complex leads to graft rejection (in the recipient mouse) in 8 to 12 days; incompatibility at one of the minor loci leads to rejection in 15 to 300 days, depending on the locus. If the donor and recipient are incompatible at many of the minor loci simultaneously, then a graft may be rejected in 8 to 12 days even though the donor and recipient match at the H-2 locus.

The genetic situation regarding transplants in humans is essentially similar to that in the mouse: one major (MHC) complex and several minor ones. The human MHC is designated **HLA**, and it is located on chromosome 6. The minor loci are scattered throughout the chromosomes. In humans, the critical period of graft rejection is the first four months after surgery; if a severe rejection crisis is to occur, it will usually occur during this period. The important point is that the immune barrier confronting tissue antigens produced by the major histocompatibility complex in both mice and humans is a much stronger barrier than confronts antigens from minor loci. Several methods have been developed for tissue typing in the laboratory to determine whether tissue from a potential donor is compatible at the major locus with the recipient. Obviously, near relatives are genetically more similar than distant relatives or unrelated people, so the search for a suitable donor usually begins with the recipient's immediate family, particularly with brothers and sisters. Once a suitable donor is identified, the transplant can be carried out with incompatibility at only the minor loci to be overcome.

Incompatibility at minor histocompatibility loci can be treated with **immunosuppressive drugs**—chemicals that interfere with the immune system. When transplants first began to be carried out on a wide scale in the 1950s, the only immunosuppressives available knocked out almost all immunity. This left the patient wide open to infection, and such minor infections as a sore throat could explode into a lethal disease. But drugs and other procedures in use at the present time are somewhat more specific in their effects; they suppress only part of the immune system. The patient is still in danger of succumbing to severe infections, but the risk is less than it would have been in times past.

In spite of tissue typing and immunosuppressive drugs, transplants of major organs are still not easy. A suitable donor must first be found. Organ banks containing organs from cadavers are of limited usefulness, because major organs cannot be maintained for long outside the human body without undergoing irreversible damage. (Certain tissues, such as eye corneas, are exceptions.) In addition, the transplant surgery can be extremely delicate. The major organ that can be transplanted most readily is a kidney. Kidney transplants, now carried out almost routinely with nearly 50 percent long-term graft acceptance, are relatively easy surgically: One major artery, one major vein, and the ureter have to be hooked up, and the organ functions well without a nerve supply. (The longest survival of a kidney transplant is over 20 years; the patient in this case was fortunate in having an identical twin donor.)

Kidney transplants are also made practical by the fact that everyone has two kidneys, and many people are willing to donate one to an otherwise doomed relative. Moreover, patients can be kept alive for years with an artificial kidney machine (if one is available and the expense can be borne) while a search for a suitable donor goes on. The importance of donor suitability in kidney transplants is illustrated in Figure 12.8, which shows the marked improvement in graft

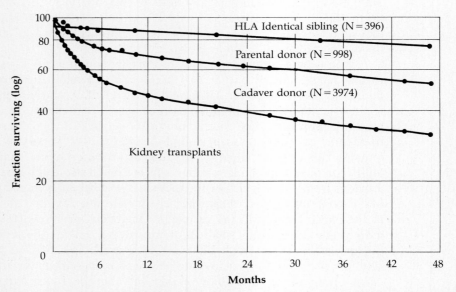

Figure 12.8 Survival time of kidney transplants using *HLA*-matched sibling donors, parents, or unrelated individuals (cadavers). N is the number of transplants in each category. Note that the scale of the vertical axis is logarithmic.

survival when *HLA*-matched sibling donors are used rather than parents or cadavers.

The advantages of kidney transplants do not apply to such organs as the liver, lungs, and heart. Many transplants of these organs have been performed, some successfully, but countless practical problems stand in the way of their widespread use. A relatively new approach to transplants is to use mechanical organs to substitute for the real thing, most notably in the case of the heart. While promising in the long run, this approach is still in its earliest experimental stages.

Ethical concerns about organ transplants seem to have been somewhat resolved. It is now taken for granted that a man or woman is free to bequeath any body parts (e.g., eye corneas) for immediate transplantation after death or for storage in an organ bank. It is also generally agreed that "death" means the complete cessation of brain function rather than heart failure. This eliminates the possibility of such favorite fictional fantasies as brain transplants (which the intricacies of surgery would have made impossible anyway). It is also generally agreed that adults can freely donate such nonvital organs as one kidney for transplantation, provided only that they be fully apprised of the risks and agree to accept them.

HLA AND DISEASE ASSOCIATIONS

The *HLA* complex consists of at least four **regions** (presumably corresponding to loci in the conventional sense of the term), which are denoted A, B, C, and D and which are ordered on the chromosome D-B-C-A. (See Figure 12.9; the "DR" region may represent a fifth locus or may, instead, be part of the D locus.) Each

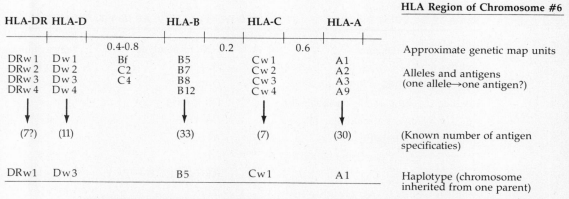

Figure 12.9 Organization of the *HLA* region in humans showing the major subdivisions (*HLA*-DR and *HLA*-D may represent a single gene instead of the two shown here) and the known number of "alleles" (antigen specificities) in each region.

region is involved in regulating different aspects of the immune system, and each region has its own series of alleles that can be identified by the appropriate laboratory tests. Each chromosome carries one allele for each region, and the collection of such alleles on a chromosome constitutes the chromosome's **haplotype**. For example, a chromosome carrying *HLA*-Dw2, *HLA*-B7, *HLA*-Cw4, and *HLA*-A2 has the haplotype Dw2, B7, Cw4, and A2. Since every normal individual has two copies of chromosome 6, every normal individual carries two *HLA* haplotypes, one inherited from each parent.

The number of possible haplotypes is enormous because of the large number of D, B, C, and A alleles that exist. Among European Caucasians, for example, there are at least 11 forms of the D region, at least 33 of B, at least 7 of C, and at least 30 of A (see Figure 12.9)—and more are being rapidly discovered. Theoretically, any of these alleles can be found in association with any of the others, so there are at least $11 \times 33 \times 7 \times 30 = 76,230$ possible haplotypes. However, certain haplotypes are much more common than others, and some are exceedingly rare or nonexistent. Nevertheless, there is a great deal of haplotype variation in the human population.

As might be expected of a complex that plays a fundamental role in regulating the immune response, there is an association between certain *HLA* haplotypes and certain diseases. Several examples are listed in Table 12.2, where it can be seen that the strongest association is between **ankylosing spondylitis** (a disease of the joints) and B27; individuals who have a haplotype carrying B27 have an 88-fold increased risk of developing this condition as compared with non-B27 individuals. The other associations in Table 12.2 are significant but weaker than the ankylosing spondylitis–B27 association.

THE ABO BLOOD GROUPS

The most widespread kind of tissue transplant is an ordinary blood transfusion. Transfusions are not usually thought of as transplants, perhaps partly because

Table 12.2   SOME DISORDERS ASSOCIATED WITH PARTICULAR *HLA* HAPLOTYPES

Disease	*HLA* haplotype	Relative risk (approximate)
Ankylosing spondylitis (joint disease)	B27	88
Rheumatoid arthritis (joint disease)	Dw4; Cw3	16
Juvenile diabetes (sugar metabolism)	DR3; DR4	4
Addison disease (adrenal glands)	B8; Dw3	4–11
Chronic active hepatitis (liver)	A1; B8	2–3
Ulcerative colitis (gastrointestinal tract)	B5	9
Psoriasis (skin)	B13; Bw17; Bw37	4–5
Anterior uveitis (eye disease)	B27	15
Hodgkin disease (malignancy)	A1, B5, B8, Bw18	1–2

Source: F. Vogel and A. G. Motulsky, 1979, Human Genetics: Problems and Approaches, Springer-Verlag, New York.

they are so commonplace. But in addition, the immune barrier to transfusions does not involve the major histocompatibility locus (*HLA*), since histocompatibility antigens are not found on red blood cells. Other antigens predominate on these cells, and these antigens define the familiar ABO and Rh blood groups as well as dozens of other blood groups that are less widely known. The ABO and Rh blood groups were discussed briefly in Chapter 3 as examples of simple Mendelian inheritance in humans. Here a somewhat more detailed discussion is in order.

The **ABO blood groups** are the most important ones in transfusions. The antigens are determined by one locus on chromosome 9 at which there are three alleles: I^A, I^B, and I^O. Individuals who are genetically I^A/I^A have on their red blood cells a particular antigen called the A antigen. (Remember that an antigen represents a *group* of antigenic determinants.) People who are genetically I^B/I^B have on their red blood cells an antigen called B. Heterozygotes of genotype I^A/I^B have both the A and B antigens on their red blood cells. The I^O allele does not produce a specific antigen, and the red blood cells of I^O/I^O individuals carry neither the A nor the B antigen. A person's ABO blood type is determined by which antigens are present on the red blood cells (Table 12.3). I^A/I^A and I^A/I^O individuals have **blood type A** because only A antigens are present on their red blood cells; I^B/I^B and I^B/I^O individuals have **blood type B**; I^A/I^B individuals have **blood type AB**; and I^O/I^O individuals are said to have **blood type O**, which means the absence of both A and B antigens. The frequencies of the ABO blood types vary from population to population, as illustrated in the maps in Figure 12.10. Among Caucasians (Londoners), the frequencies are O (48%), A (42%), B (8%), and AB (1%). Among Chinese, the frequencies are O (34%), A (31%), B (28%), and AB (7%). As with *HLA*, certain disease associations involve the ABO blood groups. Examples are stomach cancer (20 percent greater risk in type A than in type O), duodenal ulcers (30 percent greater risk in type O than in other blood

Table 12.3 IMPORTANT PROPERTIES OF ABO, Rh, AND MN BLOOD GROUPS

Blood group and genotype	Antigens on red blood cells	Antibodies present in blood	Antigens present in saliva and other body fluids?
ABO blood group:			
I^A/I^A or I^A/I^O	A	Anti-B	Yes, if secretor (*Se/Se* or *Se/se*)
I^B/I^B or I^B/I^O	B	Anti-A	Yes, if secretor (*Se/Se* or *Se/se*)
I^A/I^B	A and B	None	Yes, if secretor (*Se/Se* or *Se/se*)
I^O/I^O	None	Anti-A and anti-B	No
Rh blood group:			
DD	Rh$^+$	None	
Dd	Rh$^+$	None	
dd	None	Anti-Rh$^+$ if exposed to Rh$^+$ antigen	
MN blood group:			
MM	M	None	
MN	M and N	None	
NN	N	None	

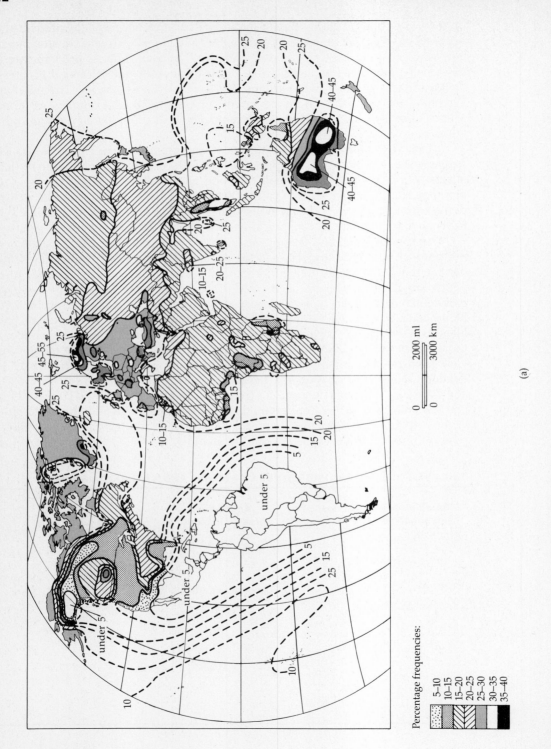

2000 ml
3000 km

(a)

Percentage frequencies:

5–10
10–15
15–20
20–25
25–30
30–35
35–40

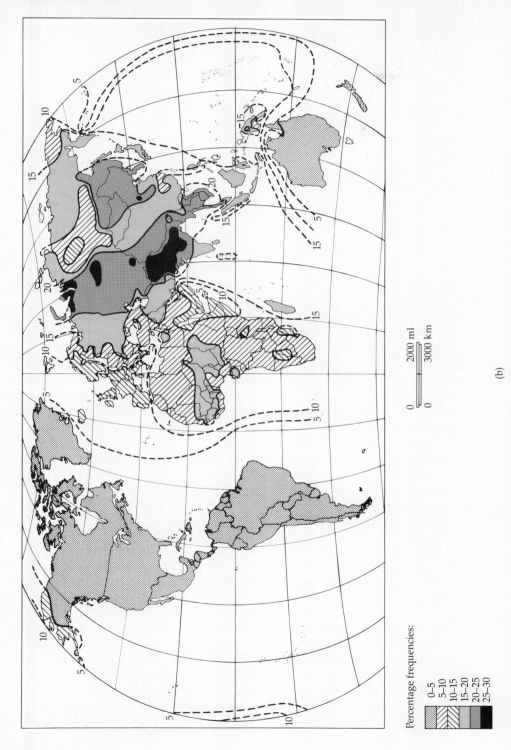

Percentage frequencies:

0–5
5–10
10–15
15–20
20–25
25–30

Figure 12.10 Maps showing the frequencies of (a) the I^A allele and (b) the I^B allele among aboriginal populations of the world. The frequencies of both I^A and I^B are low among the aboriginal Indians of southern North America and South America.

(b)

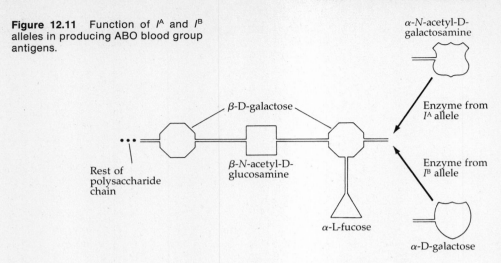

Figure 12.11 Function of I^A and I^B alleles in producing ABO blood group antigens.

groups), and obstructive blood clots (thromboembolic disease—60 percent greater risk in types A, B, and AB than in type O).

From a chemical point of view, the A and B antigens correspond to allele-specific modifications of a complex carbohydrate found on the surface of all red blood cells (Figure 12.11). The I^A and I^B alleles code for alternative forms of an enzyme that attaches one or another sugar to the end of this molecule; the I^A form of the enzyme attaches one sugar, and the I^B form of the enzyme attaches the alternative sugar. The I^O form of the enzyme causes no extra sugar to be attached to the basic carbohydrate. Under certain genetic conditions, the A and B antigens not only are present on red blood cells but are also secreted into the saliva, sweat, semen, and other body fluids. The secretion of these antigens is controlled by another locus—the **secretor locus**. The secretor locus has two alleles, *Se* and *se*, and the allele producing the secretion, *Se*, is dominant. *Se/Se* and *Se/se* genotypes are therefore secretors, whereas *se/se* genotypes are not. Thus, in secretors, the ABO phenotype of a person can be determined from the body fluids alone (see Table 12.3). Among Caucasians, about 77 percent of all individuals are secretors (i.e., *Se/Se* or *Se/se*).

The A and B antigens are highly antigenic; the antibodies against them are very potent. Unlike the situation with most antigens, an individual need not have been exposed to A or B antigens on red blood cells to produce antibodies. That is, a person not carrying the A antigen (and therefore of blood type B or O) will have so-called naturally occurring antibodies against A even though he or she was never exposed to red blood cells carrying the A antigen. How this happens is not clear, but it may result from the "A-like" or "B-like" antigenicity exhibited by several classes of microorganisms. Therefore, although a person may never have been exposed to type A blood, repeated exposure to substances that have antigens similar to A will induce the production of anti-A antibodies. The same goes for the B antigen. Remember, however, that individuals do not make antibodies against their own antigens because of immunological tolerance. The result of all this is that a person will have circulating in the blood serum (the fluid part of the blood) antibodies against the A or B antigens not present on his or her own red blood cells. Therefore, individuals with type A red blood cells will possess anti-B antibody in the serum; people with type B blood will possess anti-A antibodies; people with type O blood will have both anti-A and anti-B antibodies; and those with blood type AB will have neither anti-A nor anti-B antibodies. (These naturally occurring antibodies are of the IgM type.)

If a blood transfusion is performed using red blood cells against which the recipient has antibodies, the antibodies in the recipient will immediately attack the introduced cells and cause them to clump together (**agglutinate**) in various parts of the body (see Figure 12.12). These clumps block the tiny capillary vessels, and parts of the brain, heart, and other vital organs are deprived of their oxygen supply. The patient goes into shock and may die, so blood to be transfused must not carry any major antigens that the recipient does not also possess. Conversely, antibodies in the donor blood may clump the cells of the recipient. Therefore, best medical practice requires a two-way crossmatch of donor and recipient. However, one can transfuse limited amounts of blood that carries antibodies against the recipient's antigens; for example, people with type AB blood have A and B antigens but they may receive blood from type O people even though it contains antibodies against both A and B antigens. The reason is that the small volume of transfused antibody is diluted so rapidly in the large volume of the recipient's blood that it does no harm.

The main transfusion rule that must be observed, then, is that transfused cells can have no antigen unless the recipient has the same antigen. Therefore, individuals of blood type AB can receive blood of type A, B, AB, or O; type A people can receive A or O blood; type B people can receive B or O blood; and type O individuals can receive only type O blood. (Type AB people were formerly

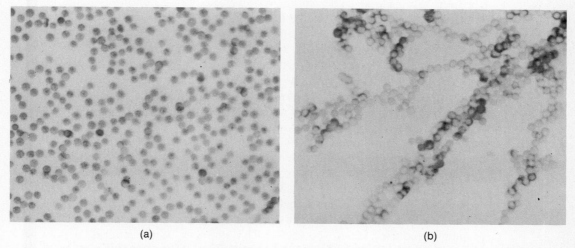

(a)   (b)

Figure 12.12 (*a*) Normal red blood cells as viewed through a microscope. (*b*) Red blood cells that have been agglutinated as a result of an antigen-antibody reaction.

called **universal recipients** because they can receive blood of any type; type O individuals were known as **universal donors** because they can donate blood to anyone. Nowadays it is realized that this terminology is misleading and simplistic in that it ignores the possible effects of other blood groups.)

THE Rh BLOOD GROUPS

The one other blood group system that is of medical importance is the **Rh system**. The genetic basis of the Rh system is rather complex, and it is not clear whether it involves a single locus on chromosome 1 with many alleles or a tightly linked cluster of loci. For our purposes, we may treat it as a single locus with two alleles, usually called D and d, although the actual situation regarding the number of loci and the number of alleles is still unknown. The D allele produces on the surface of red blood cells an antigen, chemically unrelated to the A and B antigens, called the Rh^+ antigen; the d allele does not produce a distinct antigen. Therefore, the red blood cells of both D/D and D/d genotypes carry the Rh^+ antigen; they are said to be Rh^+, or **Rh positive**. The red blood cells of d/d genotypes do not carry the Rh^+ antigen; they are said to be Rh^-, or **Rh negative** (see Table 12.3). Individuals who are Rh^- are capable of producing antibodies against Rh^+ antigen. Unlike the ABO system, however, Rh^- individuals will not produce anti-Rh^+ antibodies unless they have previously been exposed to Rh^+ cells.

The medical significance of the Rh system arises in those cases in which an Rh^- mother who is producing anti-Rh^+ antibody due to previous exposure to Rh^+ antigen becomes pregnant with an Rh^+ fetus. (This can occur when the mother is d/d and the father is D/D or D/d.) In such a case, some of the anti-Rh^+ antibodies from the mother can traverse the placenta, since these antibodies are of the IgG class. The antibodies that invade the fetus's blood begin to attack the fetus's red blood cells, leading to a serious blood condition called **erythroblastosis fetalis**, sometimes called **hemolytic disease of the newborn**.

In the most severe cases, the fetus cannot replace its red cells quickly enough and dies. Less severely affected infants may be born in apparent good health but soon become jaundiced due to toxic by-products from damaged red cells. If the infant lives, brain damage and mental retardation may result. Emergency treatment of the child may require an exchange transfusion—removing virtually all the child's blood and replacing it with Rh^- blood. Gradually the Rh^- red blood cells break down and are replaced with new Rh^+ cells made by the child, but the child does not produce anti-Rh^+ antibodies and therefore no subsequent problems arise. The time factor in these transfusions is critical because the untransfused child suffers from a shortage of oxygen. In the absence of treatment, three-quarters of the affected infants would die.

Actually, an Rh^- woman will not produce anti-Rh^+ antibodies unless she has been exposed to Rh^+ antigens. This exposure usually requires one or more pregnancies during which she carries Rh^+ embryos. With each pregnancy, some of the Rh^+ cells from the fetus seep across the placenta into the mother's bloodstream, especially at the time of birth. The mother produces antibodies against these cells. Because of the need for prior exposure, the first Rh^+ child is often not at risk, but it serves to immunize the mother. Often the mother does not produce sufficient antibody in response to the first exposure to cause problems with even the second Rh$^+$ child. But third and subsequent Rh$^+$ children incur a substantial risk of erythroblastosis fetalis. (Only 5 percent of infants with Rh hemolytic disease are first-born, roughly 45 percent are second-born, and about half are the third or later births. At one time the frequency of the condition in the United States was 1 in 150 births, but today's smaller families have reduced the incidence.) In matings of d/d mothers with D/d fathers, half the children will be d/d; these d/d children do not have the Rh$^+$ antigen and they are never at risk, nor do they sensitize the mother.

The incidence of erythroblastosis fetalis in a population depends on the frequency of marriages of Rh$^-$ women with Rh$^+$ men. Among U.S. whites, the frequency of Rh$^-$ blood is about 15 percent and that of Rh$^+$ blood is 85 percent. Thus the type of mating in question occurs in about 13 percent of white marriages (i.e., $0.15 \times 0.85 = 0.13$). Rh hemolytic disease is virtually nonexistent in Japan, where more than 99 percent of the people are Rh$^+$. Among the Basques, a genetically semi-isolated group that lives in the Pyrenees Mountains between France and Spain, the incidence of Rh$^-$ is exceptionally high—about 43 percent. (A similar situation exists among Basque descendants who live in northern Nevada.) When Rh$^-$ women from these semi-isolates marry outside their own cultural group, in the majority of cases they marry Rh$^+$ men and therefore they run the risk of erythroblastosis fetalis among their children.

For Rh$^-$ women who are currently coming into their reproductive years and have never been transfused with Rh$^+$ blood, the risk of Rh hemoloytic disease has all but been eliminated by a preventive treatment. The treatment involves injecting anti-Rh$^+$ antibodies into the Rh$^-$ mother soon after the birth of her first Rh$^+$ child (Figure 12.13). Any Rh$^+$ cells from the child that make it into the mother's blood are destroyed by these antibodies before they can immunize the mother. By the time the next pregnancy occurs, the injected antibodies have disappeared. (All proteins, including antibodies, break down or are destroyed in time.) Neither can a second Rh$^+$ child immunize the mother if, as before, she is injected with anti-Rh$^+$ antibody. This treatment is very effective provided that an Rh$^-$ woman is not already producing anti-Rh$^+$ antibody. The injected anti-Rh$^+$, by destroying any Rh$^+$ cells that might be present, prevents the mother's immune system from being stimulated to produce its own anti-Rh$^+$ antibody. Strangely enough, a similar kind of prevention sometimes occurs naturally. If red blood cells from the fetus carry an A or B antigen not present in the mother in addition to the Rh$^+$ antigen, then the mother's anti-A or anti-B antibodies will destroy the cells before they can stimulate the production of anti-Rh + antibody. In this way ABO incompatibility prevents Rh disease.

One may wonder why hemolytic disease of the newborn can come about from Rh incompatibility but not from ABO incompatibility. For example, why is it that hemolytic disease of the newborn does not ordinarily occur in type A or AB children whose mothers produce anti-A antibody? One part of the answer is that a type of ABO hemolytic disease *does* occur, but rarely—affecting perhaps 0.1 percent of all newborns. One reason the condition is so rare is that the antibodies responsible are not the same ones involved in ABO transfusion incom-

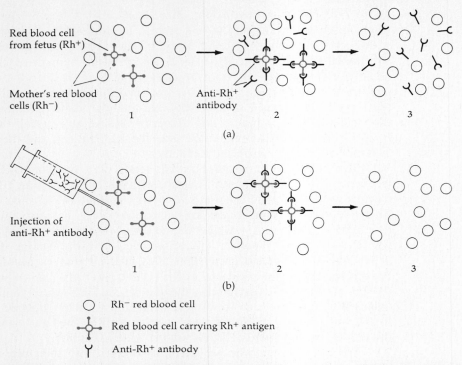

Figure 12.13 Preventive treatment of Rh hemolytic disease. (a) A portion of the Rh⁻ mother's bloodstream in the absence of treatment. Rh⁺ red blood cells from an Rh⁺ fetus seep into the mother's bloodstream (1). These cells stimulate production of anti-Rh⁺ antibody and are destroyed (2). However, the mother continues to produce antibody (3), and these antibodies may cross the placenta in a subsequent pregnancy. If the fetus in the subsequent pregnancy is Rh⁺, the antibodies will cause Rh hemolytic disease. (b) The preventive treatment: anti-Rh⁺ antibody is injected into the mother (1), and this injected antibody destroys the Rh⁺ cells (2). The Rh⁺ cells are destroyed before they can stimulate production of anti-Rh⁺ antibody by the mother. Thus, after the injected antibody has broken down, the mother's blood remains free of anti-Rh⁺ (3).

patibility. The usual anti-A and anti-B antibodies important in transfusion are of the IgM class, which are unable to cross the placenta. The antibodies responsible for ABO hemolytic disease are of the IgG class, which can traverse the placental barrier. Under ordinary circumstances, this IgG class of ABO antibody is not produced. Nevertheless, there is a smaller than expected frequency of I^A/I^O children arising from matings of I^O/I^O mothers with I^A/I^O fathers. The amount of the deficiency varies from study to study. (In one Japanese study the deficiency was more than 10 percent of the number of I^A/I^O children expected.) This deficiency has been attributed to very early spontaneous abortion brought about by ABO incompatibility between mother and fetus.

OTHER BLOOD GROUPS

Blood groups have application extending beyond genetics and medicine. Besides the ABO system and the Rh system, there are literally dozens of blood group systems. (A **blood group system** refers to the several antigens specified by the different alleles at a single locus.) Other than the ABO and Rh systems, most of these antigens are not recognized as being very antigenic by human lymphocytes; they therefore do not usually play a significant role in blood transfusions or pregnancy. They are, however, antigenic in other animals, such as rabbits. Red blood cells from one person can be injected into an animal and at some later time antibodies can be isolated. These antibodies are then tested against the red blood cells from other people, and in some instances there will be an antigen-antibody reaction (the red blood cells will clump) but not in other instances. The people whose blood cells are recognized by the antibody and clump must have the same antigen as the person whose cells were originally used to immunize the animal, but those whose red cells are not recognized by the antibody must lack this antigen. Studies of this kind are then combined with family studies to determine the mode of inheritance of the antigenic difference.

One blood group that was discovered in almost the way just described is the **MN system**. The MN blood group antigens are due to two alleles at a single locus on the long arm of chromosome number 4. By injecting red blood cells of the appropriate kind into rabbits, anti-M and anti-N antibodies can be obtained. Anti-M antibody reacts only with cells carrying the M antigen; anti-N antibody reacts only with cells carrying the N antigen. The alleles that produce the M and N antigens are also called *M* and *N*. Three genotypes at the *MN* locus can be found (see Table 12.3). Some people are genetically *M/M* (their red blood cells react only with anti-M), others are *N/N* (their red blood cells react only with anti-N), and still others are *M/N* (their red blood cells react with both anti-M and anti-N). Thus, in the MN system, the antigens carried on the red blood cells can be used to identify the genotype of an individual at this locus.

APPLICATIONS OF BLOOD GROUPS

The dozens of different blood group systems used collectively can identify a person almost as uniquely as fingerprints. Although many people have the same ABO blood type in common, or the same Rh blood type in common, or the same MN blood type in common, when these blood groups are considered separately, many fewer people have the same ABO, Rh, and MN blood types simultaneously. This is simply to say that there are fewer people with the precise blood type O;Rh⁻;MN than there are people with blood type O with no regard to their Rh and MN types. Now if one considers, say, 30 blood groups instead of only three, then the odds would be exceedingly high that any two people would have a different genotype with respect to one or more of the 30 groups. Every human being is genetically unique, different from all who have come before and all who will follow; the only exceptions are identical twins, triplets, and quadruplets.

The property of blood groups to characterize individuals and groups of people has been used widely by anthropologists to study the origin and migrations of various human populations. In these kinds of studies, blood group alleles that are common in one population but less common or rare in others are very important. For example, the B type antigen is virtually absent in American Indians (with a few exceptions, such as the Blackfeet). Another example: Among the many alleles of the Rh locus (the previous discussion divided these alleles into only two classes, *D* and *d*), one allele is highly prevalent in African Negroes but rare elsewhere; a different allele is frequent in European Caucasians but rare elsewhere. Provided sufficient supporting information can be gained from, for example, similarities in language, alleles of this sort can sometimes be used as markers of ancestry. In a similar vein, comparisons of humans and other primate species have been pursued using blood group similarities to measure genetic relationship.

The many blood group systems have found legal application in such litigations as paternity suits. The principle behind these applications is that a person falsely accused of fathering a child can sometimes be exonerated on the grounds that, based on genetic evidence, he could not possibly have fathered the child. For example, a man of blood type AB cannot have a type O child, except in the extremely unlikely event of mutation. As another example, a man of blood type O;M cannot have been the father of an A;MN child whose mother is AB;M, because although the ABO system works out satisfactorily (the child could be genotypically I^A/I^O, having received the I^O gene from the father), the MN system does not work out. Such a child would have to receive the gene for the N antigen from its father because the mother does not have it. But the accused father does not have this gene either, and therefore the real father must be somebody else. The same kind of reasoning applies to matching infants to their natural mothers in the unlikely event of a mix-up in the hospital nursery. An O;Rh$^+$;M woman could not, for instance, be the mother of an A;Rh$^-$;N child.

The ABO antigens are important in certain kinds of police work. These antigens are extremely stable and persistent, so minute quantities of dried blood (or body fluids, in the case of secretors) can be typed. Indeed, the ABO blood types of even some Egyptian mummies can be determined! A minute amount of an assailant's blood under a victim's fingernail can sometimes be decisive in excluding innocent suspects; knowledge that the assailant's blood type is AB will exclude 99 percent of innocent white suspects in London, for example.

It seems appropriate to end this section with an account of one of the earliest criminal cases involving the use of blood typing for evidence:

It was a Japanese who took the lead. In 1928 K. Fujiwara, Director of the Institute of Forensic Medicine in Niigata, reported a crime which had been solved by means of blood group determination from seminal stains. A sixteen-year-old girl, Yoshiko Hirai, who went from village to village selling fortune-telling slips, had been found strangled. Witnesses had seen the girl at 7:30 P.M. on her way from the village to the railroad station. She must have been raped and killed shortly afterward. Fujiwara found whitish spots which contained well-preserved semen on the girl's body. Her blood group was O. Fujiwara tested the semen spots by the absorption method, adding anti-A and anti-B serums, and thus determined that the semen had the group characteristic A. Meanwhile the police had arrested two suspects. One of them was a twenty-four-year-old mentally retarded beggar named Mochitsura Tagami, who straightway confessed that he had committed the murder. The second suspect, Iba Hoshi, denied all guilt. The police were on the point of accepting Tagami's confession and releasing

Hoshi when Fujiwara asked for time to determine the blood groups of the prisoners. The results proved that Tagami, who had made the confession, could not possibly be the criminal; his blood group was O. Hoshi, on the other hand, had group A. Confronted with the data, Hoshi confessed; that Fujiwara knew his blood and his semen had something in common impressed [him as] witchcraft.*

SUMMARY

1. The **immune response** is the body's defense against invading substances. The response is mediated by two types of **lymphocytes** (white blood cells): **thymus-dependent** or **T cells**, which undergo differentiation in the thymus gland and include **killer** T cells and other agents of **cellular immunity**; and **bone-marrow-derived** or **B cells**, which differentiate primarily in the bone marrow and which produce circulating **antibodies** and are thus agents of **humoral immunity**.

2. An **antigen** is any substance that is able to elicit an immune response by virtue of certain three-dimensional configurations of atoms on its surface; these surface landmarks are known as **antigenic determinants**, and a typical antigen carries many different antigenic determinants. An antigenic determinant elicits an immune response by combining with a complementary **receptor** molecule on the surface of a B cell or a T cell. Any B cell or T cell has only one type of receptor, which can combine with only one type of antigenic determinant. The antigenic determinant-receptor complex is essential for B-cell or T-cell stimulation, but proper stimulation also requires **cell cooperation** involving **helper** T cells and a class of white blood cells called **antigen-presenting macrophages**.

3. Stimulated lymphocytes undergo successive mitotic divisions to give rise to a **clone** of cells; each is capable of recognizing the antigenic determinant that caused the stimulation. Since stimulation is selective in the sense that only those lymphocytes that have the appropriate receptor molecules can respond to an antigenic determinant, the mitosis-stimulating process is called **clonal selection**. In the case of B cells, certain members of the clone differentiate into **plasma cells**, which are the actual antibody-producing cells. Other members of the clone do not immediately differentiate into plasma cells but are held in readiness to mount an antibody response against reinvasion by the same antigenic determinant. This readiness to mount a secondary response is referred to as **immunity**.

4. Five principal **classes** of antibody can be distinguished: IgG, IgM, IgA, IgD, and IgE, where the symbol Ig stands for **immunoglobulin**. Each class consists of a number of **light** polypeptide chains, which may be of either the κ type or the λ type but not both, and a number of **heavy** polypeptide chains, which may be of five types: γ (found in IgG), μ (IgM), α (IgA), δ (IgD), or ϵ (IgE); the type of heavy chain thus determines the class of antibody. However, there are alternative genes for some of the heavy chains—four for the γ chain, for example. The

* J. Thorwald, 1966, Crime and Science, Harcourt, Brace & World, New York, p. 90.

particular heavy-chain gene used in an antibody determines the antibody's **subclass**.

5. IgG is the prevalent type of antibody in the blood serum, and it is the only class able to cross the placenta. IgM is a large and powerful antibody that, in addition to other functions, constitutes the B-cell receptor. IgA is the class of antibody found in the gut, respiratory tract, urogenital tract, and other mucous membranes. IgE seems to be involved in allergic reactions. The functions of IgD are not well understood, but it may also serve as a receptor molecule.

6. All antibodies consist of one or more Y-shaped units. IgG, IgD, and IgE each have one such unit, IgM has five, and IgA has a variable number. Each Y-shaped unit is actually a double Y, like this: $\backslash\!\!/\!\!/$. The short lines in the tines of the Y represent the light chains, the long lines the heavy chains. The entire structure is held together by disulfide (-S—S-) bridges, and each class of antibody has its own type of carbohydrate attached to the heavy chains in the stalk of the Y. The tips of the tines of the Y are the **antigen-binding sites**—sites able to combine with an antigenic determinant because of their complementary surface landscape. The surface landscapes of the antigen-binding sites are determined by the folding and interaction of the light and heavy chains in the upper half of the tines of the Y. The amino acid sequences in this portion of the antibody are said to be **variable** because they differ in each particular type of antibody. The lower half of the light chains and the lower three-quarters of the heavy chains have a **constant** amino acid sequence; that is, their amino acid sequence is determined by the specific light-chain gene and the specific heavy-chain gene used in coding for the antibody.

7. A normal person is capable of producing at least 1 million different antibodies. Two broad hypotheses have been proposed regarding the origin of so much variability. According to the **germ-line theory**, a large number of genes code for different light chains and a large number of genes code for different heavy chains; this theory requires at least 1000 genes of each type. According to the **somatic theory**, there are relatively few antibody-coding genes; the great diversity of antibody types is generated anew in each individual by mutation and recombination. Both views have turned out to be partly correct. Antibody variability is generated by **DNA splicing** of alternative versions of various parts of the completed antibody gene. The variable part of the heavy chain consists of three regions, a V region (approximately 250 alternative forms exist in germ-line DNA), a D region (about 10 alternative forms), and a J region (about 4 alternatives). In DNA splicing, one of each of these regions is spliced together and attached to a DNA sequence corresponding to the constant portion of the heavy chain. Light chains are produced similarly by DNA splicing, although light chains lack a region corresponding to D. This process of DNA splicing is called **combinatorial joining**, and it is a principal source of antibody diversity, along with somatic mutation in the V, D, and J regions of heavy chains and the V and J regions of light chains.

8. Breakdowns of immunity include **immunodeficiency** diseases, in which patients lack B-cell function or T-cell function or both, and **autoimmune** diseases, in which the immune system attacks one's own antigens. **Rheumatoid arthritis** is an example of an autoimmune disease, as is the

heart damage that sometimes accompanies **rheumatic fever**. Autoimmunity results from a breakdown of **immunological tolerance**, the process whereby the body comes to recognize and to tolerate "self" antigens while retaining the ability to attack "nonself" antigens. Nonself antigens include abnormal antigens on the surface of malignant cells, which provide the basis of **immune surveillance** against malignancies.

9. Rejection of transplanted tissues is primarily a T-cell function directed against antigens produced by more than a dozen loci called **histocompatibility loci**. The general transplantation rule is that a recipient individual will reject any tissue that carries histocompatibility antigens that the recipient does not also possess. However, some antigens are much more potent than others, and the cluster of loci responsible for the most potent antigens is called the **major histocompatibility complex (MHC)**. In the mouse the MHC is designated **H-2**. In humans it is designated **HLA**, and it is present on chromosome 6. The success of transplants is greatly increased if donor and recipient are compatible for *HLA* (i.e., if the donor has no *HLA* antigens not also present in the recipient). With *HLA* compatibility, rejection reactions due to the minor loci can often be controlled by **immunosuppressive drugs**.

10. *HLA* plays a key role in the immune response, and the complex can be subdivided into four functionally distinct regions (order D-B-C-A on the chromosome), with each region having many alternative alleles. The **haplotype** of a chromosome refers to its particular combination of alleles corresponding to the D, B, C, and A regions. Certain diseases have an increased prevalence among individuals who have a particular haplotype. One example is **ankylosing spondylitis**, which tends to be associated with haplotypes containing B27.

11. Clinical problems in blood transfusions do not involve *HLA* but primarily two other loci: one responsible for the **ABO blood groups** and the other for the **Rh blood groups**. An individual's ABO blood group is determined by the presence of particular carbohydrate antigens on the surface of the red blood cells. Presence of these antigens is, in turn, controlled by a single locus on chromosome 9 at which there are three alleles: I^A, I^B, and I^O. Genotypes I^A/I^A and I^A/I^O produce the A antigen and are said to have **blood type A**; genotypes I^B/I^B and I^B/I^O produce the B antigen and are said to have **blood type B**; genotype I^A/I^B produces both A and B antigens and has **blood type AB**; and genotype I^O/I^O produces neither antigen and has **blood type O**. Because the A and B antigens are similar to certain antigens of microorganisms, virtually all individuals are stimulated to produce antibodies (usually IgM) against antigens not present on their own red cells. Thus, type A individuals will also have anti-B antibodies in their serum (the fluid portion of the blood), type B individuals will have anti-A, type O will have both anti-A and anti-B, and type AB will have neither anti-A nor anti-B. Difficulties in transfusion arise when donor red blood cells carry antigens against which the recipient has antibodies, for then the antibodies will cause **agglutination** of the red blood cells and blockage of capillaries. As with *HLA*, there is an elevated risk of particular diseases in persons with certain blood types, such as stomach cancer with type A and duodenal ulcers with type O. Secretion of the A and B antigens into the body fluids is controlled by a dominant allele at another locus called the **secretor locus**.

12. The **Rh blood groups** are determined by a red cell antigen distinct from the A and B antigens. The Rh antigen is controlled by a single locus on chromosome 1 (or perhaps by a cluster of loci) at which there are two principal alleles (*D* and *d*). Genotypes *D/D* and *D/d* produce the Rh antigen and are said to be Rh^+ (**positive**); genotype *d/d* does not produce the antigen and is said to be Rh^- (**negative**). Unlike the situation with ABO, Rh^- individuals do not usually produce anti-Rh^+ antibody unless stimulated with Rh^+ antigen. However, anti-Rh^+ antibodies are typically of the IgG type and can cross the placenta. Thus, antibodies from an Rh^- mother can cross the placenta and attack the red cells of an Rh^+ fetus, causing a blood condition known as **erythroblastosis fetalis** or **hemolytic disease of the newborn**. On the other hand, Rh^- mothers do not usually produce anti-Rh^+ antibody unless they have previously carried one or two Rh^+ fetuses; seepage of Rh^+ cells from these prior fetuses is necessary for stimulation of the immune system. Modern preventive treatment for newborn hemolytic disease involves injecting the mother with anti-Rh^+ antibodies near the time of birth to destroy the otherwise stimulatory Rh^+ cells from the fetus. This sort of protection sometimes occurs naturally; matings of Rh^-;O mothers with Rh^+;AB fathers never lead to Rh hemolytic disease, because the maternal ABO antibodies destroy the fetal Rh^+ cells before they can stimulate production of anti-Rh^+ antibodies.

13. Dozens of different blood group systems analogous to ABO and Rh are known. Most involve substances that are not antigenic in humans but are antigenic in another organism. For example, the MN system involves two red cell antigens (M and N) that are antigenic in rabbits. The M antigen is determined by a single allele (*M*); the N antigen is determined by an alternative allele (*N*). Thus, *MM* genotypes have blood type M, genotype *NN* has blood type N, and genotype *MN* has blood type MN (because both antigens are expressed). The simple Mendelian inheritance of ABO, Rh, MN, and the other blood groups makes these loci particularly useful in anthropological studies, paternity exclusion, criminology, and other applications.

KEY WORDS

ABO blood groups	DNA splicing	Lymphocyte
Acquired immunodeficiency disease (AIDS)	Germ-line theory	Macrophage
	Haplotype	MHC
Agglutinate	Heavy chain	MN blood groups
Ankylosing spondylitis	Hemolytic disease of the newborn	Monoclonal antibody
Antibody		Plasma cell
Antigen	Histocompatibility loci	Receptor site
Antigenic determinant	*HLA*	Rh blood groups
Autoimmune disease	Immune surveillance	Secretor
B cell	Immunity	Somatic theory
Cell cooperation	Immunodeficiency disease	T cell
Clonal selection	Immunoglobulin	Variable region
Combinatorial joining	Immunosuppressive	
Constant region	Light chain	

PROBLEMS

12.1. What is immunity?

12.2. What is an autoimmune disease?

12.3. What is an antigen?

12.4. Which of the following is not normally antigenic? Protein; Red blood cell; Virus; Bacterium; Amino acid.

12.5. What does the T stand for in T cells? What does the B stand for in B cells?

12.6. What would be the major immunological effect of a total lack of B cells in an individual, and what would be the principal consequence of such a defect?

12.7. Is tissue rejection in transplants primarily a T-cell or a B-cell function?

12.8. What is the principal dangerous side effect of immunosuppressives used to increase transplant success?

12.9. Ignoring the small possibility of crossing over in the major histocompatibility complex, what is the probability that a mother will share an *HLA* haplotype with her son?

12.10. Assuming that each *HLA* haplotype in two parents is different, as will almost always be the case, and ignoring rare crossing over within the *HLA* complex, what is the probability that two siblings will have identical *HLA* haplotypes?

12.11. If there were 10 histocompatibility loci with 4 alleles at each, how many histocompatibility genotypes would be possible?

12.12. Why are the MN blood groups not normally important in transfusions?

12.13. To which blood groups can a type A individual donate blood? From which blood groups can a type A individual receive blood?

12.14. A couple has four children, one A, one B, one AB, and one O. What are the genotypes of the parents?

12.15. If there were a fourth major allele in the ABO blood groups, called I^C, which acted like I^A and I^B in being dominant to I^O but codominant with other alleles, how many ABO blood groups would there be?

12.16. How would it be possible for a type A secretor male and a nonsecretor female to have a type B secretor child?

12.17. An "incompatible" mating with respect to the ABO blood groups is a mating in which the fetus could potentially possess ABO antigens not present in the mother. With regard to the ABO blood groups, which of the possible types of matings are incompatible?

12.18. A woman of blood type O;Rh$^+$;N has an O;Rh$^+$;MN child and claims a man of type AB;Rh$^-$;M is the father. Are the blood types consistent with her claim?

12.19. Considering the following data, could Couple 1 be the parents of baby 1? Of baby 2? Could Couple 2 be the parents of baby 1? Of baby 2?

A;Rh$^+$;MN
B;Rh$^+$;N } Couple 1

AB;Rh$^-$;M
O;Rh$^+$;N } Couple 2

AB;Rh$^-$;MN Baby 1
A;Rh$^+$;MN Baby 2

FURTHER READING

Baltimore, D. 1981. Somatic mutation gains its place among the generators of diversity. Cell 26: 295–96. Although DNA splicing is elegant, a great deal of antibody diversity is generated by somatic mutation.

Bernard, O., N. Hozumi, and S. Tonegawa. 1978. Sequences of mouse immunoglobulin light genes before and after somatic changes. Cell 15: 1133–44. Direct evidence of V-J joining.

Buisseret, P. D. 1982 (August). Allergy. Scientific American 247: 86–95. The immune system gone wrong.

Dausset, J. 1981. The major histocompatibility complex in man. Science 213: 1469–74. Nobel-prize lecture reviewing past, present, and future concepts regarding HLA.

Davis, M. M., S. K. Kim, and L. Hood. 1980. Immunoglobulin class switching: Developmentally regulated DNA rearrangements during differentiation. Cell 22: 1–2. A review of DNA splicing and some unanswered questions.

Golub, E. S. 1980. Know thyself: Autoreactivity in the immune response. Cell 21: 603–04. Some degree of reaction against "self" seems essential in order to mount a normal immune response.

———. 1981. The Cellular Basis of the Immune Response. 2d ed. Sinauer Associates, Sunderland, Massachusetts. A leading textbook introducing cellular immunology.

Herberman, R. B., and J. R. Ortals. 1981. Natural killer cells: Their role in defense against disease. Science 214: 24–30. Special types of T cells and their functions.

Hood, L., M. Steinmetz, and R. Goodenow. 1982. Genes of the major histocompatibility complex. Cell 28: 685–87. Transplantation antigens and other features encoded by the MHC.

Kindt, T. J., and A. Coutinho. 1983. More sources of antibody diversity. Nature 304: 306–07. Briefly reviews a multiplicity of mechanisms.

Koffler, D. 1980 (July). Systemic lupus erythematosus. Scientific American 243: 52–61. Discusses a serious autoimmune disease found mainly in women.

Leder, P. 1982 (May). The genetics of antibody diversity. Scientific American 246: 102–15. Review of combinatorial joining and class switching.

Lee, J., and J. Trowsdale. 1983. Molecular biology of the major histocompatibility complex. Nature 304: 214–15. Cloning reveals even more diversity than expected.

Mourant, A. E., A. C. Kopec, and K. Domaniewska-Sobczak. 1976. The Distribution of the Human Blood Groups and Other Polymorphisms. 2d ed. Oxford University Press, New York. A compendium of information on the varying incidences of blood groups in different branches of humanity.

Potash, M. J. 1981. B lymphocyte stimulation. Cell 23: 7–8. What induces B lymphocytes to secrete antibody.

Reinherz, E. L., and S. F. Schlossman. 1980. The differentiation and function of human T lymphocytes. Cell 19: 821–27. A good review of a highly complex population of cells.

Robertson, M. 1982. The evolutionary past of the major histocompatibility complex and the future of cellular immunology. Nature 297: 629–32. Report of a meeting on the cloning of HLA and H-2.

Rose, N. R. 1980 (February). Autoimmune diseases. Scientific American 244: 80–103. Malfunctions of the immune system and strategies of treatment.

Ryder, L. P., A. Svejgaard, and J. Dausset. 1981. Genetics of HLA disease association. Ann. Rev. Genet. 15: 169–87. Much more data on HLA and disease.

Seidman, J. G., et al. 1978. Antibody diversity. Science 202: 11–17. The role of recombination in the origin of antibody diversity.

Snell, G. D. 1981. Studies in histocompatibility. Science 213: 172–78. Nobel-prize lecture reviewing the mouse H-2 system from its earliest studies.

chapter ***13***

Population Genetics

- Genetic Variation
- Allele Frequencies and Genotype Frequencies
- Implications of the Hardy-Weinberg Rule
- Differentiation of Populations: Race
- Inbreeding
- Mutation
- Migration
- Selection
- Mutation-Selection Balance
- Random Genetic Drift
- Founder Effects
- Summary

This and the next chapter constitute a brief summary of **population genetics**—the study of how Mendel's laws and other principles of genetics apply to entire populations (interbreeding groups) of organisms. On the one hand, population genetics includes studies of contemporary populations, such as tribes of Indians or major human races, in which the focus is on the genetic relationships among groups and on the reasons that underlie differences in the incidence of inherited disorders or other traits. On the other hand, population genetics attempts to clarify the major forces involved in the process of **evolution**, by which we mean cumulative change in the genetic makeup of a population through time, genetic change that is associated with increasing adaptation of a population to its environment.

GENETIC VARIATION

Populations of humans and most other organisms contain substantial amounts of genetic variation, so much so that no two individuals, with the exception of identical twins, are ever genetically identical. Widespread genetic variation is not obvious from casual inspection of phenotype. Although there is great phenotypic diversity in such traits as facial appearance, eye color, skin color, height, weight, growth rate, athletic skills, and certain mental abilities, some of this phenotypic variation is caused, not by genes, but by environmental differences among individuals, as evidenced by the fact that even identical twins may differ in such traits. Much of the genetic variation occurring among individuals is **hidden** in the sense that it is not expressed as obvious differences in phenotype. For example, the ABO and Rh blood groups differ among individuals because there are different genotypes in the population, yet without special tests the blood groups cannot be identified. In studying such hidden genetic variation, one must use procedures that study gene products—proteins—directly, or, better yet, that study the DNA itself.

A particularly useful technique for studying hidden genetic variation is **electrophoresis**, a technique applicable to the separation of nucleic acid fragments as discussed in Chapter 8, but that can also be used to separate proteins. Whereas separation of nucleic acid fragments is based on their size, protein separation is based on size and the net electric charge of the molecule, which is determined by the amino acid composition of the protein. That is to say, two otherwise

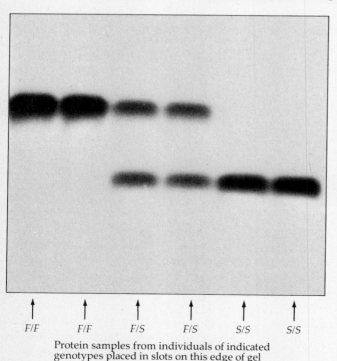

Direction of current flow and of migration of proteins

F/F   F/F   F/S   F/S   S/S   S/S

Protein samples from individuals of indicated genotypes placed in slots on this edge of gel

Figure 13.1 Gel slab used in electrophoresis showing dark bands that mark the position to which a particular enzyme has migrated in response to the electric field. The enzyme here is amylase in *Drosophila melanogaster,* and the electrophoretic pattern of three genotypes (*F/F, F/S,* and *S/S*) is shown. (Negative print of gel photo courtesy of J. Coyne and D. Hickey.)

identical proteins differing in a single amino acid that alters the charge of the molecule will be separated by protein electrophoresis. Since nucleotide substitutions in the DNA can lead to charge-changing amino acid substitutions in the corresponding protein, electrophoresis is a convenient method by which to study hidden genetic variation.

Enzymes separated by electrophoresis can be revealed by treating the gel with a staining solution that changes color wherever the enzyme-catalyzed reaction occurs. Figure 13.1 is an example involving a starch-degrading enzyme in *Drosophila*. Two alleles (here called simply *F* and *S*) and three genotypes (*F/F, F/S,* and *S/S*) are represented among the six individuals. The *F* allele codes for a fast-migrating enzyme, and *F/F* homozygotes exhibit only the fast-moving enzyme band; the *S* allele codes for a slow-migrating enzyme, and *S/S* homozygotes exhibit only the slow-moving enzyme band. *F/S* heterozygotes (in the middle of Figure 13.1) exhibit both the fast-enzyme form (corresponding to the *F* allele) and the slow-enzyme form (corresponding to the *S* allele). The *F* and *S* alleles are said to be **codominant**, which means that they are both expressed in heterozygotes. Enzymes that differ in electrophoretic mobility because of genetic differences at the corresponding locus are called **allozymes**, and codominance is the usual situation regarding allozyme-associated alleles. With electrophoresis, many different enzymes from hundreds of individuals can be examined for hidden genetic variation affecting allozymes.

In the largest electrophoretic survey of humans yet carried out, involving 71 enzymes in Europeans, hidden genetic variation was found at about 30 percent of the loci (Table 13.1). (Rare alleles, which are found at most loci, are excluded

Table 13.1 FREQUENCIES OF MOST COMMON HOMOZYGOTES AND HETEROZYGOTES AT VARIOUS ENZYME LOCI IN EUROPEANS

Enzyme	Locus	Frequency of most common homozygote	Frequency of heterozygotes
Peptidase D	*PEP-D*	0.986	0.014
Peptidase C	*PEP-C*	0.978	0.016
Glutamic oxaloacetate transaminase	*GOT-M*	0.966	0.033
Phosphogluconate dehydrogenase	*PGD*	0.962	0.037
Alcohol dehydrogenase-2	*ADH*$_2$	0.941	0.058
Adenylate kinase	*AK*	0.920	0.077
Adenosine deaminase	*ADA*	0.897	0.100
Esterase D	*ES-D*	0.815	0.175
Phosphoglucomutase-1	*PGM*$_1$	0.573	0.367
Phosphoglucomutase-3	*PGM*$_3$	0.550	0.383
Alkaline phosphatase (placental)	*PL*	0.394	0.502
Acid phosphatase (red cell)	*ACP*$_1$	0.358	0.519
Alcohol dehydrogenase-3	*ADH*$_3$	0.352	0.482
Glutamic pyruvic transaminase	*GPT*	0.255	0.495

Proportion of polymorphic loci $\frac{8}{14} = 0.57$

Average heterozygosity   0.23

Source: H. Harris, D. A. Hopkinson, and F. B. Robson, 1974. The incidence of rare alleles determining electrophoretic variants: Data on 43 enzyme loci in man, Ann. Hum. Genet: Lond. 37: 237–53.

from consideration. Hidden genetic variation is generally considered to be present if the frequency of the most common homozygote at a locus is less than 90 percent.) Among Europeans, an average individual was found to be heterozygous at about 7 percent of the loci studied. This amount of genetic variation, while somewhat greater than that observed among mammals in general, is actually less than that found among plants, invertebrates, and even *E. coli*.

Hidden genetic variation can also be studied directly at the DNA level using nucleotide sequencing techniques or restriction enzymes. For example, the positions of restriction sites for the restriction enzyme *HpaI* that flank the β-globin gene are found to vary from chromosome to chromosome. Some *HpaI* restriction fragments containing the β-globin gene are 7.0 kilobase pairs (kb) in length, others are 7.6 kb, still others are 13.0 kb. Since the β^S allele is frequently found on the 13.0-kb fragment whereas the normal β allele rarely is, this difference can be used for detecting β^S heterozygotes or for diagnosing sickle cell anemia in fetal cells obtained by amniocentesis. Similarly, a restriction-fragment difference associated with the gene for Huntington disease has been discovered and used to localize the gene to chromosome 4. Such differences in the length of DNA restriction fragments are becoming increasingly important in human genetics.

ALLELE FREQUENCIES AND GENOTYPE FREQUENCIES

This section concerns the relationship between the frequency of individual alleles at a locus and the frequency of genotypes at the locus. Genetic variation in populations is, of course, organized into genotypes. **Genotype frequencies** (the relative proportions of the various genotypes) are determined by the frequencies of various kinds of matings, by the rate of mutation, by differences among genotypes in survival or fertility, by migration of individuals into or out of the population, and by other factors as well. Predicting genotype frequencies might therefore seem a hopeless task, but it is not. In spite of the complexities, there is a remarkably simple rule, called the **Hardy-Weinberg rule**, that can often be used to calculate the genotype frequencies at a locus. The rule works when the population is large and when the effects of mutation, differential survival or fertility, and migration are small. It also requires that there be **random mating**, which means that mates be chosen without regard to the locus in question. In the cases of many loci in humans and other organisms, these requirements are satisfactorily met; the Hardy-Weinberg rule for finding genotype frequencies is therefore of great practical utility.

The Hardy-Weinberg rule may be illustrated using the surname analogy of Chapter 3. For this purpose, imagine a population having only two surnames, Smith and Jones, and consisting of 425 individuals with the following names: 137 Smith-Smith, 207 Smith-Jones (here we include the Jones-Smiths along with the Smith-Jones for purposes of name counting), and 81 Jones-Jones. The 425 individuals represent $2 \times 425 = 850$ surnames because each individual has two surnames. Among these surnames the 137 Smith-Smiths represent $2 \times 137 = 274$ occurrences of the name Smith, and the 207 Smith-Jones represent 207 occurrences of the name Smith. Consequently, the overall frequency of the name Smith in this population is

$$\frac{2 \times 137 + 207}{2 \times 425} = \frac{274 + 207}{850} = \frac{481}{850} = 0.5659$$

Similarly, the frequency of Jones among surnames is

$$\frac{2 \times 81 + 207}{2 \times 425} = \frac{162 + 207}{850} = \frac{369}{850} = 0.4341$$

(Notice that the frequency of Smith plus the frequency of Jones equals 1.0; this is a mathematical consequence of the fact that there are just two surnames in the population.)

Of course, instead of dealing in terms of surnames we could just as well consider genotypes. For example, in one study of the MN blood group (recall from Chapter 12 that this blood group has two codominant alleles, M and N), 425 individuals were found to have the following genotypes: 137 *MM*, 207 *MN*, and 81 *NN*. Here the numbers are exactly the same as above, but we are using M in place of Smith and N in place of Jones. Consequently, the frequency of M among all alleles studied in this population is 0.5659, and the frequency of N is 0.4341. (The calculations are summarized in Table 13.2.) These numbers are called the **allele frequencies** of M and N, respectively. More generally, the allele frequency of any prescribed allele among a group of individuals is the proportion of all alleles at the locus that are of the prescribed type. Allele frequencies are sometimes called **gene frequencies** if the allele in question is clear from context.

The Hardy-Weinberg rule connects allele frequencies with genotype frequencies when there is random mating, and here again it will be convenient to invoke the Smith and Jones surname example. With random mating, the surnames in the parental generation are combined at random according to their frequencies. The consequences of random combination of surnames can be worked out using the same sort of cross-multiplication squares as were used in Chapters 3 through 5, but here the frequencies along the margins are not simply 1's and $\frac{1}{2}$'s but can be any numbers, depending on the surname frequencies.

Table 13.2   EXAMPLE OF ALLELE FREQUENCY CALCULATIONS FOR THE *MN* LOCUS

Type of observation	Sample data			
Genotype	*MM*	*MN*	*NN*	
Number of individuals	137	207	81	(total = 425)
Number of alleles at *MN* locus	274 *M*	207 *M* + 207 *N*	162 *N*	(total = 850)
Total *M* (or *N*) alleles	481 *M*		369 *N*	
	Calculation of gene frequency			
	Frequency of *M* allele = 481/850 = 0.5659			
	Frequency of *N* allele = 369/850 = 0.4341			

Father's contribution

	Smith (0.5659)	Jones (0.4341)
Smith (0.5659)	Smith-Smith $(0.5659 \times 0.5659 = 0.3202)$	Smith-Jones $(0.5659 \times 0.4341 = 0.2457)$
Jones (0.4341)	Jones-Smith $(0.4341 \times 0.5659 = 0.2457)$	Jones-Jones $(0.4341 \times 0.4341 = 0.1884)$

(Left side label: Mother's contribution)

Figure 13.2 Random mating illustrated with two surnames.

An example with the numbers calculated earlier is shown in Figure 13.2, where across the top are the surnames contributed by the males in the population and along the side are the surnames contributed by the females. The expected offspring surnames with random mating are obtained by the cross-multiplications indicated, and lumping the Smith-Jones and Jones-Smiths together, the overall frequency of unpaired surnames in the population will be $0.2457 + 0.2457 = 0.4914$. Since these numbers are frequencies (i.e., proportions), we can obtain the expected numbers among 425 individuals by multiplying each frequency by 425. The expected number of each kind of surname is therefore

Smith-Smith	$0.3202 \times 425 =$	136.1
Smith-Jones or Jones-Smith	$0.4914 \times 425 =$	208.8
Jones-Jones	$0.1884 \times 425 =$	80.1

These are the numbers that would be expected were the surnames in perfect random-mating proportions, and they are to be compared with the observed numbers 137, 207, and 81, respectively. It is apparent that the agreement between the observed and expected numbers is very good indeed. However, the Smith and Jones surnames have merely been used as stand-ins for the M and N alleles of the MN blood group, and the observed numbers are those actually obtained for MM, MN, and NN genotypes. We can therefore conclude that the population of 425 individuals has genotypes at the MN locus that are very close to those that would be expected with random mating with the allele frequencies calculated.

Although the allele frequencies of M and N in the population being discussed are 0.5659 and 0.4341, respectively, other populations might very well have different allele frequencies at this same locus. Population geneticists usually use the symbols p and q to represent allele frequencies, so we could use p to represent the allele frequency of M and q that of N. In the example being discussed, $p = 0.5659$ and $q = 0.4341$. (Note that $p + q = 1.0$, which will always be the case when there are just two alleles at a locus.) For other populations, p and q could very well be different (though $p + q$ will nevertheless equal 1.0), but the same kind of reasoning as used in Figure 13.2 will still apply. Specifically, the genotype frequencies with random mating will be as given in Figure 13.3, which is to say

Frequency of $MM = p^2$
Frequency of $MN = 2pq$ (i.e., $pq + qp$)
Frequency of $NN = q^2$

These genotype frequencies constitute the **Hardy-Weinberg rule**, which plays a prominent role in many aspects of population genetics.

As an example of the use of the Hardy-Weinberg rule, we may consider a population with allele frequencies $p = 0.6$ of M and $q = 0.4$ of N. Then the expected genotype frequencies with random mating will be

MM: $p^2 = (0.6)^2 = 0.36$
MN: $2pq = 2(0.6)(0.4) = 0.48$
NN: $q^2 = (0.4)^2 = 0.16$

Conversely, if the genotype frequencies (or numbers) are known, then allele frequencies can be calculated as before. For example, if the genotype frequencies in a population are 0.49 (for MM), 0.42 (MN), and 0.09 (NN), then the allele frequencies will be

$$p = \frac{2 \times 0.49 + 0.42}{2} = 0.7$$

$$q = \frac{2 \times 0.09 + 0.42}{2} = 0.3$$

and in this case the expected genotype frequencies will be $p^2 = (0.7)^2 = 0.49$ (for MM), $2pq = 2(0.7)(0.3) = 0.42$ (MN), and $q^2 = (0.3)^2 = 0.09$ (NN). (This is a contrived example in which genotype frequencies fit the Hardy-Weinberg rule exactly. Perfect fits will not normally occur in actual data because of chance variation in the numbers.)

Recall now that we have made certain assumptions: large population size; negligible effects of mutation, migration, and selection; and random mating. It is important to realize that random mating with respect to one trait—for example, the MN blood groups—does not mean indiscriminate mating with regard to other

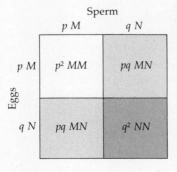

Figure 13.3 Punnett square showing genotype frequencies expected with random mating at the *MN* locus. Allele frequencies of *M* and *N* are denoted *p* and *q*, respectively.

traits. Mating can be random with respect to some traits and at the same time nonrandom with respect to others. Mating in humans is random or nearly random with respect to such traits as blood groups and allozymes, but among people in the United States who have European or African ancestry, mating tends to be nonrandom with respect to such traits as height, IQ score, and skin color. In these latter cases, individuals tend to mate with others more similar to themselves than would be expected by chance; tall individuals tend to choose tall mates, and so on. Such a tendency toward like-with-like mating is called **assortative mating**. Another principal departure from random mating is called **inbreeding**, in which there is mating between relatives such as first cousins. The consequences of inbreeding in regard to genotype frequencies will be discussed shortly.

IMPLICATIONS OF THE HARDY-WEINBERG RULE

Allele Frequencies Remain Constant

One important implication of the Hardy-Weinberg rule is that the allele frequencies stay the same from generation to generation. This can be seen from Figure 13.3, which provides genotype frequencies among offspring in terms of allele frequencies among the parents. The allele frequencies of M and N among the offspring (call them p' and q', respectively) will therefore be

$$p' = \frac{2 \times p^2 + 2pq}{2} = p(p + q) = p \text{ (because } p + q = 1)$$

$$q' = \frac{2 \times q^2 + 2pq}{2} = q(q + p) = q \text{ (because } q + p = 1)$$

That is to say, the allele frequencies *among the offspring* are the same as they were *among the parents,* which is just another way of saying that the allele frequencies remain constant. Since the allele frequencies remain the same, the genotype frequencies will also remain the same generation after generation, which is to say that the population will retain its genetic variation at the locus.

This genetic-variation-preserving quality of Mendelian inheritance seems obvious and straightforward today, but it stands in marked contrast to prevailing views of heredity in the last century, which implied that genetic variation in a population would gradually decrease unless replenished by mutation. The preservation of genetic variation is one of the most important consequences of Mendel's laws and one of the most important implications of the Hardy-Weinberg rule. Of course, we have assumed a large population size and negligible effects of mutation, migration, and differences in survival or fertility. If these assumptions are violated, then allele frequencies can indeed change from generation to generation.

Case of Complete Recessive Alleles

So far we have dealt with cases in which the alleles are codominant so that all three genotypes can be identified. When one of the alleles is recessive, then the dominant homozygotes cannot be distinguished from the heterozygotes. However, if there is random mating, then the Hardy-Weinberg rule can still be used to calculate the genotype frequencies. The allele frequency of the recessive can be obtained as follows: If R equals the frequency of the recessive homozygote, then

$$q = \sqrt{R}$$

The reason is that with random mating the frequency of the homozygote must equal q^2. But we know that this frequency equals R. Consequently it must be the case that $q^2 = R$, or, taking square roots of both sides, $q = \sqrt{R}$.

As an example, consider the Rh blood groups among the Basques. In one study, a proportion 0.575 was found to be Rh$^+$ (genotypes DD or Dd) and a proportion 0.425 (genotype dd) was found to be Rh$^-$. Since d is the recessive allele, R corresponds to 0.425. Consequently $q = \sqrt{0.425} = 0.652$ and $p = 1 - q = 1 - 0.652 = 0.348$. The expected genotype frequencies are therefore

$$DD: \quad p^2 = (0.348)^2 = 0.121$$
$$Dd: \quad 2pq = 2(0.348)(0.652) = 0.454$$
$$dd: \quad q^2 = (0.652)^2 = 0.425$$

Frequency of Heterozygotes

In Chapter 4 we emphasized that rare recessive alleles occur much more frequently in heterozygotes than in recessive homozygotes. The reason is that when q is small, $2pq$ will be larger than q^2. In the case of phenylketonuria, for example (see Chapter 9), the incidence of homozygous recessives is about $R = 0.0001$ (i.e., 1 per 10,000). Therefore $q = \sqrt{0.0001} = 0.01$ and $p = 1 - 0.01 = 0.99$. The corresponding genotype frequencies are therefore

$$\text{Homozygous normal:} \quad p^2 = (0.99)^2 = 0.9801$$
$$\text{Heterozygous:} \quad 2pq = 2(0.99)(0.01) = 0.0198$$
$$\text{Homozygous recessive:} \quad q^2 = (0.01)^2 = 0.0001$$

The frequency of carriers is consequently about one person in 50 (i.e., 1/0.0198), and there are about 198 carriers for each homozygous recessive.

X-linked Loci

The Hardy-Weinberg rule also holds for genes on the X chromosome, provided we keep separate account of males and females because males have only one X chromosome whereas females have two. The reasoning is the same as before, but the single X in males creates a slight complication. Random mating for an X-linked locus is outlined in Figure 13.4. Here we are considering two X-linked alleles, designated C and c, with respective allele frequencies p and q. When

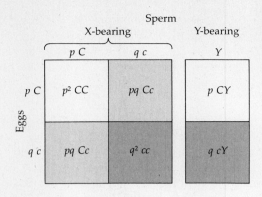

Figure 13.4 Punnett square showing genotype frequencies expected with random mating at an X-linked locus. Note that the genotype frequencies in males correspond to the allele frequencies.

mating is random, X-bearing sperm join randomly with eggs, so, *among the female offspring,* the expected genotype frequencies will be

$$CC: \quad p^2$$
$$Cc: \quad 2pq$$
$$cc: \quad q^2$$

On the other hand, males have only one X chromosome, received from their mothers, so the frequencies of CY and cY males (Y represents the Y chromosome) will be the same as the frequency of the C and the c alleles in eggs. Hence, *among male offspring,* the genotype frequencies from Figure 13.4 are

$$C: p$$
$$c: q$$

Figure 13.4 can be applied to the human Xg blood groups, which are due to a pair of alleles (designated Xg^a and Xg) on the X chromosome. Among the British population, the allele frequency of Xg^a is 0.675 (p) and that of Xg is 0.325 (q). Expected genotype frequencies are therefore

Among *females*:
Xg^a/Xg^a: $p^2 = (0.675)^2 = 0.456$
Xg^a/Xg: $2pq = 2(0.675)(0.325) = 0.439$
Xg/Xg: $q^2 = (0.325)^2 = 0.105$

Among *males*:
Xg^a: $p = 0.675$
Xg: $q = 0.325$

In Chapter 5 we emphasized that traits caused by rare X-linked recessive alleles will be more common in males than females and stated the rule

Freq of affected females = (Freq of affected males)^2

This rule comes directly from Figure 13.4, since the frequency of affected males is q and that of affected females is q^2. We can carry calculations one step further and arrive at another useful conclusion. For an X-linked allele the frequency of carrier females will be $2pq$. If the recessive allele is rare, however, q will be close to 0 and p will be close to 1. Thus $2pq$, the frequency of carriers, will be approximately $2(1.0)q = 2q$, or, in words, *for a rare X-linked allele, the frequency of carrier females will be approximately equal to twice the frequency of affected males*. For example, the frequency of the X-linked green form of color blindness among European males is about 0.05, which corresponds to q. The approximate frequency of carrier females would then be $2 \times 0.05 = 0.10$. The exact frequency of carriers in this case is $2(0.95)(0.05) = 0.095$, so the rough approximation is quite good even though this form of color blindness is not particularly rare.

DIFFERENTIATION OF POPULATIONS: RACE

Populations spread over a wide geographic area rarely undergo random mating over their entire range, if for no other reason than that all organisms are limited in their ability to disperse. Consequently, most populations are effectively split up into **local populations** consisting of individuals living within a sufficiently restricted geographical area that matings usually occur within the group. Within-group mating preferences need not be based on geography, however. In humans, cultural similarity is important, and individuals tend to mate with others of the same or a similar cultural background. Although there is some intermating and gene exchange among all large human groups, the entire human population cannot be considered as a single, large, randomly mating entity. The limitation of the exchange of genes among populations is known as **genetic isolation**. Genetic isolation among most human populations is by no means complete, but partial isolation does occur because of the tendency for within-group matings.

Over the course of hundreds of thousands of years, and especially in the distant past, when populations were small and migration was more difficult than it is at present, isolated or semi-isolated human populations came to have different allele frequencies. Some differences in allele frequency probably arose by chance, as a result of historical accident. Allele frequencies can change by chance from generation to generation, and small populations are especially susceptible to such effects. Other differences in allele frequency probably occurred as each population became adapted to its own physical environment; genotypes better adapted to a particular environment survive and reproduce more successfully, and their alleles come to be represented in greater frequency in future generations. Although all humans share essentially the same chromosomal complement, the frequency of individual alleles at many loci varies from population to population.

By way of analogy, the human gene pool is composed of a number of smaller pools—puddles—like the water puddles on a gravel street after a heavy rain, all interconnected in tiny rivulets through which material flows from one to the next. If one examines the puddles closely, one sees that they are not identical. Some are large, some small; some have more rotifers, nematodes, bacteria, or algae than others. But the differences are quantitative, not qualitative. The puddles are distinguished by how much or how little they have of each component and not by whether or not they possess the constituent. No puddle is completely isolated, so whatever one puddle has in abundance travels through the rivulets to the others. Because of this structure, the puddles form one large, interconnected unit, but a unit with local differences. The puddles in the analogy represent **races**—groups whose allele frequencies differ from those of other groups.

How many human races are to be distinguished is a matter of where one draws the lines. The purpose of distinguishing human races genetically is to trace the origin and history of the human population and to anticipate its evolutionary future. To draw the line at differences in allele frequency that are very small is simply not useful; the people in virtually every city or town of every country would then represent a distinct race, because allele frequencies do vary even within a single country. In practice, an arbitrary judgment must be made about how different two sets of allele frequencies must be to consider them suffciently distinct to define separate races. Where the line is drawn depends on the particular application, on whether large groups or small ones are to be considered. Anthropologists, using various physical and cultural criteria, have defined the number of races as anywhere from three—usually called Caucasoid, Negroid, and Mongoloid—to more than 30. A few anthropologists deny that the concept of race has any relevance at all, quite apart from the danger of misuse of the concept (such as in the nonsensical idea of "racial purity"). In any case, there is no "correct" number of human races. The number depends on how finely one makes distinctions and on the purpose, which makes the concept of "race" an arbitrary but sometimes useful tool—nothing more.

Genetic distinctions among human populations must be based on allele frequencies at many loci simultaneously. Some loci have allele frequencies that are virtually the same in all populations; others have allele frequencies that are somewhat different in all populations. Genetic distinctions must be based on the whole constellation of allele frequencies; no one locus or small group of loci gives an adequate picture. (This is why skin color is not an acceptable trait for distinguishing human populations; it puts too much emphasis on the one small group of loci that determines skin pigmentation and ignores everything else; it also gives a false impression of the amount of genetic differentiation among populations.)

The concept of race as a constellation of allele frequencies is illustrated with several blood group loci in three large human groups in Figure 13.5. Allele frequencies at some loci (e.g., MN, Lutheran) do not vary a great deal among populations, whereas at other loci (e.g., P, Rh, Lewis) the variation is substantial.

A convenient quantitative measure of genetic differentiation among populations is called the **fixation index** (conventionally represented with the symbol F). The fixation index is calculated from the allele frequencies among a group of populations, and it is a measure of how different the allele frequencies are. If all populations have the same allele frequencies, then $F = 0$. As the allele frequencies become more different, the value of F increases and reaches a maximum of 1.0 when each population is fixed for a different allele. (An allele that is **fixed** is

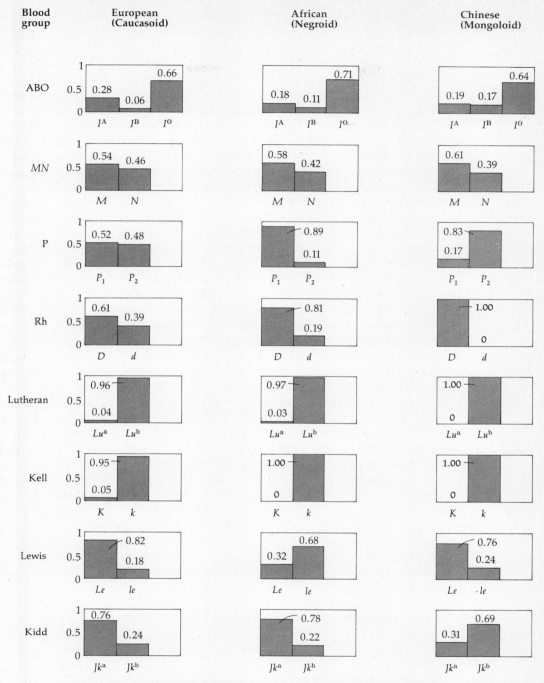

Figure 13.5 Allele frequencies at eight blood group loci in three large human populations.

an allele with a frequency of 1.0.) Generally speaking, a value of F between 0 and 0.05 indicates little genetic differentiation among the populations, F between 0.05 and 0.15 indicates moderate differentiation, and F above 0.15 indicates great or very great differentiation. For the MN locus in Figure 13.5, for exmaple, $F = 0.003$, which reflects the similarity of allele frequencies in the populations. For the P locus, on the other hand, where the allele frequencies are very different, $F = 0.35$, which corresponds to very great genetic differentiation at this locus.

No one locus adequately portrays genetic differentiation among populations, of course, because some loci (e.g., MN) will have little differentiation whereas others (e.g., P) will have much. An average value of F for many loci is therefore necessary to gain a reliable picture, and relevant data are summarized in Table 13.3. The upper row pertains to the three major races (Caucasoid, Negroid, and Mongoloid), and it indicates an average fixation index of 0.069, corresponding to moderate amounts of differentiation (i.e., allele-frequency differences). The second row pertains to the native Yanomama Indians of Venezuela and Brazil—an exceptionally well-studied group from the standpoint of both anthropology and genetics. Here again the picture is similar. The fixation index among populations (in this case Yanomama villages) is 0.077, indicating moderate genetic differentiation. To summarize the situation in a slightly different way, we could say that there is so much genetic diversity *within* human populations that the genetic differences *among* populations are relatively small in comparison.

The modern concept of race is based on how the human gene pool is structured. This is quite unlike the traditional view, which was based on a few easily identifiable traits such as skin color, facial features, and body build, and which used these traits to define certain human "types"—a Caucasoid type, a Negroid type, a Mongoloid type, and so on. The traditional concept of race does not correspond to anything real in terms of genetic characteristics. There is no such thing as a "perfect specimen" of a Caucasian, a black, or anything else. There is tremendous genetic diversity *within* races, and to define arbitrarily a perfect "type" to which everyone in a group should conform is to imagine a genetic uniformity that does not exist. Because the traditional concept of race was mixed up with the fanciful notion of a "pure race"—one with little or no genetic variation—the concept led to fearful imaginings of the biological consequences of interracial mating, or "mongrelization" of the races. The modern viewpoint emphasizes similarities among races as much as it recognizes the differences in allele frequencies that occur. The modern view also emphasizes that genetic variation *within* races is so enormous that it all but swamps genetic differences *among* races.

Table 13.3 **FIXATION INDEX AMONG VARIOUS SUBPOPULATIONS**

	Number of populations	Number of loci	Fixation index
Major races	3	35	0.069
Yanomama Indians	37	15	0.077

Source: M. Nei, 1975, Molecular Population Genetics and Evolution, American Elsevier, New York.

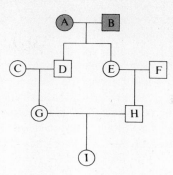

Figure 13.6 Pedigree showing a mating between first cousins (G and H). The individual I is the result of inbreeding, as A and B (shaded) are common ancestors of G and H.

INBREEDING

Inbreeding refers to matings between relatives. In Chapter 4 we used the term **consanguineous matings** to refer to matings between relatives, so the two terms— *consanguineous mating* and *inbreeding*—can be used interchangeably. The most common form of inbreeding in most human populations is mating between first cousins, designated by the letters G and H in Figure 13.6. Individual I in the figure, the offspring of the consanguineous mating between G and H, is said to be **inbred,** and a characteristic feature of all forms of inbreeding is the closed loop that occurs in the pedigree. At the top of Figure 13.6 are two individuals (A and B) who are called **common ancestors** of G and H. One of the principal consequences of inbreeding is that a particular allele in one of the common ancestors can be transmitted to both of the consanguineous partners, and their mating can thus lead to this allele becoming homozygous in the inbred offspring.

The genetics involved in inbreeding can again be illustrated using surnames, and the considerations are outlined in Figure 13.7. In this case one of the common

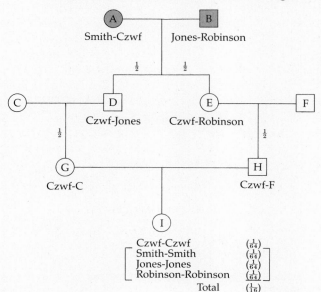

Figure 13.7 The principal consequence of inbreeding is that an allele (here represented as a surname) in a common ancestor can be transmitted down both sides of the pedigree and become homozygous in the inbred individual.

ancestors (A) has a rare surname (Czwf), but this surname, like an allele at a locus, can be transmitted down both sides of the pedigree and come together in the inbred individual I. Because of the inbreeding, individual I could have the name Czwf-Czwf, although the name might be so rare in the population that Czwf-Czwfs would virtually never be formed by random mating. The Czwf surnames in individual I are not only identical, they are identical in a very special way—namely, by being transmitted from the same common ancestor. In a genetic context, identical alleles that have been transmitted from a common ancestor are said to be **identical by descent**, because they arise by DNA replication of the same ancestral allele.

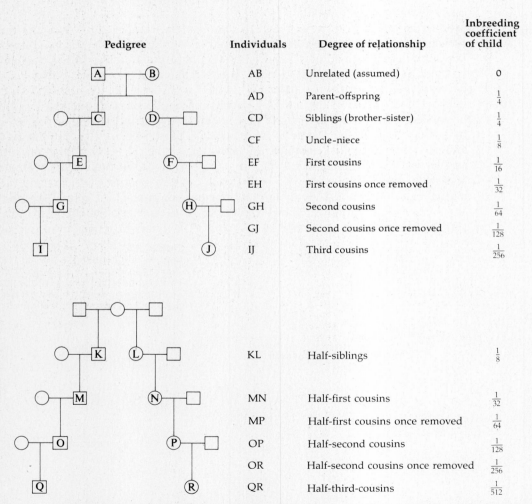

Pedigree	Individuals	Degree of relationship	Inbreeding coefficient of child
	AB	Unrelated (assumed)	0
	AD	Parent-offspring	$\frac{1}{4}$
	CD	Siblings (brother-sister)	$\frac{1}{4}$
	CF	Uncle-niece	$\frac{1}{8}$
	EF	First cousins	$\frac{1}{16}$
	EH	First cousins once removed	$\frac{1}{32}$
	GH	Second cousins	$\frac{1}{64}$
	GJ	Second cousins once removed	$\frac{1}{128}$
	IJ	Third cousins	$\frac{1}{256}$
	KL	Half-siblings	$\frac{1}{8}$
	MN	Half-first cousins	$\frac{1}{32}$
	MP	Half-first cousins once removed	$\frac{1}{64}$
	OP	Half-second cousins	$\frac{1}{128}$
	OR	Half-second cousins once removed	$\frac{1}{256}$
	QR	Half-third-cousins	$\frac{1}{512}$

Figure 13.8 Inbreeding coefficients of potential offspring of matings between relatives who have the pedigree relationships shown.

The effects of inbreeding are measured in terms of the **inbreeding coefficient** (conventionally designated f), which is the probability that two alleles in an inbred individual are identical by descent. In the case of Figure 13.7 (a first-cousin mating), the probability of individual I being named Czwf-Czwf is $\frac{1}{64}$, and these surnames are identical by descent. This probability can be calculated from simple Mendelian considerations. The probability that D receives the name Czwf from A is $\frac{1}{2}$. Once D has the name, the probability that G receives it from D is also $\frac{1}{2}$. Once G has the name, the probability that I receives it from G is also $\frac{1}{2}$. The same reasoning applies to the other side of the pedigree as well. The probability that E receives Czwf from A is $\frac{1}{2}$. Once E has the name, the probability that H receives it from E is $\frac{1}{2}$. Once H has the name, the probability that I receives it from H is $\frac{1}{2}$. Consequently, for I to have the surname Czwf-Czwf, all six of these transmission events must occur, so the overall probability that I will be named Czwf-Czwf is $\frac{1}{2} \times \frac{1}{2} \times \frac{1}{2} \times \frac{1}{2} \times \frac{1}{2} \times \frac{1}{2} = (\frac{1}{2})^6 = \frac{1}{64}$.

However, individual I could as well have been named Smith-Smith, Jones-Jones, or Robinson-Robinson, and these names, too, would be identical by descent. Each of these other possibilities also has probability $\frac{1}{64}$, so the overall probability of identity by descent in individual I is $\frac{1}{64}$ (Czwf-Czwf) + $\frac{1}{64}$ (Smith-Smith) + $\frac{1}{64}$ (Jones-Jones) + $\frac{1}{64}$ (Robinson-Robinson) = $\frac{1}{16}$. Consequently, $\frac{1}{16}$ is the inbreeding coefficient of the offspring of a first-cousin mating. Similar considerations apply to other types of pedigrees as well, and the inbreeding coefficients of the offspring of various other types of consanguineous matings are summarized in Figure 13.8.

When there is inbreeding, the genotype frequencies in a population are no longer given by the Hardy-Weinberg rule because of the increased homozygosity brought about by inbred individuals having alleles that are identical by descent. Genotype frequencies with inbreeding are summarized in Table 13.4. The parts of the formulas involving $1 - f$ pertain to individuals who, in spite of the inbreeding, have alleles that are not identical by descent, and within this class of individuals, the genotype frequencies are the familiar p^2, $2pq$, and q^2. The parts of the formulas involving f pertain to inbred individuals whose alleles are identical by descent, and it is to be noted that the effect of this portion is to increase the frequency of homozygous genotypes at the expense of heterozygous genotypes.

To illustrate use of the formulas, we may calculate the genotype frequencies at the phenylketonuria locus among the offspring of first cousins. In a previous example we have calculated $p = 0.99$ and $q = 0.01$ for this locus, and for the

Table 13.4 GENOTYPE FREQUENCIES IN INBRED POPULATIONS

Genotype	Frequency in inbred population	Frequency with random mating ($f = 0$)
AA	$p^2(1 - f) + pf$	p^2
Aa	$2pq(1 - f)$	$2pq$
aa	$q^2(1 - f) + qf$	q^2

offspring of first cousins, $f = \frac{1}{16} = 0.0625$. The quantity $1 - f$ is therefore 0.9375, and the required genotype frequencies are

$$AA:\ p^2(1 - f) + pf = (0.99)^2(0.9375) + (0.99)(0.0625)$$
$$= 0.9188 + 0.0619 = 0.9807$$
$$Aa:\ 2pq(1 - f) = 2(0.99)(0.01)(0.9375) = 0.0186$$
$$aa:\ q^2(1 - f) + qf = (0.01)^2(0.9375) + (0.01)(0.0625)$$
$$= 0.0001 + 0.0006 = 0.0007$$

These numbers are to be compared with the random-mating frequencies calculated earlier (i.e., 0.9801, 0.0198, and 0.0001 for AA, Aa, and aa). Note that the genotype frequencies among the normal homozygotes and heterozygotes are not strikingly different than they would be with random mating. The large discrepancy occurs among the homozygous recessives, whose frequency increases from 0.0001 with random mating to 0.0007 among the offspring of first cousins. This increase represents a sevenfold greater risk of phenylketonuria homozygotes among the offspring of first cousins, and it comes about because of the identity by descent of alleles that can occur in such pedigrees.

The phenylketonuria example illustrates that the principal effect of inbreeding is to increase the frequency of rare homozygous recessive genotypes. Consequently, although first-cousin matings account for only a minority of all matings (about 1 percent of all matings in the United States are between first cousins; the corresponding figure for Japan is 6 percent), first-cousin matings can nevertheless account for a disproportionate share of matings that give rise to rare homozygous

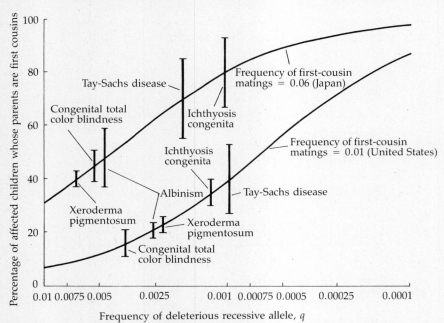

Figure 13.9 Proportion of homozygous recessive individuals who have first-cousin parents in Japan (upper curve) and in the United States (lower curve).

recessives. This aspect of inbreeding is illustrated in Figure 13.9 with data from the United States and Japan. Taking Tay-Sachs disease in the United States as an example, although first-cousin matings account for 1 percent of all matings, they account for almost 40 percent of matings that give rise to Tay-Sachs off-spring. Inbreeding thus exerts its most important effect by bringing rare alleles together in the inbred offspring—alleles that are identical by descent with an allele in a common ancestor—just as inbreeding in a pedigree like that of Figure 13.7 can give rise to an offspring named Czwf-Czwf.

Most human societies discourage or prohibit marriages between close relatives. That this might be due to the generally harmful effects of close inbreeding seems doubtful. More than likely, the marriage prohibitions were based on sociological considerations, although perhaps reinforced by the ill effects of inbreeding that were sometimes observed in the offspring of close relatives. Possibly the harmful effects of inbreeding were first noted among domesticated animals, which are usually monitored carefully. Indeed, William Penn observed that people ''are generally more careful of the breed of their horses and dogs than of their own children.''*

On the other hand, some societies, such as the ancient Egyptians and the Incas, encouraged marriages between brothers and sisters of the reigning dynasty, partly for political purposes and partly in the belief that ''royal blood'' was fit to mix only with ''royal blood.'' Figure 13.10 illustrates the intense inbreeding that occurred in the Ptolemaic dynasty in Egypt in the centuries just before the birth of Christ. The inbreeding coefficient of Cleopatra-Berenike III is 0.427, which is almost as large as would be obtained in four consecutive generations of brother-sister mating. Cleopatra-Berenike III, by the way, was an aunt of the famous Cleopatra of Antony and Cleopatra.

Although the types of matings that occur in a population do not generally change allele frequency, allele frequencies can change as a result of mutation, migration, differences in ability to survive or reproduce among the genotypes, and other factors as well. Such changes in allele frequency are one aspect of the evolutionary process, and the causes of allele-frequency change therefore warrant consideration.

MUTATION

The ultimate and original source of all genetic variation is mutation, the variety and types of which were discussed in Chapter 10. *New alleles arise only by mutation.* Of course, a population may receive a new allele by migration, but this simply means that the original mutation occurred in another population somewhere else at an earlier time.

Notwithstanding its importance in creating new genetic variation, mutation by itself is a decidedly weak evolutionary force. Imagine a large population that is fixed for some allele, say A, at a particular locus. Suppose that in every generation some fraction of the A alleles mutates to a. Although mutation rates

---

* William Penn, from *Some Fruits of Solitude.*

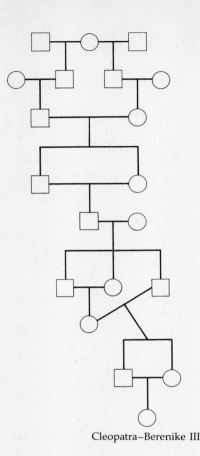

Cleopatra–Berenike III

vary from locus to locus, a rate of mutation from A to a of 10^{-5} (i.e., 0.00001) may be taken as a realistic example. In the absence of evolutionary forces other than mutation, what will happen as time goes on is that in every generation the frequency of the A allele will be reduced somewhat because a few more A alleles will have mutated to a. Therefore the a allele steadily accumulates in the population, but the magnitude of the effect is very small. Starting with a population in which A is fixed, in the next generation the frequency of the A allele will be 0.99999, in the generation after that 0.99998, then 0.99997, and so on, decreasing very gradually. After 40 generations the frequency of A will be 0.9996; after 80 generations it will be 0.9992. Even after 8,000 generations—which corresponds in humans to about 200,000 years—the frequency of A will still be 0.92. (See Figure 13.11.) Nevertheless, as time goes on, the frequency of A will continue to decrease very slowly until a comes to predominate. (The frequency of A will be 0.50 after 69,315 generations; thereafter a will become more frequent than A.)

As the a allele becomes more frequent, however, reverse mutation from a to A will become increasingly important as an evolutionary force because there will be a greater number of a alleles available for mutation; the a allele therefore

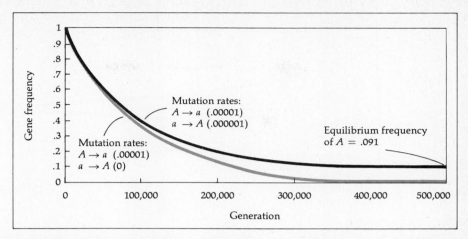

Figure 13.11 Mutation is capable of changing gene frequencies, but the force is very weak. Depicted here are changes in the frequency of an allele (A) when mutation is the only evolutionary force at work. The gray curve shows how the frequency of A changes when mutation from A to a occurs at a rate of .00001 but when reverse mutation (from a to A) does not occur; eventually the frequency of A goes to zero (all the alleles in the population will then be a), but the length of time involved is enormous—extending to hundreds of thousands of generations. The black curve shows what happens when mutation of A to a is opposed by reverse mutation of a to A. (Here the mutation rate of a to A is assumed to be ten times smaller than that of A to a.) Eventually the gene frequencies come into a balance, an equilibrium, at which the number of A alleles that mutate to a in any generation is exactly balanced by the number of a alleles that mutate to A.

cannot completely replace A. Over a very long time, the A and a alleles will reach frequencies from which they will no longer change in the absence of other forces because the number of A alleles lost by mutation of A to a in every generation will be exactly counterbalanced by the number of new A alleles that arise by mutation of a to A. At this stage the allele frequencies are said to be in **equilibrium**. In the example in Figure 13.11, in which the forward (A to a) mutation rate is 10^{-5} and the reverse (a to A) mutation rate is 10^{-6}, the equilibrium frequencies of A and a are 0.091 and 0.909, respectively, although this equilibrium requires a long time to be achieved because the force of mutation on allele frequencies is very weak. (In general, the ratio of the equilibrium allele frequencies of A and a will equal the ratio of the reverse to forward mutation rates, and the equilibrium frequencies will not depend on the allele frequencies in the starting population.)

MIGRATION

Migration refers to the movement of individuals among populations. Since migrants can bring new alleles into a population, the effects of migration are similar to those of mutation. However, the effects of migration on allele frequency are generally much greater than those of mutation because migration rates are typically much greater than mutation rates. Figure 13.12 provides an example of migration between two populations, one starting with an allele frequency of 0.75

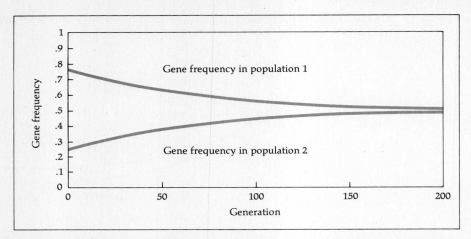

Figure 13.12   The effect of migration is to equalize the gene frequencies in the populations involved. Population 1 starts with an initial frequency of a given allele of .75; population 2 starts with a frequency of .25. The curves show the changes in gene frequency over time when the amount of migration between the two populations is 1 percent. The gene frequencies in both populations converge toward .5, and the change is relatively rapid. (After 100 generations the frequencies are .53 and .47.)

and the other with 0.25. In each generation, 1 percent of the alleles in each population derives from individuals in the other population (so the migration rate is 1 percent per generation). After one generation of migration, the allele frequency in population 1 will be $(0.75 \times 0.99) + (0.25 \times 0.01) = 0.745$. The first term in this expression accounts for the nonmigrants, and the second term accounts for the migrants. Similarly, the allele frequency in population 2 after one generation will be $(0.25 \times 0.99) + (0.75 \times 0.01) = 0.255$. Note that the allele frequency in population 1 has decreased, whereas that in population 2 has increased. The populations therefore become progressively more similar.

Calculations like those above can be carried out on a generation-by-generation basis (most easily with a computer or a programmable calculator), and these are the results plotted in Figure 13.12. After 40 generations, which corresponds to about 1000 years in humans, the allele frequencies are 0.61 and 0.39; after 80 generations they are 0.55 and 0.45. Of course, the speed at which populations become similar because of migration depends on the migration rate. In Figure 13.12 the difference in allele frequency between the two populations decreases by 2 percent each generation. With a greater rate of migration the allele frequencies would approach equality more rapidly. In an extreme case in which 50 percent of the two populations migrate every generation, the allele frequencies would become equal after only one generation.

Migration in humans is much more complex than the example above would indicate. In the first place, although humans tend to be clustered into groups in towns and cities, there are few habitable areas completely devoid of people; this sometimes makes it difficult to say precisely where one ''population'' ends and the other one begins (where would you draw the distinction between Los Angeles and San Diego, or between Tokyo and Yokohama?). Moreover, men may migrate more than women in one instance, or women more than men in another; and people in their middle years are more likely to migrate than the very young or the very old. (In the massive immigration into the United States between 1901 and 1910, for example, 70 percent of the immigrants were men and 83 percent were in the early adult ages, but over the last several decades more women than men have migrated to the United States.) In addition, migration is partly determined by local population density and social and economic conditions; people have their own reasons for migrating, but they seldom do it for fun. Then, too, most populations do not receive immigrants from only one nearby population; immigrants usually come from many populations, some near and some far away, although the rate of migration certainly falls off with distance. Even "distance" between populations is hard to define because humans do not usually travel "as the crow flies" but by roads and waterways.

Another major complication is that patterns and rates of migration are not constant from generation to generation; they change with time. The flood of immigrants that entered the United States in 1901 to 1910—nearly nine million people in all, over 90 percent from Europe—was by far the largest number ever for a comparable period of time. To complicate the issue further, immigrants may retain their own cultural identity and marry assortatively within their own groups for a number of generations. Thus a completely realistic picture of human migration is difficult to achieve, and to assess the exact magnitude of the evolutionary effect of migration as compared to other evolutionary forces is impossible at the present time. Nevertheless, despite the complications, it remains true in sum that *migration tends to make populations genetically more similar than they would otherwise be.*

SELECTION

Evolution refers to the processes whereby populations become progressively more adapted to survival and reproduction in their environment. Fundamentally, evolution can occur because certain genotypes are better able to survive and reproduce than are others, and these favored genotypes contribute more offspring to succeeding generations. The favored alleles in these genotypes thus increase in allele frequency: From a genetic point of view, evolution can be looked upon as changes in the allele frequency of favored alleles.

The relationship between evolution and genotypic differences in survival and reproduction among individuals was first emphasized by Charles Darwin in his monumental book *The Origin of Species,* first published in 1859, in which he argued that the genetic characteristics of populations are molded by external forces of the environment acting on preexisting hereditary variants that occur among members of the population. (Similar ideas were proposed by Alfred Russel Wallace at about the same time.)

Darwin called his theory of the mechanism of evolutionary change **natural selection**, and the idea rests on three observations:

1. All populations produce more young than can possibly survive and reproduce.
2. Since not all individuals are equally likely to survive and reproduce, there will be phenotypic variation in this ability from individual to individual.
3. At least some of the phenotypic variation in ability to survive and reproduce is inherited; that is, it is due to differences in genotype among individuals.

Now, the argument continues, since there will certainly be variability in the ability of individuals to survive and reproduce, those most able to survive and reproduce in a given environment will do so, thereby contributing their genes disproportionately to individuals of the next generation. Since part of the parents' ability to survive and reproduce is heritable, their offspring will receive the beneficial genes. Thus, genes that enhance the ability of an organism to survive and reproduce in its environment will increase in frequency in the population as time goes on. The population thereby becomes progressively more adapted to its environment because of the action of natural selection. A key element in natural selection is the environment, and perhaps it would be more precise to say that a population becomes adapted to its environment because the environment selects the adaptive traits. The arguments leading to the theory of natural selection seem rather obvious to biologists today, even self-evident. But Darwin perceived them in the middle of a century when most people disputed the very occurrence of evolution, let alone worried about its mechanism.

In population genetics, the ability of a genotype to survive and reproduce is called the genotype's **fitness**, and fitness is usually expressed in relative terms. For example, if one genotype's fitness is 5 percent smaller than another's, then this means that the first genotype leaves, on the average, 0.95 offspring for every offspring left by the second genotype. Once the fitnesses of all relevant genotypes are known, then it becomes possible to calculate the allele frequencies in successive generations by taking the fitnesses into account, although the details of such calculations will not be necessary for our purposes.

Figure 13.13 shows three examples of such calculations involving two alleles at an autosomal locus. In each case the disfavored homozygote has a fitness 5 percent less than that of the favored homozygote, and the three cases pertain to a favored dominant allele (black curve), a favored recessive allele (gray curve), and additive alleles. (**Additive** alleles are alleles in which the fitness of the heterozygote is exactly the average of the fitnesses of the homozygotes.) In all cases, of course, the favored allele increases in frequency until it ultimately becomes fixed. The curve for a favored dominant has an important feature that warrants emphasis, however. Note that when the favored dominant is common (i.e., frequency above about 0.80), further changes in its allele frequency become slow. To say exactly the same thing in another way, selection against a *disfavored*

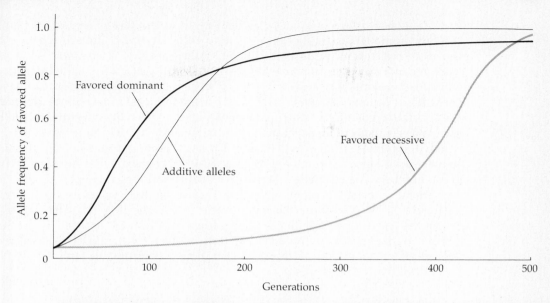

Figure 13.13 Pattern of increase in the allele frequency of a favored dominant (thick black line), a favored recessive (thick gray line), and an additive (thin black line) allele. In each case selection against the disfavored homozygote is 5 percent and the initial frequency of the favored allele is 0.05.

recessive allele will be inefficient when the recessive allele is rare. This inefficiency occurs because when a recessive allele is rare, most of the alleles will be found in heterozygotes, and these are equal in fitness to the normal homozygotes.

Indeed, selection against a rare recessive is inefficient whether the recessive homozygotes are mildly or severely affected. For this reason, eugenic proposals to sterilize those affected with rare recessive diseases are misguided. Such individuals have a low fitness to begin with, and eliminating their reproduction altogether will have little effect on allele frequency because selection against rare recessives is so inefficient anyhow. The reservoir of rare harmful recessives is not in the affected homozygotes; it is in the phenotypically normal heterozygotes.

Figure 13.13 typifies those cases of selection in which the fitness of the heterozygote equals that of one of the homozygotes or is between the fitnesses of the homozygotes. An important additional case occurs when the fitness of the heterozygote is greater than that of the homozygotes. This case is known as **overdominance**, and a characteristic of overdominance is that selection will not ultimately lead to the elimination of one of the alleles. Since selection against one homozygote will tend to decrease the frequency of a particular allele, whereas selection against the other homozygote will tend to increase it, overdominance leads to an equilibrium situation in which both alleles are maintained in the population.

Although overdominance appears to be relatively rare in populations, one well-established case in humans concerns sickle cell anemia. Homozygotes for the sickle-cell hemoglobin allele, β^S/β^S, are severely anemic. The condition is essentially lethal since most affected individuals do not reproduce, and the fitness of β^S/β^S homozygotes is therefore close to zero. However, heterozygotes, β^S/β, are more resistant to malarial infections by the parasite *Plasmodium falciparum* than are normal homozygotes β/β. Although the occurrence of mild malarial infections is about the same in β/β and β^S/β genotypes, the occurrence of severe malarial infections is about twice as high among β/β homozygotes as among β^S/β heterozygotes. In some African populations the overall survival of β/β genotypes is reduced about 15 percent as compared with β/β^S heterozygotes. The relative survivals of β^S/β^S, β^S/β, and β/β are therefore about 0, 1.0, and 0.85, respectively. Since the genotypes β^S/β and β/β are approximately equal in fertility, these numbers correspond to the relative fitnesses of the genotypes. Generation-by-generation calculations of allele frequency using these fitnesses are presented in Figure 13.14. Whatever the initial allele frequency of β^S may be, the allele frequency moves very rapidly toward a frequency of 0.13, and this is the equilibrium frequency. A β^S frequency of 0.13 is close to the average actually observed over most of west Africa, although there is considerable variation in allele

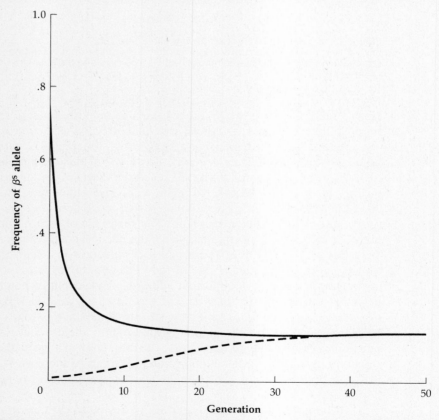

Figure 13.14 Expected change in frequency of β^S allele in populations starting at different frequencies. With the fitnesses given in the text, all populations eventually attain a frequency of β^S of 0.13, which is close to the actual frequency found in parts of West Africa.

frequency among populations, ranging from about 0.05 to 0.15. (For a map, see Figure 4.4 in Chapter 4.)

MUTATION-SELECTION BALANCE

In a large population, selection cannot completely eliminate all harmful alleles, because in each generation some new harmful alleles arise by mutation. Eventually there is established an equilibrium in which the number of old harmful alleles eliminated in each generation by selection is exactly counterbalanced by the number of new harmful alleles introduced into the population by mutation.

A mechanical analogy of this balance between mutation and selection is illustrated in Figure 13.15, in which the spring pulling up on the weight represents mutation and gravity pulling down on the weight represents selection. (For realistic values of mutation rates and selection, the equilibrium frequency of the harmful recessive will usually be smaller than 0.01.) The mechanical analogy is adequate in implying that an increase in the mutation rate will increase the equilibrium frequency of the harmful allele, or that an increase in selection against genotypes carrying the harmful allele will decrease the equilibrium frequency. The analogy is inadequate in that it is not quantitative. On a scale such as the one shown in Figure 13.15, the response of the pointer to changes in the weight would be virtually instantaneous. With mutation-selection balance, response to changes in mutation or selection are extremely slow—requiring thousands of generations—particularly with regard to rare recessive alleles, against which selection is inefficient in any case.

The analogy also fails to take into account possible effects of the harmful

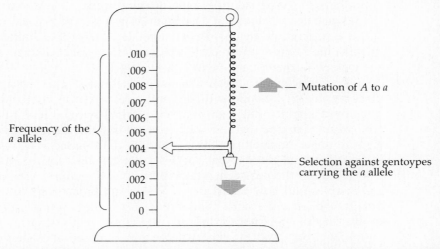

Figure 13.15 Mechanical analogy showing how the force of selection against an allele (weight pulling down) is opposed by mutation (spring pulling up), eventually leading to an equilibrium gene frequency indicated on the scale by the pointer. The scale of gene frequencies runs only from 0 to .010 because the force of mutation is generally too weak to maintain equilibrium gene frequencies above .01.

allele on the fitness of heterozygotes. In general, because harmful alleles are rare, there will be many more heterozygotes than homozygotes, and a small amount of selection against heterozygotes will therefore have a disproportionately large effect on allele frequency, because the small amount of selection is spread over so many individuals. For example, the equilibrium in Figure 13.15 (allele frequency 0.004) would occur in the case of a completely recessive allele causing a 50 percent reduction in fitness of homozygotes and in which the mutation rate to the recessive allele is 10^{-5}. Rendering the homozygotes sterile or lethal would have little effect on the equilibrium frequency; if the recessive homozygotes had a fitness of 0, the equilibrium frequency would be 0.003. However, a small heterozygous effect of the "recessive" allele will bring about a major change in the equilibrium frequency. For example, if the allele in Figure 13.15 produced a 1-percent reduction in the fitness of the heterozygotes, its equilibrium frequency would be reduced from 0.004 to 0.001. The take-home lesson is that a small amount of selection against the carriers of harmful recessives can have a major effect on the equilibrium allele frequency.

RANDOM GENETIC DRIFT

So far we have considered three evolutionary forces that can change allele frequency—mutation, migration, and selection. The fourth principal force for changing allele frequency is **random genetic drift**, which refers to chance changes in allele frequency that occur in populations because of their limited size. Unlike mutation, migration, and selection, in which allele-frequency changes tend to occur in one direction, random genetic drift causes random changes; in some generations the allele frequency may increase, in others it may decrease, and in still others it may remain the same. Random genetic drift comes about because not all gametes produced in a population give rise to zygotes and so appear in the next generation. Indeed, from among the potentially infinite pool of gametes, only the lucky ones appear in the next generation, and the smaller the population, the luckier a gamete must be to be successful.

The principal features of random genetic drift can perhaps best be made clear by analogy. Suppose there were a large barrel containing an enormous number of marbles (the gametes), exactly half red and half green, and all very well mixed. Imagine that one were to reach into this barrel and, without looking, remove some number (an even number). This process represents successful fertilization, and the chosen marbles represent the lucky gametes that will be transmitted to the next generation. With such a random withdrawal, how many red marbles and how many green ones would be chosen? In particular, would one draw exactly the same number of reds as greens? As it happens, the chance of drawing an equal number of reds and greens is not large. If two marbles are withdrawn, only half the time will there be one red and one green; if four are withdrawn, only 37.5 percent of the time will there be two red and two green; if ten are withdrawn, only 24.6 percent of the time will there be five red and five green; and if 50 are withdrawn, only 11.2 percent of the time will there by 25 of each color. In short, the larger the number withdrawn, the smaller will be the chance that exactly half will be of each color. So, too, with random genetic drift—the larger the population size, the smaller the chance that allele frequencies in one generation will be *exactly* the same as they were the generation before.

On the other hand, the larger the number of marbles withdrawn, the smaller will be the average deviation from exact equality. Some 96 percent of withdrawals of 50 marbles will result in a proportion of reds between 0.36 and 0.64; 96 percent of withdrawals of 100 will result in a proportion between 0.40 and 0.60; and 96 percent of withdrawals of 1000 will result in a proportion between 0.47 and 0.53. That is to say, the larger the population, the smaller will be the random fluctuations in allele frequency from generation to generation because of random genetic drift. Of course, the same sort of random fluctuations would occur with any other proportion of red and green marbles, and these proportions represent the allele frequencies in the original population.

Random genetic drift is therefore especially important in small populations, "small" in this context meaning populations of size 100 or less. In small populations allele frequencies will fluctuate randomly from generation to generation, and a particular allele, after a series of lucky increases in frequency, could very well become fixed. On the other hand, this very same allele could as well have undergone a series of unlucky decreases in frequency and have become lost. Large populations undergo random genetic drift too, but, as noted above, the magnitude of the changes in allele frequency in large populations is small compared to those that occur in small populations.

It might seem on first consideration that the human population is too large for random genetic drift to have had appreciable effects on allele frequency, but further consideration will revise this perspective. The human population is not a single large random-mating unit but is split up into smaller units based on geographical proximity, cultural similarity, and other factors, and matings usually occur within these smaller units. These smaller mating units, corresponding to local populations, are the evolving populations, and random genetic drift will exert an effect because of their small size. In addition, random genetic drift must have had significant effects in hunting-and-gathering societies when the population density was much smaller than at present and individuals lived and mated in small tribal groups. Consequently, much of the variation in allele frequency among populations seen today could have originated as a result of random genetic drift in times past.

FOUNDER EFFECTS

Certain aspects of random genetic drift are related to inbreeding because in a small population some mating between relatives is unavoidable. It is not that individuals in such groups actively seek out and mate with relatives, but that almost everyone is related to everyone else to one extent or another. For example, in a small religious community (the Hutterites) in South Dakota and Minnesota, the average inbreeding coefficient of individuals is 0.04. Almost none of this inbreeding is caused by individuals mating with relatives over and above that which would necessarily occur in a population of such restricted size.

Figure 13.16   Several present-day members of the Kel Kummer Tuareg.

Another example of inbreeding due to small population size is the Kel Kummer Tuareg (Figure 13.16), a group that lives in the southern Sahara Desert. Complete genealogies of all 2420 living and dead members of the Kel Kummer tribe have been compiled (a task that took the French ethnologist Chaventré 15 years). These genealogies trace the ancestry of every member of the group back to 156 people who founded the group some 300 years ago. (Figure 13.17 shows part of the complex pedigree of the Kel Kummer.) As of 1970 the size of the Kel Kummer Tuareg was 367 people. Their ancestry was extraordinarily complex; each pair of members of the tribe had at least 15 ancestors in common. Furthermore, the 156 original founders did not contribute equally to the genetic com-

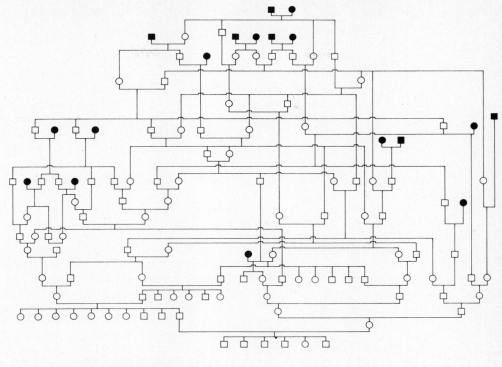

■ Male founders
● Female founders
□ Other males
○ Other females

Figure 13.17 Partial pedigree of Kel Kummer population showing the extensive interbreeding made necessary in part by the small size of the population. Marriages in this pedigree are denoted by horizontal lines somewhat below the level of the marriage partners.

position of the present population. The parents of the great warrior Kari-Denna contributed more than 10 percent of the genes in the present population, and Kari-Denna's wife's parents were the origin of another 20 percent of the genes in the present population. Indeed, 40 percent of the genes in the tribe today trace back to just six individuals among the founders, and virtually all the present-day genes trace back to a mere 25 individuals. These 25 individuals were the true genetic founders of the Kel Kummer tribe, even though 156 individuals were present at the outset.

This founder aspect of the history of the Kel Kummer illustrates an important principle known as the **founder effect**. When a small group of individuals moves away from a large group and establishes a genetically semi-isolated population, then the allele frequencies in the new population may be very different from those in the original population, simply because the founders of the new population may not be perfectly representative of the population from which they came. (Founder effects are thus special varieties of random genetic drift.)

Founder effects are especially important if the founders are few in number, because the smaller the group, the less genetically representative it is likely to be. Being nonrepresentative genetically may come about because of chance differences in allele frequency among the founders, but it may also occur because founders of a new population are sometimes groups of relatives. One striking example of the founder effect concerns the allele causing Huntington disease. In the Australian state of Tasmania there are about 350,000 people, about 120 of them affected with Huntington disease. Every single affected individuals has an ancestry that traces back to a Huguenot woman who left England in 1848 to settle in the islands off New Zealand. Such founder effects explain in part why there is variation among populations in the incidence of such inherited conditions as cystic fibrosis, Tay-Sachs disease, albinism, blood groups, allozymes, and so on.

SUMMARY

1. **Population genetics** deals with the application of genetic principles to entire **populations** (interbreeding groups) of organisms. Population genetics thus deals with **genetic variation** in populations and with the principles governing changes in the genetic makeup of populations through time. Genetic variation is widespread in most natural populations. The most widely studied form of genetic variation involves alleles associated with easily identified phenotypes, such as those determining various blood groups, those influencing the rate of movement of particular enzymes in response to an electric field (**electrophoretic mobility**), or, more recently, those associated with the presence or absence of a particular restriction site in a fragment of DNA. Proteins that have an altered electrophoretic mobility as a result of mutation in the corresponding gene are said to be **allozymes**.

2. The frequencies of genotypes in a population are determined in part by the frequencies of the various types of matings that can occur. The simplest situation regarding types of matings is known as **random mating**, in which mating pairs occur randomly with respect to the locus in question. With random mating, and in the absence of such evolutionary forces as mutation, migration, and selection, genotype frequencies are related in a simple manner to the **allele frequencies** at the locus in question. (The allele frequency of a prescribed allele among a group of individuals is equal to the proportion of all alleles at the locus that are of the prescribed type.) For two alleles A and a at respective allele frequencies p and q (where $p + q = 1$), the genotype frequencies with random mating are given by the **Hardy-Weinberg rule**:

$$
\begin{array}{ll}
AA: & p^2 \\
Aa: & 2pq \\
aa: & q^2
\end{array}
$$

3. The Hardy-Weinberg rule has several important implications: (*a*) The allele frequencies remain constant from generation to generation, provided that the assumptions of no mutation, selection, and so on, are

valid. (*b*) In the case of recessive alleles, the allele frequency of the recessive is given by the square root of the frequency of the recessive homozygote. (*c*) When a recessive allele is rare, there will always be many more heterozygotes than recessive homozygotes. (*d*) For X-linked loci, the genotype frequencies in females are given by the Hardy-Weinberg rule, whereas the genotype frequencies in males are given by the corresponding allele frequencies.

4. Most geographically widespread populations are effectively split up into **local populations**, with matings usually occurring within the local population. This structure hinders genetic exchange among local populations and creates partial **genetic isolation** among them. In time and for a variety of reasons, local populations may come to have different allele frequencies at many loci. Populations with such differing constellations of allele frequencies constitute **races**. In terms of this allele-frequency concept of race, the human species can be said to have anywhere from three races to hundreds of races, depending on how finely one wishes to make distinctions. In terms of measures of genetic variation, however, there is much more genetic variation *within* human races than there are genetic differences *among* races.

5. **Inbreeding** refers to matings between relatives (i.e., consanguineous matings). The consanguineous parents of an inbred individual thus have one or more **common ancestors**, and the principal effect of inbreeding is that the inbred individual can inherit the same allele from one of these common ancestors. Alleles that originate by replication of a common ancestral allele are said to be **identical by descent**, and the **inbreeding coefficient**, used to measure the extent of inbreeding, is the probability that an inbred individual carries alleles that are identical by descent.

6. When there is inbreeding, the genotype frequencies are no longer given by the Hardy-Weinberg rule. Genotype frequencies among a group of individuals who have inbreeding coefficient *f* are given by

$$
\begin{aligned}
AA&: \quad p^2(1 - f) + pf \\
Aa&: \quad 2pq(1 - f) \\
aa&: \quad q^2(1 - f) + qf
\end{aligned}
$$

where *p* and *q* are the allele frequencies of *A* and *a*, respectively. From these genotype frequencies, it is clear that inbreeding increases the frequency of homozygous genotypes. Consequently, although consanguineous matings (mainly first-cousin matings, for which the inbreeding coefficient of the offspring is $f = \frac{1}{16}$) account for only a minority of matings in most populations, they account for a much larger proportion of matings that give rise to offspring that are homozygous for rare harmful recessive alleles.

7. The principal forces that can change allele frequency are (*a*) **mutation**, which refers to spontaneous changes in genes, (*b*) **migration**, which refers to movement of individuals into or out of populations, (*c*) **selection**, which refers to genetically caused differences in survival or reproduction among individuals, and (*d*) **random genetic drift**, which refers

to chance fluctuations in allele frequency that occur in small populations.

8. **Mutation** in its widest sense includes all changes in the genetic material, from gross chromosomal aberrations, to changes in the locations of transposable elements, to DNA rearrangements, to simple nucleotide substitutions. Thus, mutation is the ultimate source of all evolutionary novelty. Notwithstanding its importance in generating new genetic variation, mutation is a very weak force for changing allele frequencies because mutation rates are typically small.

9. **Migration** is similar to mutation in that it introduces new alleles into a population. However, migration generally has larger effects on allele frequency than does mutation because migration rates are typically much larger than mutation rates. Migration limits the amount of genetic differentiation that can occur among local populations, so migration is a sort of ''glue'' that holds populations together.

10. **Selection** is the principal process whereby populations become progressively more adapted to survive and reproduce in their environment, and it is the process emphasized by Charles Darwin in his 1859 book *The Origin of Species*. A genotype's ability to survive and reproduce is known as its **fitness**, and differences in fitness among genotypes can cause allele frequencies to change through time. In the simplest case of selection involving two alleles at a single locus when a favored allele is dominant or recessive (or when the heterozygote has a fitness intermediate between the homozygotes), the favored allele will ultimately become **fixed** (i.e., allele frequency of 1.0) in the population. However, selection involving recessive alleles is highly inefficient when the recessive allele is rare because most of the recessive alleles will be hidden in heterozygotes and therefore not exposed to selection.

11. **Overdominance** refers to a situation in which the heterozygote has a greater fitness than both homozygotes. With overdominance, selection will ultimately lead to an **equilibrium** in which both alleles are maintained in the population at frequencies that remain the same generation after generation. One well-documented case of overdominance in humans involves sickle cell anemia, in which the normal homozygotes are not anemic but are susceptible to malaria, the sickle cell homozygotes are severely anemic, and the heterozygotes are mildly anemic but resistant to malaria. The heterozygotes are therefore the fittest genotype, and over most of west Africa the frequency of the sickle cell allele is about 0.10.

12. Selection cannot completely rid a population of harmful recessive alleles because new mutations recreate the allele. Eventually there is established a balance between mutation and selection in which the old alleles eliminated by selection are exactly balanced by the new alleles arising from mutation. This equilibrium frequency is highly dependent on the heterozygous effects of the allele. Alleles that have small harmful effects in heterozygotes have much lower equilibrium frequencies than do completely recessive alleles.

13. **Random genetic drift** is a process whereby allele frequencies can change in random fashion from one generation to the next. Random genetic

drift comes about because populations are not infinite in size; thus, the allele frequencies among successful gametes that give rise to each new generation may not be perfectly representative of allele frequencies in the previous generation. These random effects tend to be larger in small populations, so random genetic drift is particularly important in populations that are small. In populations of restricted size, matings will necessarily often occur between individuals who are related to one extent or another, so one aspect of random genetic drift is the inbreeding that tends to occur in small populations.

14. **Founder effects** refer to the random genetic drift that occurs when a relatively small group of individuals founds a new population, as the allele frequencies among the founders may not be perfectly representative of those in the population of origin. Founder effects may be particularly important in humans because founders of new populations are often groups of relatives. Founder effects and random genetic drift are responsible for an unknown but possibly major fraction of the allele frequency variation among human populations.

KEY WORDS

Additive alleles	Fitness	Local population
Allele frequency	Fixed allele	Mutation-selection balance
Allozymes	Founder effect	Overdominance
Assortative mating	Genotype frequencies	$p^2 : 2pq : q^2$
Codominant alleles	Hardy-Weinberg rule	Race
Common ancestor	Identity by descent	Random genetic drift
Equilibrium	Inbreeding	Random mating
Evolution	Inbreeding coefficient	Selection

PROBLEMS

13.1. Considering the molecular basis of mutation, what types of mutations are detected by electrophoresis? Why can the procedure not detect silent nucleotide substitutions?

13.2. Why is mutation normally a negligible contributor to changes in allele frequency over a few generations?

13.3. What are the three premises upon which the modern theory of natural selection is based?

13.4. What does overdominance imply about the relative fitnesses of homozygotes and heterozygotes? What are the long-term consequences of overdominance?

13.5. A balance between which two evolutionary forces will account for the maintenance of a low frequency of harmful alleles in a population?

13.6. What is a random genetic drift?

13.7. Would random genetic drift be a more important process in a hunting-and-gathering society than in a large industrialized society? Why?

13.8. Among Swiss living in Zurich, the genotype frequencies at the ABO blood group locus are approximately

$I^O I^O$	0.450
$I^O I^A$	0.363
$I^A I^A$	0.073
$I^O I^B$	0.079
$I^B I^B$	0.003
$I^A I^B$	0.032

What are the allele frequencies of I^O, I^A, and I^B?

13.9. A certain population has the following observed numbers of the genotypes in the MN blood groups. What is the allele frequency of M? Of N?

MN	354
MN	492
NN	154

13.10. If the population in Problem 9 had perfect Hardy-Weinberg proportions of the three genotypes, what genotype frequencies would be expected? Do the observed numbers seem to agree with these? What do you conclude from the comparison?

13.11. If a random-mating population has allele frequencies 0.56 of M and 0.44 of N in the MN blood groups, what is the probability that an MM mother will have an MM child? An MN child?

13.12. In Problem 11, what is the probability that an MN mother will have an MN child?

13.13. A certain population has the following observed numbers of Rh^+ and Rh^-. Assuming random mating, what is the allele frequency of d? Of D? How many individuals in this group would be expected to be heterozygous?

Rh^+	9856
Rh^-	144

13.14. A mating that is incompatible in the Rh blood group has an Rh^- female and an Rh^+ male. If a random-mating population has an allele frequency of d of 0.20, what is the frequency of incompatible matings?

13.15. If an X-linked recessive affects 1 in 30 males, what is the expected random-mating frequency of affected females? What is the expected frequency of heterozygotes?

13.16. If an autosomal recessive disorder affects 1 individual in 4000, what is the expected frequency of carriers?

13.17. For the trait in Problem 16, what is the expected incidence among the offspring of first cousins?

FURTHER READING

Cavalli-Sforza, L. L., and W. F. Bodmer. 1971. The Genetics of Human Populations. W. H. Freeman, San Francisco. Excellent textbook of human population genetics.

Dickerson, R. E. 1980 (March). Cytochrome c and the evolution of energy metabolism. Scientific American 242: 136–53. Traces the evolution of this important protein from bacteria to humans.

Dobzhansky, T., F. J. Ayala, G. L. Stebbins, and J. W. Valentine. 1977. Evolution. W. H. Freeman, San Francisco. An introduction to evolution.

Friedman, M. J., and W. Trager. 1980 (March). The biochemistry of resistance to malaria. Scientific American 244: 154–64. The biochemical basis of resistance in sickle-cell and thalassemia heterozygotes.

Hartl, D. L. 1980. Principles of Population Genetics. Sinauer Associates, Sunderland, Mass. A nonmathematical introduction to population genetics.

––––––. 1981. A Primer of Population Genetics. Sinauer Associates, Sunderland, Mass. A brief introduction to various aspects of the field for those wishing an overview.

Jacquard, A. 1978. Genetics of Human Populations. (Trans. by D. M. Yermanos.) Excellent introduction to elementary population genetics as applied to humans.

Lewontin, R. C. 1974. The Genetic Basis of Evolutionary Change. Columbia University Press, New York. An easy-to-read discussion of the problems in studying the importance of genetic polymorphisms.

Lovejoy, C. O. 1981. The origin of man. Science 211: 341–50. What makes humans unique.

Milkman, R. (ed.). 1982. Perspectives on Evolution. Sinauer Associates, Sunderland, Mass. Good collection of readings covering the range from molecular evolution to species formation.

Neel, J. V. 1978. The population structure of an Amerindian tribe, the Yanomama. Ann. Rev. Genet. 12: 365–413. The principal features of population structure that have a pronounced impact on human evolution.

Nei, M., and R. K. Koehn (eds.). 1983. Evolution of Genes and Proteins. Sinauer Associates, Sunderland, Mass. Excellent introduction to molecular evolution.

Scientific American, September 1978 (Vol. 239, No. 3). An entire issue devoted to the subject of evolution.

Stansfield, W. D. 1977. Science of Evolution. Macmillan, New York. A good introductory textbook of evolution.

chapter 14

Quantitative and Behavior Genetics

- Multifactorial Inheritance
- Multifactorial Traits: An Example
- Total Fingerprint Ridge Count
- Heritability
- Another Type of Heritability
- Twins
- MZ and DZ Twins in the Study of Threshold Traits
- Heritabilities of Threshold Traits
- Genetic Counseling
- Genetics of Human Behavior
- Male and Female Behavior
- Race and IQ
- Summary

Many human traits, including many inherited disorders, can be traced to the effects of individual genes, and the inheritance of these traits follows a simple Mendelian pattern brought about by the segregation of alleles at the relevant locus. Simple Mendelian traits include the common inherited traits mentioned in Table 3.1 in Chapter 3, such as common baldness, type of ear cerumen, or dimpled chin. Among the simple Mendelian disorders are familial hypercholesterolemia, achondroplasia, Huntington disease, cystic fibrosis, Tay-Sachs disease, sickle cell anemia, albinism, phenylketonuria, and many others. In the latest compendium of Mendelian inheritance in humans, 934 traits are considered as definitely attributable to autosomal dominants, 588 traits are considered as definitely attributable to autosomal recessives, and 115 traits are considered as definitely attributable to X-linked alleles. If probable mode of inheritance is included along with confirmed cases, then the numbers become 1827 dominants, 1298 recessives, and 243 X-linked, totaling 3368 traits whose simple Mendelian inheritance is probable or confirmed. Judging from these numbers, simple Mendelian inheritance is an important source of genetic variation among humans, even though many of the known simple Mendelian disorders are individually rather rare.

Simple Mendelian traits, particularly inherited disorders, are frequently emphasized in human genetics precisely because their inheritance is simple Mendelian. With such a trait and the appropriate pedigrees, the mode of inheritance of the trait can be inferred, and even linkage with other loci can sometimes be detected. Biochemical studies of such traits have often proven successful in identifying the primary biochemical defect associated with the trait (i.e., the "inborn error of metabolism"). In some cases, notably phenylketonuria, this biochemical knowledge has been employed to devise a successful therapy for the condition.

MULTIFACTORIAL INHERITANCE

There is another class of traits whose inheritance is not simple Mendelian, however. These are called **multifactorial traits**, which, as the name implies, are determined by the interactions of "many factors." The "factors" in question may be multiple genetic loci, as in cases in which the alleles at several or many loci interact together to produce an aggregate effect on phenotype. The factors may also be environmental influences, such as the amount of exposure to sunlight in the case of skin color. Most multifactorial traits are influenced jointly by the effects of multiple environmental factors and the genotype at multiple loci. Because of the complexity of their determination, multifactorial traits do not exhibit simple Mendelian patterns of inheritance.

Most multifactorial traits can be classified as being of one of three types. **Quantitative traits** are traits that can be measured on a continuous scale in single individuals. Traits such as height, weight, IQ score, intensity of skin pigmentation, and blood pressure are all examples of quantitative traits. Such traits are supremely important in animal and plant breeding. The growth rates of beef cattle and pigs, milk production in dairy cattle, egg production in poultry, and yields of corn, wheat, rice, and soybeans are all important quantitative traits.

The second category of multifactorial traits consists of **meristic traits**—traits that are measured by counting. Examples of meristic traits include number of petals on a flower, number of kernels on an ear of corn, number of bristles on the abdomen of a fruit fly, and so on.

The third major class of multifactorial traits consists of **threshold traits**, which are traits that are either present or absent in an individual, such as schizophrenia or diabetes, but which are determined by an underlying **liability** (or predisposition) toward the trait: Individuals with a high liability (i.e., above a certain **threshold**) will develop the trait; those with liabilities below the threshold will not. Liability toward a threshold trait is thus a quantitative trait and may be determined by multiple factors but, unlike typical quantitative traits, the liability of an individual cannot be measured directly. All that can be observed is whether the individual and his or her relatives have the condition or not (i.e., whether their liability is above the threshold), but this information can be used to make inferences about the underlying liability.

Determining the genetic basis of multifactorial traits is no easy matter. In the first place, one cannot investigate them by compiling elaborate pedigrees and making inferences about how individual genes have trickled down through the generations. This pedigree method works well for traits determined by single genes, of course, because the presence of a trait in an individual is indicative of genotype. But in the case of multifactorial traits, each individual gene may have such a small effect that its hereditary transmission cannot be discerned. Moreover, multifactorial traits are influenced by genes at several loci, in many cases tens of loci. So even if the segregation and transmission of a single gene were discernible, the effects caused by simultaneous segregation and transmission of all other genes affecting the same trait would hopelessly complicate the picture.

Another complicating factor in the study of multifactorial traits is that they are often influenced by environment. The strength of the environmental influence on the expression of a trait as compared with the genetic effects varies from trait to trait. In some cases, the environmental effects are rather small; in others the environmental and genetic factors may be about equally important; in still other cases environmental factors may be so important as to render insignificant whatever genetic contributions there may be. It is worth bearing in mind that all traits are influenced to some extent by environment. If this were not the case, then no inherited disorders could be successfully treated by surgery, special diets, drugs, or any other medical practices. But it should also be kept in mind that the influence of the usual environments to which individuals are exposed on such traits as cystic fibrosis and sickle cell anemia are small compared with the effects of the single mutant gene. In the case of multifactorial traits, however, the environment may be as important as or even more important than hereditary factors.

MULTIFACTORIAL TRAITS: AN EXAMPLE

Figure 14.1 is a hypothetical example showing how the alleles at several loci can interact to influence a trait, in this case a meristic trait with possible phenotypes ranging from 0 through 6. The example assumes three loci arranged from left to right with two alleles (denoted 0 and 1) at each, and the phenotypic expression of the trait is assumed to be equal to the number of type 1 alleles in the genotype. (That is, each "1" allele adds one unit to a phenotype.) At the top of the figure are two genotypes, $\frac{000}{000}$ (phenotype 0) and $\frac{111}{111}$ (phenotype 6), which might represent, for example, the genotypes of two different inbred lines with respect to the relevant loci. When the lines are crossed, the hybrids will have genotype $\frac{000}{111}$ and will have a phenotype equal to 3. If the hybrids are now crossed among them-

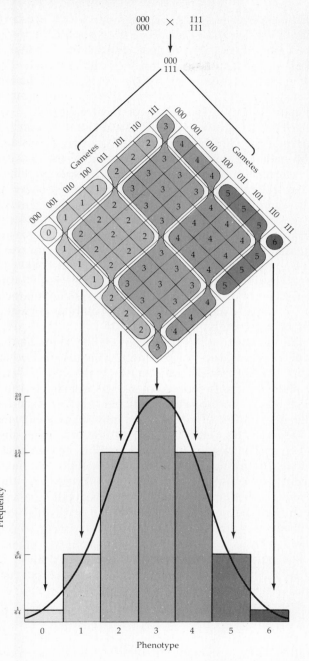

Figure 14.1 Hypothetical example illustrating a multifactorial trait influenced by two alleles at each of three unlinked loci. Each "1" allele is assumed to add one unit to phenotype. With ordinary Mendelian segregation in the crosses shown, there will be seven phenotypic classes, and the genotype of an individual cannot be ascertained by examination of its phenotype. Note that the distribution of phenotypes approximates a smooth, bell-shaped normal distribution.

selves, the resulting gametes in sperm and eggs can be represented along the sides of a large Punnett square as shown. The phenotype (i.e., number of 1's) in the zygote resulting from the union of each pair of gametes is shown within the square. If the loci are unlinked, then each gamete will have a frequency of $\frac{1}{8}$, and each of the 64 possible zygote genotypes within the square will have a frequency of $\frac{1}{64}$.

The bottom part of Figure 14.1 is a bar graph of the frequency of the phenotypes that occurs in the Punnett square above it. The relative frequencies of the phenotypes are 1:6:15:20:15:6:1, which is as expected in a cross involving three unlinked loci. Were the phenotype in question a quantitative trait instead of a meristic trait, then the distribution of phenotypes would correspond to a smooth curve instead of a bar graph. The smooth, bell-shaped curve superimposed on the bar graph in the figure closely approximates the bar graph. Such a bell-shaped curve is called a *normal distribution*, and many quantitative traits are normally distributed. This type of distribution will be discussed shortly, but for now it is sufficient to note that the good agreement between the smooth curve in the bar graph in Figure 14.1 implies that the original meristic trait could just as well be treated as a quantitative trait having the normal distribution shown. Many meristic traits are conveniently dealt with in this manner. Consequently, the term "quantitative trait" is usually meant to include not only quantitative traits in the strict sense but meristic traits as well.

Unfortunately, examining the distribution of phenotypes in a population tells nothing about the mode of inheritance of the trait. The distribution of phenotypes in Figure 14.1 could come about in an almost limitless variety of ways. At one extreme, the phenotypic differences among individuals could be entirely genetic in origin, as illustrated in the figure, but there would be no way of inferring merely from the distribution of phenotypes how many loci were involved, whether individual alleles were dominant or recessive or something in between, how the alleles in a genotype may interact to determine phenotype, or what were the allele frequencies at the relevant loci. At the other extreme, with only the distribution of phenotypes known, the phenotypic differences among individuals could be entirely environmental in origin. The "phenotype" in Figure 14.1 could represent, for example, the number of mistakes on a driver's road test. Indeed, a distribution of phenotypes like that in Figure 14.1 is not incompatible with simple Mendelian inheritance if there are enough environmental influences on the trait to blur phenotypic distinctions among the genotypes. Because of these uncertainties, determining the genetic basis of a multifactorial trait can become complex and difficult.

TOTAL FINGERPRINT RIDGE COUNT

As an example of a quantitative trait in humans we may use total fingerprint ridge count. Fingerprint ridges refer to the raised skin ridges that carry sweat glands and that vary in pattern from individual to individual so much that each individual, including members of identical twin pairs, has unique fingerprints suitable for identification. Each finger may have a ridge pattern forming an arch, a loop, or a whorl, and certain conventions are used to identify the center and edge of the pattern. The number of ridges between the center and edge constitutes the ridge count for any finger, and the sum of the ridge counts for all ten fingers constitutes the total fingerprint ridge count.

Total fingerprint ridge count is a convenient human quantitative trait because it is expressed at all ages, is stable throughout life, and the phenotype of a large number of individuals can be obtained relatively easily without having to deal with blood or tissue samples. Moreover, the trait is virtually free of emotional overtones like those encountered in the study of weight, IQ, or potentially heriditary disorders. In addition, the trait is highly heritable with no complications due to social or other environmental factors. (Studies of traits like weight and IQ are complicated by familial but nongenetic causes of resemblance among relatives, such as learned eating habits in the case of weight or the availability of books in the home in the case of IQ.) This high degree of hereditary determination is reflected in the close similarity of total ridge count in identical twins as compared with siblings.

An extreme example of such phenotypic similarity is illustrated in Figure 14.2, which gives the probable mode of origin of the famous Dionne quintuplets from a single zygote. The girls had total ridge counts in the narrow range from 99 to 102, as compared with three older siblings who had ridge counts of 69, 78, and 139. (The Dionne quintuplets had a tragic history. Born in Canada on May 28, 1934, they achieved celebrity status immediately and remained objects of public attention throughout their lives. The press and the public were insistent and often obnoxious, and in an attempt to protect the girls they were reared in a special compound with little contact outside their nurses and immediate family. Perhaps because of this strange upbringing, the girls later showed signs of social maladjustment and ultimately became recluses.) In spite of the virtually identical ridge counts of the Dionne quintuplets, the details of their fingerprints were sufficiently different to distinguish them. In any event, the close resemblance among the girls, contrasted with the smaller similarity among their siblings, indicates a strong genetic determination of the trait.

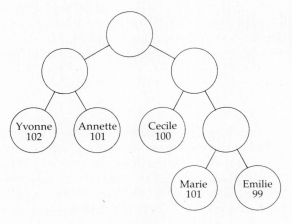

Figure 14.2 Likely origin of the Dionne MZ quintuplets from a single zygote. The numbers are the total fingerprint ridge count of the girls.

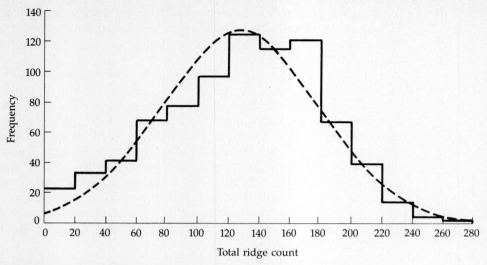

Figure 14.3 Total fingerprint ridge count among 825 British women.

Figure 14.3 shows the distribution of total ridge count in a large sample of British females. (Note that some of the scores are 0 or close to 0. This does not imply a lack of fingerprints; fingers with an arch pattern lack convenient landmarks from which to count the ridges, so arches are arbitrarily assigned a ridge count of 0. Thus, an individual with a total ridge count of 0 simply has an arch on each finger.) The broken-line curve in Figure 14.3 is the normal distribution that best fits the observed data. Its **mean** (average) is 127 and its standard deviation is 52. (The **standard deviation** of a distribution is a measure of its breadth. In a normal distribution, approximately 68 percent of the individuals will lie within one standard deviation from the mean, and approximately 95 percent will lie within 2 standard deviations from the mean.) For the data in Figure 14.3, therefore, approximately 68 percent of the women would be expected to have a ridge count between 127 ± 52 (i.e., between 75 and 179), and approximately 95 percent would be expected to have a ridge count between $127 \pm (2 \times 52)$—i.e., between 23 and 231. Another often used measure of the breadth of a distribution is called the **variance**, which corresponds to the square of the standard deviation. In this example, the variance of the distribution is $52^2 = 2704$.

HERITABILITY

As emphasized earlier, the mere fact that the distribution of phenotypes in a population is normal implies nothing about the mode of inheritance of the trait. It bears neither on the number of loci influencing the trait, nor on the effects of the alleles, nor on their dominance or recessiveness, nor on possible interactions among the alleles at different loci. To know these things would be to understand the genetic basis of the trait completely, but, unfortunately, this depth of understanding can rarely be achieved in quantitative genetics; even in experimental organisms most quantitative traits are too complex to enable the various possibilities to be sorted out. It is therefore necessary to adopt an alternative approach for discussing phenotypic resemblance among relatives, an approach that does not require detailed understanding of the genetic basis of the trait. It is in this context that the concept of **heritability** is useful.

This alternative approach is illustrated in Figure 14.4. The distribution of phenotypes in the entire population is shown in the top panel, and it may represent, for example, total fingerprint ridge count. We focus upon a restricted group of parents whose phenotypes are delimited by the shaded bar. The offspring of these parents have the phenotypic distribution shown in the bottom panel. Typically, these offspring will have a mean phenotype that lies between the mean of the population in the previous generation and the mean of their parents. The offspring mean will be greater than the population mean to the extent that phenotypic variation in the trait is inherited, and it will be smaller than the parental mean to the extent that phenotypic variation in the trait is environmentally determined.

When observations like that in Figure 14.4 are carried out for parents whose

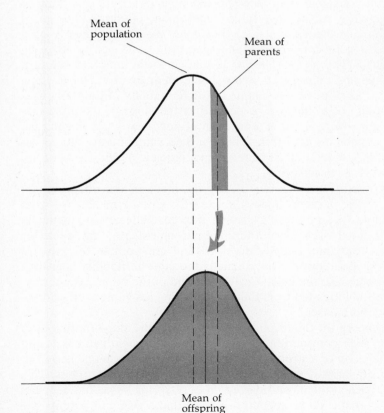

Figure 14.4 With quantitative traits, the mean of the offspring is determined by the mean of the parents, the mean of the population from which the parents derive, and the heritability of the trait.

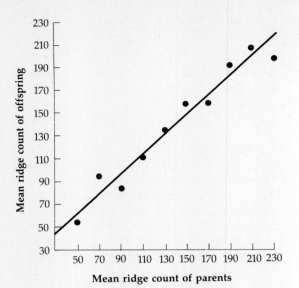

Figure 14.5 Relationship between offspring mean and parental mean (midparent) for the trait total fingerprint ridge count in the British population.

phenotype spans a broad range, then a graph like that in Figure 14.5 can be produced. This graph gives the average ridge count among offspring in terms of the average ridge count of their parents. As indicated, the data fall nearly on a straight line, and the slope of this line is then a measure of how nearly offspring resemble their parents. (The slope of a line is related to the angle at which it rises. A horizontal line has a slope of 0, one rising at an angle of 45° has a slope of 1.) In Figure 14.5, this slope is an estimate of the heritability of the trait. That is to say, the **heritability** of a trait can be defined as the slope of the offspring-parent line, as in Figure 14.5, and it is a measure of how nearly the phenotypic mean of the offspring matches that of the parents. Said another way, the heritability is a measure of how much of the phenotypic variation in a trait is genetically transmissible from parents to offspring.

Notice that heritability tells us how nearly the mean of the offspring will resemble the mean of the parents. It does not tell us *why* the means are what they are. For example, in considering growth rate in cattle, two genetically identical populations may be reared in different environments, one favorable for growth rate and the other nutritionally deprived and unfavorable for growth. Nevertheless the two populations can have exactly the same heritability of growth rate, even though in one case the calves grow well and in the other case poorly. All that the equal heritabilities tell us is that the offspring resemble their parents to the same extent in both cases.

In the case of fingerprint ridge count in Figure 14.5, the slope (heritability) is 0.86, which is very high considering that its maximum possible value is 1, and it reinforces the conclusion of a strong genetic component that was reached from consideration of the similarity of ridge count among the Dionne quintuplets. This value can be used to predict the mean ridge count of offspring based on the mean ridge count of their parents. To be specific, suppose that a group of parents has a mean of P and that they derive from a population with a mean of M. Let us represent the heritability of the trait with the symbol h^2, which is the symbol conventionally used. To calculate the mean phenotype expected among the children (C), one uses the formula

$$C = M + (P - M)h^2$$

For example, in a population with a mean finger ridge count of 145 (which corresponds to M), what ridge count would be expected among children of parents whose average ridge count is 170? In this case $P = 170$ is given, and we use the value of $h^2 = 0.86$ obtained from Figure 14.5, so the offspring mean (C) is calculated as

$$
\begin{aligned}
C &= 145 + [(170 - 145) \times 0.86] \\
&= 145 + (25 \times 0.86) \\
&= 166.5
\end{aligned}
$$

This kind of formula for prediction is frequently used in plant and animal breeding when a breeder wishes to predict how much improvement in a trait could be expected by using only superior animals for breeding. However, breeders typically must deal with traits that have much lower heritabilities than 0.86. Some examples are given in Table 14.1. Using the principles of quantitative genetics and heritability, breeders have made substantial improvements in breeds of cattle, sheep, poultry, swine, and other animals, and in varieties of corn, wheat, rice, rye, soybeans, potatoes, and other crops.

ANOTHER TYPE OF HERITABILITY

The heritability discussed in connection with Figure 14.5 and Table 14.1 is sometimes called the **narrow-sense heritability** because it relates to inheritance in the narrow sense of transmission from parents to offspring. In some quantitative traits, the resemblance between identical twins or siblings is greater than would be expected on the basis of the value of the narrow-sense heritability. This increased resemblance can occur because there are certain sources of genetic variation that are not transmissible from parents to offspring, so these sources of variation do not appear in the narrow-sense heritability. An important example is in the contribution to variation in phenotype due to dominance. Since siblings (and, of course, identical twins) can have the same genotype at a locus and therefore share two alleles, siblings can resemble one another more closely than a parent and an offspring, which can share only one allele at each locus because the other allele must come from the other parent.

This greater resemblance between sibs than between parents and offspring is illustrated in the greater survival rate of kidney transplants when a sibling is the donor than when a parent is the donor (see Figure 12.8). Thus, part of the contribution to phenotype due to dominance that occurs in a parent may not be transmissible because the favorable combination of alleles in the parent will be disrupted by segregation. Consequently, there are two effects of dominance in a quantitative trait, one part of which appears in the narrow-sense heritability and

Table 14.1   RANGES OF NARROW-SENSE HERITABILITIES OBSERVED FOR VARIOUS TRAITS IN FARM ANIMALS

Animal	Trait	Heritability (narrow sense)
Cattle	Milk yield	0.2–0.4
	Percent fat in milk	0.4–0.8
	Percent protein in milk	0.4–0.7
	Feed efficiency	0.3–0.4
Sheep	Fleece weight	0.3–0.6
	Fiber diameter	0.2–0.5
	Body weight	0.2–0.4
Poultry	Eggs per hen	0.05–0.15
	Rate of egg production	0.15–0.30
	Body weight	0.3–0.7
	Egg weight	0.4–0.7
Swine	Daily gain in weight	0.1–0.5
	Feed efficiency	0.15–0.60
	Back-fat thickness	0.50–0.70
	Litter size	0.10–0.20

Source: F. Pirchner, 1969, Population Genetics in Animal Breeding, W. H. Freeman, San Francisco.
Note: Narrow-sense heritability differs from population to population. Nevertheless, the narrow-sense heritability is a useful concept to animal breeders because it provides a measure of the speed with which a particular population can be improved by selective breeding.

the other part of which does not. A similar duality applies to genetic variation due to the interaction among alleles at different loci.

The phenotypes of siblings and other types of relatives can be compared in a graph like those in Figures 14.8 and 14.9 but because of the possibility of nontransmissible genetic variation, the slope will no longer correspond to narrow-sense heritability. When the relatives being compared can resemble one another because of nontransmissible genetic factors, as in the case of identical twins or siblings, the slope of the line is called the **broad-sense heritability** (symbolized H^2). The broad-sense heritability takes into account all genetic sources of resemblance between relatives, not just the tranmissible sources, and therefore H^2 will be larger than h^2 to the extent that nontransmissible sources of variation are important. Said in another way, the broad-sense heritability refers to the proportion of the phenotypic variation in a population that can be attributed to all genetic differences among individuals; the narrow-sense heritability refers only to the proportion of the phenotypic variation that is transmissible and so is reflected in the offspring-midparent similarity. (Midparent refers to the average of the phenotypes of the parents.) Consequently, for purposes of partitioning phenotypic variation into genetic (including both transmissible and nontransmissible sources) and environmental components, the broad-sense heritability is important. On the other hand, for purposes of predicting the offspring mean based on the parental mean, the narrow-sense heritability is important.

In human genetics, the broad-sense heritability is often the one of interest, and values for various traits are listed in Table 14.2. As the table illustrates, the values vary widely among traits. On the whole, the values of H^2 are likely to be

Table 14.2   BROAD-SENSE HERITABILITIES OF SOME TRAITS IN CAUCASIANS

Trait	Heritability*
Fertility	10–20
Birth weight	16
Longevity	(29)
Handedness	31
Serum lipid levels	(44)
Diastolic blood pressure	44
Systolic blood pressure	57
Body weight	63
Amino acid excretion	(72)
Stature	85
Total fingerprint ridge count	80–95

* Estimates in parentheses are considered tentative.
Source: C. Smith, 1975, Quantitative inheritance, in G. Fraser and O. Mayo (eds.), Textbook of Human Genetics, Blackwell Scientific, Oxford.

overestimates of the true values because nongenetic similarities in environment among members of a family have not been corrected for. Nevertheless, it is clear from Table 14.2 that some of the phenotypic variation in quantitative traits in humans is due to genotypic differences among individuals, but the precise magnitude of the genotypic contribution is variable in magnitude and uncertain in amount.

TWINS

The discussion of fingerprint ridge count among the Dionne quintuplets, depicted in Figure 14.2, illustrates the potential value of multiple identical births in the study of quantitative and threshold traits. Identical quintuplets are, of course, exceedingly rare, but the same principles apply to considerations of twins. Two types of twins must be distinguished. One type consists of **identical** twins (also called **monozygotic** or just **MZ** twins). The other type consists of **fraternal** twins (also called **dizygotic** or **DZ** twins).

As the name implies, MZ twins originate from a single fertilized egg (zygote) that gives rise to two separate embryos, as would happen if an embryo in an early stage of development were to fall apart into two distinct clumps of cells. The falling apart of embryonic cells can occur at almost any early stage of development. If it occurs early enough, each twin may have its own amnion, chorion, and placenta. (See Figure 14.6 for diagrams of the several intrauterine relationships of twins.) If the separation occurs later, the twins may have separate amnions but will share the same chorion. If still later, the twins will be in the same amnion.

Sometimes it happens that the separation of the two embryos is not complete or that parts of the embryos fuse together. Such fused embryos give rise to **Siamese** twins—twins who are joined together. Siamese twins can often be separated surgically if, for example, they are joined by the arms or legs. In other

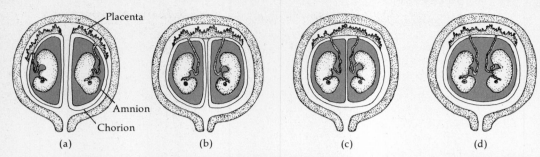

Figure 14.6 The several possible intrauterine relationships of twins. (*a*) The twins (either MZ or DZ) have separate amnions, chorions, and placentas. (*b*) The twins (again either MZ or DZ) have separate amnions and chorions but a fused placenta. (*c*) MZ twins may have separate amnions but the same chorion and placenta. (*d*) MZ twins may share the same amnion, chorion, and placenta.

cases the twins are so fused that they share one vital organ, such as the heart; the life of one twin must then be sacrificed for the sake of the other. Fortunately, this dilemma arises only infrequently because Siamese twins are extremely rare.

Dizygotic twins arise from a double ovulation, with each egg being fertilized by a different sperm. DZ twins almost always have their own amnion and chorion, and they very often have their own placentas. DZ twins are genetically related as siblings and can be alike or different in sex. MZ twins, because they originate from a single egg and a single sperm, are genetically identical and must therefore always have the same sex.

The frequency of twinning varies from population to population. Among whites in the United States, about 1 in 85 births is twins; among Japanese, about 1 in 145 births is twins. This rather large difference is due almost entirely to differences in the rate of DZ twinning. The rates of DZ twinning in the two populations are about 1 in 135 births and 1 in 370 births, respectively; the rates of MZ twinning are 1 in 256 and 1 in 238—almost the same. There does seem to be an inherited tendency toward two-egg twinning, but the tendency is not strong and the mode of inheritance is complex.

The importance of twins in the study of threshold traits derives from the genetic identity of one-egg twins. Whereas differences between two-egg twins with regard to any trait may arise because of environmental or genetic factors—or combinations of the two—differences between one-egg twins must be due only to environmental factors because the twins are genetically identical. Consequently, if the occurrence or nonoccurrence of a trait has a genetic component, then identical twins will more often be alike with regard to the trait in question than will fraternal twins. Conversely, if the genetic component is small and unimportant, then one-egg twins will be no more similar with respect to the trait then two-egg twins. The importance of genetic factors in influencing such traits as facial features is illustrated in the sometimes striking resemblance between MZ twins (Figure 14.7).

One difficulty that arises in twin studies—and it is a serious one—is that the environment of one-egg twins may be more similar than that of two-egg twins.

Figure 14.7 Monozygotic twins.

Within the uterus, one-egg twins frequently share the same embryonic membranes and placenta; two-egg twins rarely do. After their birth, the striking resemblances in facial features, body conformation, and so on between identical twins probably lead their parents, teachers, and friends to treat them more similarly than if they were two-egg twins. To the extent that these factors are important relative to the trait in question, resemblances between identical twins due to similarities in their environments will mistakenly be attributed to their genetic identity. This difficulty can be avoided in part by studying identical twins who were separated from each other shortly after birth and raised in different environments; such cases are not numerous, though. Even when reared separately, however, MZ twins may have resemblances resulting, not from their identical genes, but from their special and intimate intrauterine association. The overall conclusion to keep in mind is that the results of twin studies are to be interpreted with caution.

MZ AND DZ TWINS IN THE STUDY OF THRESHOLD TRAITS

The extent to which twins are alike with respect to any threshold trait is expressed as a number called the **concordance**. A concordance is simply the percentage of cases in which both twins of a pair have the trait when it is known that at least one has it. Suppose, for example, that 70 sets of identical twins were found in which only one member of the pair was affected with some disorder and that 30 sets were found in which both were affected; then the concordance would be $\frac{30}{(30 + 70)} = 30$ percent.

The concordance varies according to the degree of genetic determination of a trait and according to the mode of inheritance. A trait that is completely determined by heredity will have a concordance in one-egg twins of 100 percent. The concordance of the same trait in two-egg twins will depend on how many genes are involved and on how these interact. To take a simple case, the concordance in fraternal twins of a trait determined by a single dominant gene with complete penetrance will be 50 percent (if only one parent is affected); if the trait is determined by a single recessive gene, then the concordance in fraternal twins will be 25 percent (if neither parent is affected). If penetrance is incomplete or if several genes are involved in the inheritance, then the concordance cannot easily be calculated. It will depend on the number of loci and where they are on the chromosomes, on the allele frequencies in the population, on whether any or all of the loci are dominant, and on whether the loci interact in a complicated fashion. However, the difference in the concordance rates of one-egg and two-egg twins will reveal, in a general way, the importance of genetic factors in the causation of a condition.

With respect to traits known to be almost totally genetic in origin, the concordance rates are as expected. Concordance in one-egg twins with respect to blood groups is 100 percent. Concordance in fraternal twins depends on the particular blood group under consideration, but it is as anticipated. (Because of the 100-percent concordance in one-egg twins, the various blood groups are an easy and reliable way to determine whether newborn twins of the same sex are identical or fraternal.) The concordance in one-egg twins with respect to whether eye color is blue or brown is nearly 100 percent; among two-egg twins it is 55 percent. A finer classification of eye color by shade reveals slight differences in some one-egg twins; these differences must be environmental in origin. With respect to light versus dark hair color, the concordance in one-egg twins is again nearly 100 percent, although a finer classification of color by shade reveals environmentally caused differences in some one-egg twins.

Table 14.3 lists the concordance rates in MZ and DZ twins for 14 threshold traits. The two traits set off at the bottom of the table—cancer at any site and death from acute infection—do not have significantly higher concordances in one-egg than in two-egg twins. Thus there is not a strong genetic component in these traits, at least insofar as can be ascertained from twins. All the other traits in the table do have a significant genetic component as judged by the concordance rates.

Several traits in Table 14.3 are worth special mention because they are so widespread. Diabetes affects about 1 in 250 people and so is very common. Epilepsy and manic-depressive psychosis each affect about 1 in 250 people. Mental deficiency affects about 1 in 130 people. Schizophrenia affects about 1 in 1000 people (about half of all persons committed to mental hospitals are schizophrenics). Altogether, some 2 percent of all people in the United States—1 in 50—suffer from one of these five traits. And all these traits have a significant genetic component, although each is also influenced by an individual's environment. The concordance in identical twins for these traits averages around 50 percent. This means that when a person has a genotype with a strong liability or

Table 14.3 CONCORDANCES OF MZ AND DZ TWINS FOR SEVERAL TRAITS

Trait	Concordance in MZ twins, percent*	Concordance in DZ twins, percent*
Schizophrenia	34 (203)	12 (222)
Tuberculosis	37 (135)	15 (513)
Manic-depressive psychosis	67 (15)	5 (40)
Hypertension	25 (80)	7 (212)
Mental deficiency	67 (18)	0 (49)
Rheumatic fever	20 (148)	6 (428)
Rheumatoid arthritis	34 (47)	7 (141)
Bronchial asthma	47 (64)	24 (192)
Epilepsy	37 (27)	10 (100)
Diabetes mellitus	47 (76)	10 (238)
Smoking habits (females only)	83 (53)	50 (18)
Cancer (at same site)	7 (207)	3 (767)
Cancer (at any site)	16 (207)	13 (212)
Death from acute infection	8 (127)	9 (454)

* Numbers in parentheses denote number of twin pairs studied.
Source: L. L. Cavalli-Sforza and W. F. Bodmer, 1971, The Genetics of Human Populations, W. H. Freeman, San Francisco.
Note: The *concordance* of a trait in a group of twins is the percentage of cases in which when one twin is affected the other twin is also affected. The difference in concordance rates between one-egg and two-egg twins provides a very crude and often difficult-to-interpret measure of the importance of hereditary factors in the determination of a trait. In the two examples set off at the bottom of the table, the concordance rates between one-egg and two-egg twins are not significantly different. The concordance rates for all the other traits are significantly different, thus providing evidence of some degree of genetic involvement in susceptibility to the various conditions.

predisposition toward one of these threshold conditions, the chance is still not 100 percent that the person will actually become affected; indeed, the chance is roughly half of what one would expect if the occurrence or nonoccurrence of the trait were entirely determined by genes.

HERITABILITIES OF THRESHOLD TRAITS

Threshold traits are those that are either present or absent, and in this sense they are unlike quantitative traits. Threshold traits are like quantitative traits, however, in that the **liability** or risk of becoming affected with the trait is controlled by many genes acting together with environmental influences. Examples of threshold traits in humans are diabetes and schizophrenia, as mentioned earlier, but they also include such traits as hypertension (high blood pressure) and bronchial asthma, as well as such frequent birth defects as anencephaly (smallness or absence of the brain), spina bifida (exposed spinal cord), congenital heart malformation, harelip and cleft palate, clubfoot, pyloric stenosis (obstructed digestive tract), and congenital dislocation of the hip.

Although the importance of genetic factors in the causation of threshold traits can be evaluated in studies of twins (see Table 14.3), the results of such studies are sometimes questionable because, among other objections, the environments of identical twins may be more similar than the environments of fraternal twins. Another approach to the study of threshold traits involves the use of conceptual tools developed for dealing with quantitative traits.

These applications make special use of the idea that every person has a certain liability toward a threshold disease, and they assume that the liabilities are distributed in the population approximately according to the normal curve. Any person whose liability is above a certain value—the **threshold**—will actually develop the trait; those whose liabilities are below this value will not. Recall that the liability of a person cannot be observed directly; the only information available is the incidence of the condition among related people. However, this information can be used to make inferences about the underlying distribution of liability, and by special statistical procedures the heritability of liability can be estimated.

Using these methods, the heritability of liability to schizophrenia has been estimated as about 80 percent, for example, and the heritability of liability to diabetes is 30 to 40 percent. (These are broad-sense heritabilities.) The heritability of liability toward schizophrenia being about 80 percent means that within the population studied (in this case Caucasians in the United States), about 80 percent of the variation in liability to schizophrenia is genetic in origin. This number does *not* mean that schizophrenia is genetically determined and that environment is unimportant; it means only that genotypic differences among individuals can account for about 80 percent of the variation in liability among individuals.

GENETIC COUNSELING

Genetic counseling—advising patients of the genetic risks faced by them or their relatives—is an appropriate subject to consider in the context of threshold traits because the most common birth defects are threshold traits. Genetic counseling makes available to everyone information about human heredity that would otherwise be known only to geneticists. Such knowledge is of great benefit because it prevents legitimate anxieties about one's own inheritance from deteriorating into fears based on rumor or superstition. Genetic counselors assist people in assessing the risk of hereditary illness befalling themselves, their relatives, or their children. Counselors do not make decisions for their patients. They help them evaluate risks. Final decisions are left to the patient and the patient's family.

About 85 percent of the people who request assistance at clinics that specialize in genetic counseling request counseling because they have a child affected with some disorder and wish to know the risk to a subsequent child. (This risk is known as the **recurrence risk**.) The balance of the patients is divided about equally between those who have medical problems themselves that might be transmitted to their children and those who have an affected relative and wonder whether this poses special dangers to themselves or their children. Perhaps 1 person in 10, or 1 in 20, in the general population is in serious need of genetic counseling for one of these reasons. But almost everyone could benefit from genetic counseling at one time or another because most of us have relatives with an ailment that is determined in part by heredity—ailments such as hypertension, diabetes, schizophrenia, epilepsy, or the most common kinds of birth defects. For most of us, counseling would offer comfort and reassurance.

Fortunately, about two-thirds of the people who request counseling because they have had an abnormal child and wish to know the risk of recurrence usually receive rather good news—their recurrence risk will be relatively low. Recurrence risks range essentially from 0 to 50 percent, depending on the trait. High-risk traits are generally those with simple, single-gene inheritance like phenylketonuria or cystic fibrosis; risk of recurrence in such cases is 25 percent if the gene is autosomal recessive, 50 percent if the gene is autosomal dominant. (In the present discussion, a high-risk trait will be considered as a trait with a recurrence risk of 20 percent or more; a low-risk trait will be considered as a trait with a recurrence risk of 10 percent or less.) Low-risk traits are generally those with complex inheritance such as diabetes or schizophrenia; these traits are variable in expression, heterogeneous in cause, and multifactorial in their underlying genetic basis.

As a group, high-risk traits are much less common than low-risk traits. This is shown graphically in Figure 14.8, which includes many of the conditions ordinarily encountered in Caucasians. The vertical axis on the graph represents the recurrence risk in brothers and sisters of affected individuals. Low-risk traits are denoted on the graph by dots, high-risk traits by squares. The open circles on the graph indicate traits for which the recurrence risks are intermediate (10 to 20 percent). The heavier lines represent the recurrence risks of traits inherited as simple Mendelian recessives or simple Mendelian dominants, or as polygenic traits with the narrow-sense heritabilities as noted. Of the 36 traits in the graph, 21 are low risk, 10 are high risk, and 5 are intermediate. The 36 traits have a combined incidence of 4.8 percent. This number is made up of 3.2 percent low-risk traits, 0.4 percent high-risk traits, and 1.2 percent intermediates. Thus, to repeat, about two-thirds of the families being appraised of their recurrence risk will be reassured by rather good news—their recurrence risk will be low.

Figure 14.8 makes a counselor's job look easy. The counselor merely identifies which trait is involved, finds where it is on the graph, and presto—there is the recurrence risk. Counseling is not so straightforward, however, and a graph like Figure 14.8 would not actually be used. First, the numbers graphed in this figure are mere averages. Proper counseling will take into account the racial or ethnic background of the parents (Japanese have higher frequency of harelip and cleft palate), where they live (the incidence of spina bifida is lower in the western part of the United States than in the eastern part), their social class (anencephaly is more frequent among the children of unskilled workers, at least in Ireland), how many normal children the parents have had (congenital dislocation of the hip is more frequent in firstborn babies), and the sex of the affected child (pyloric stenosis is five times more common in boys than girls; congenital hip dislocation is six times more common in girls than boys). Many other factors such as family history must also be considered.

Adding to the difficulties encountered by genetic counselors is the problem of proper diagnosis. For example, the most common variety of cleft palate is multifactorial and has a low recurrence risk, but there is a rare form, recognized by tiny indentations on the inner surface of the lip, that is inherited as a simple Mendelian dominant. Also, many traits that are inherited in a simple manner can be mimicked by nongenetic disorders, and the counselor has to sort out which is which. Genetic counseling is always a delicate business, best left to those especially trained for it.

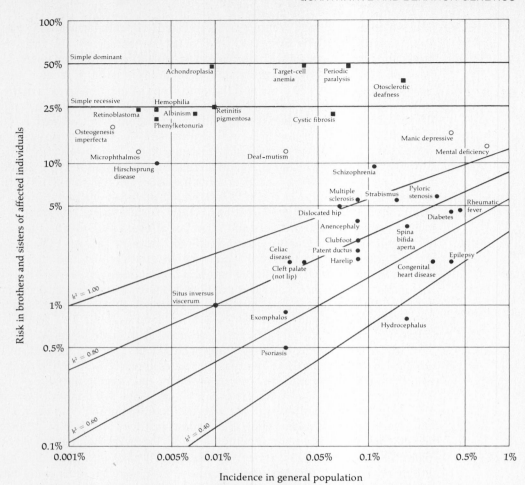

Figure 14.8 Graph of recurrence risks of common abnormalities. The numbers along the horizontal axis refer to the frequency of each trait in the population as a whole; the numbers along the vertical axis refer to the risk of each trait among brothers and sisters of affected individuals (the recurrence risk). The heavier lines on the graph denote the recurrence risks expected for traits inherited as simple dominants, simple recessives, or threshold traits with the narrow-sense heritabilities as noted. The incidences and recurrence risks plotted here are those observed in Caucasians.

At one time, the best a counselor could do was advise parents of their risks. This has all changed. Counselors now have an arsenal of medical technology at their disposal. In many cases of recessive inheritance, they can identify heterozygous carriers; counselors need no longer wait for an affected child to be born, they can alert the parents in advance. In high-risk pregnancies, amniocentesis can be used to determine whether the embryo is normal. When abnormal children are born, they are recipients of medical technology—corrective surgery for harelip, cleft palate, pyloric stenosis, and spina bifida; antibiotics for cystic fibrosis; insulin for diabetes; other drugs for schizophrenia; special diets for phenylketonuria—the list could be extended indefinitely. The arsenal of medical technology has multiplied by many times the power of counselors to advise and the power of physicians to treat.

GENETICS OF HUMAN BEHAVIOR

There is perhaps no more controversial an application of genetics than its relation to human behavior. Mistrust seems to derive from two sources: one psychological and the other scientific. On a psychological level, many people seem to believe that the finding of any significant genetic influence on behavior would be tantamount to a sort of genetic predestination, a kind of enslavement to our genes, that puts our own behavior beyond our own control. This biology-is-destiny interpretation is unjustified. From discussions throughout this chapter and, indeed, throughout the book, it ought to be clear that a genetic influence on a trait does not at all imply that the environment is unimportant. For quantitative traits especially, the environment can exert profound influences on a trait and even reverse any possible genetic predispositions. Perhaps it would be most accurate to interpret genetic influences on behavior as determining a set of potentialities, which may be very broad; the environment then determines which of these potentialities will actually be achieved. Seen from this perspective, genes do not predetermine human behavior, but they may establish limits in much the way genes put limits on our physical abilities; no human can run a two-minute mile or fly by flapping his arms.

The second objection to applying genetics to human behavior, a scientific objection, is substantive. Normal variation in behavior, if influenced by genes at all, must be analyzed as a quantitative trait. Animal and plant breeders have successfully analyzed many quantitative traits. In these fields special experiments can be carried out, but such experimental control is not available to the human geneticist. For many behavioral traits, assortative mating may be important, and environmentally relevant factors may tend to be shared among members of a family. These complications and others tend to make estimates of genotypic variation greater than the true value may actually be. Moreover, it may be argued that statistical techniques that are appropriate for animal and plant breeding because they permit prediction of response to selection are quite inappropriate in human genetics.

Heritabilities tell us nothing about how many loci may influence the trait in question, nor about the relative importance of these loci, nor about what the loci do. For these reasons, some human geneticists have adopted a more direct approach by studying genetic variation in hormones, enzymes, or other molecules thought to be involved in brain function.

In some instances, genes have an overriding influence on behavior. Examples include genetically caused abnormalities in the development of the nervous system such as Tay-Sachs disease (see Chapter 4) and phenylketonuria (see Chapter 9), genetically caused degeneration of the nervous system such as Huntington disease (see Chapter 10), and the bizarre behavioral effects of Lesch-Nyhan syndrome (see Chapter 5). At one time there was thought to be a strong association between the XYY chromosome complement and violent criminality, but this association has not been confirmed in later studies; the XYY situation, discussed in detail in Chapter 6, illustrates some of the serious methodological difficulties involved in the study of genetic influences on human behavior.

Tay-Sachs disease, phenylketonuria, and Lesch-Nyhan syndrome, while demonstrating the importance of genes in the development and maintenance of the nervous system, are individually rare disorders; thus, they leave open the question of whether genotypic variation accounts for a significant fraction of variation in behavior among *normal* individuals.

That there exists a potential for such genetic influences is illustrated in the animal experiment in Figure 14.9. Involved in these experiments are two strains of rats: a strain that learns to run mazes well (unshaded bars) and another strain that learns poorly (shaded bars). As indicated by the large difference in learning ability in the average environment, the strains clearly differ from each other and the difference is due to genotype. However, when rats are reared in a high-stimulation environment enriched with ramps, mirrors, balls, tunnels, and so on, the difference in learning ability is very much reduced; and when rats are reared in a low-stimulation environment, the difference disappears entirely. This example illustrates how the environment can strongly influence the expression of genetically based differences.

Figure 14.9 is not intended to imply that maze learning in rats is in any way comparable to learning in humans. Human learning is undoubtedly much more complex. However, since Figure 14.9 illustrates a pronounced genotype-environment interaction for the simple maze-learning situation in rats, all the more surely

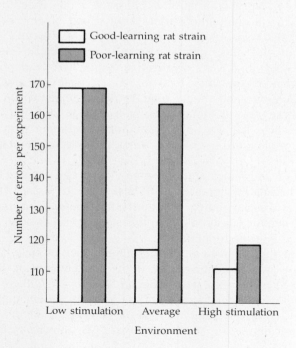

Figure 14.9 Maze-running ability among strains of rats selected for good learning and poor learning in an average environment. Strong genotype-environment interaction is indicated because the difference between strains virtually disappears in high-stimulation or low-stimulation environments.

we can expect such complications in human learning. With regard to human behavior, about all we have to go on are twin studies and measures of resemblance between relatives, and, as emphasized earlier, these approaches are fraught with difficulties when applied to humans.

In the rest of this chapter, we consider two examples in which a proposed genetic influence on human behavior is highly controversial. One case involves behavioral differences between the sexes and the other involves IQ.

MALE AND FEMALE BEHAVIOR

In Chapter 5 we considered the developmental basis of sex and discussed the importance of various hormones, particularly testosterone, in switching sexual differentiation from a preprogrammed female direction to a male direction. These early gonadal hormones are also known to influence differentiation of the central nervous system, and in experimental animals (primarily rodents), such early hormones can be demonstrated to influence nonreproductive behaviors such as overall activity, aggression, play, taste preferences, and maze learning. In these experimental organisms, there do seem to be innate behavioral differences between the sexes.

On the other hand, it may be argued that experimental studies of behavior in rodents have little or no relevance to human behavior. The human brain is vastly more complex than the rodent brain, and human family and social structures are much more sophisticated than those of rats. So, the question remains: Are there indeed inborn behavioral differences between the sexes in humans, and, if so, how important are they?

It should be emphasized at the outset that the answers to such questions are unknown. Not only are they unknown, but they may be unknowable from direct observation because innate behavioral traits become hopelessly intertwined with external environmental influences from the moment of a child's birth and sometimes prior to birth. The only way to sort out the effects of "nature and nurture" would be to isolate newborns from *all* human contact and care for them with machines of some unholy design. Such an experiment would be grotesque and immoral, unthinkable in any society.

Such experiments have, however, been carried out with monkeys; while certainly not humans, monkeys are much closer relatives than rodents. Extensive studies of behavior in rhesus monkeys have been carried out by Harry F. Harlow and collaborators at the University of Wisconsin. The observations are made possible because newborn monkeys can be raised in complete isolation from adult monkeys. They do need body contact, but this is supplied by a simple terrycloth-covered wire dummy, an inanimate "surrogate" mother (Figure 14.10). Infants raised individually with surrogate mothers are socially and psychologically abnormal. Harlow puts it this way:

> The first 47 baby monkeys were raised during the first year of life in wire cages so arranged that the infants could see and hear and call to other infants but not contact them. Now they are five to seven years old and sexually mature. As

Figure 14.10 A baby monkey with its terrycloth surrogate mother. Monkey infants raised in complete isolation from other monkeys become severely disturbed both socially and psychologically.

month after month and year after year have passed, these monkeys have appeared to be less and less normal. We have seen them sitting in their cages strangely mute, staring fixedly into space, relatively indifferent to people and other monkeys. Some clutch their heads in both hands and rock back and forth. . . . Others, when approached or even left alone, go into violent frenzies of rage, grasping and tearing at their legs with such fury that they sometimes require medical care. Eventually we realized that we had a laboratory full of neurotic monkeys.*

These monkeys are sexually incompetent. The males are unable to mate even with sexually experienced females, who react to the males' repeated failures with contempt and frustration. The motherless females, when mated with experienced and persistent males, turn out to be cruel and heartless mothers, quite devoid of maternal behavior toward their own young.

As it happens, differences in behavior between normally raised male and female monkeys can be identified in the first several months of life. For example, threat behavior, marked by a distinct facial grimace, is observed about fives times

* H. F. Harlow, 1962, The heterosexual affectional system in monkeys, The American Psychologist 17: 1–9.

as often in males as females. Passivity, observed as a turning away from an approaching animal or as a relaxed body posture, is three to four times as frequent in females as males. Rigidity, which is some four times as frequent in females as males, is a response evoked by an approaching monkey and identified by a rigid body posture with the limbs extended, often with the tail erect and the head looking back over the shoulder at the approaching animal.

There is great overlap in the behavior of male and female monkeys. Any particular animal can and does sometimes exhibit behavior found most frequently in the opposite sex, but the statistical differences between the sexes are quite marked. Behavioral patterns in the sexes not only differ on the average, but the differences become greater as the monkeys age, probably because certain kinds of behavior are reinforced by other animals. But a notable point is that at least some underlying differences in behavior seem to be innate. According to Harlow, rhesus monkeys raised in isolation exhibit behavior appropriate to their own sex with little or no alteration.

The innate behavioral patterns in monkeys can overpower even strong social influences. As mentioned before, monkeys raised in complete isolation are hopelessly neurotic. They can, however, be "rehabilitated" by "therapy," the "therapist" of a 6-month-old "patient" being a 3- or 4-month-old socially normal monkey, a monkey too young to be aggressive but old enough to cling to the patient and interact with it. Upon first seeing the therapist, the patient withdraws and huddles in a remote corner of the cage; the therapist follows along and clings to the patient. In a few days, the patient is clinging back. In weeks, the patient and therapist are playing enthusiastically with each other. After 6 months of therapy the patient's abnormal behavior has almost completely disappeared.

In one series of cases, quite by chance, all the patients were males and all the therapists were females. The male patients therefore had only female therapists after which to pattern their behavior. Yet, when identifiable sexual behavioral patterns in the patients began to emerge, they were typically male patterns. Thus some sort of innate tendency toward male behavior is not swamped out by the presence of behavioral models of the opposite sex.

The innate sex difference in behavior in rhesus monkeys is evidently due to prenatal hormone influences, as females artificially treated with male hormone during embryonic development and childhood behave in a way resembling their male counterparts. On the other hand, such conclusions must not be applied to humans without great care and thoughtfulness—and much more evidence. The biological determinants of behavior could be much stronger in rhesus monkeys than in other species of monkeys or in humans. Possibly, whatever biological differences in human behavior patterns that may exist may be too small to measure; if innate differences occur, they may easily be swamped out or altered beyond recognition by conscious or unconscious learning. On the other hand, perhaps there are innate differences in behavior between boys and girls. Among girls with the adrenogenital syndrome, a defect in hormone metabolism causing a buildup of testosterone-like substances (see Chapter 5), there is an increased incidence of "tomboyish" behavior, although this syndrome is surely rare and perhaps unrepresentative.

Whatever the case regarding innate behavioral differences may be, it is clear, even from studies of rhesus monkeys, that behavior is variable, that the sexes overlap in their behavior, and that behavioral differences between the sexes can be stated only as statistical averages. Nevertheless, the extent to which sex influences behavior is still highly debatable.

RACE AND IQ

Another application of genetics that has been highly controversial in recent years is the use of quantitative genetic techniques in the investigation of IQ scores, particularly in regard to racial differences in average IQ. The controversy arises in attempting to interpret the sort of data illustrated in Figure 14.11, which shows the distribution of IQ scores among a "standard" sample of white school children in the United States (black curve) and a sample of 1800 black school children from the southern United States (gray curve); the mean of the white children is about 100, that of the black children is about 85, and the issue is how to account for this 15-point difference. It is important to note at the outset that there is a large overlap in the distributions (shaded area). Because of this overlap, 11 percent of blacks score above the average of whites, and 18 percent of whites score below the average of blacks. Thus, the distributions in Figure 14.11 have little or no relevance when applied to particular individuals. The IQ controversy deals only with overall averages. (Incidentally, the average IQ among contemporary Japanese is about 111, which is the highest in the world.)

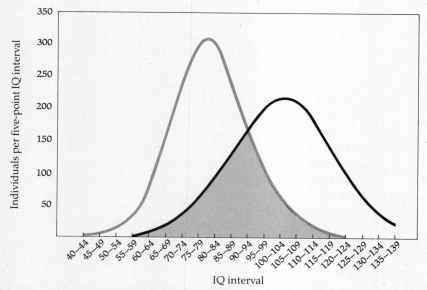

Figure 14.11 Distributions of IQ score among school children in the United States. The black curve is for a "standard" sample of white school children; the gray curve is for 1800 black school children from the southern United States. The mean of the IQ distribution of whites is about 100, that of blacks about 85; but note the extensive overlap of the distributions (shaded area). The source of the difference in the distributions is unknown, but a majority of geneticists and psychologists believe that the difference is environmental in origin and is related to the cultural bias of the IQ test itself.

One possible response to Figure 14.11 is to say that the data are completely meaningless on grounds that the IQ test is invalid. The IQ tests used today evolved from those developed by Alfred Binet and his colleagues in Paris in the early 1900s. Binet's goal was to develop a test that would distinguish mildly subnormal children from normal ones with a view to providing the mildly subnormal children with special attention in schools. Later on, the tests were revised and extended and used widely to classify *normal* individuals with regard to their IQ. Modern tests involve items that examine verbal and geometrical skills, mathematical and other abstract tasks, and memory.

Relative to middle-class U.S. whites, IQ scores are fairly reliable predictors of performance in schools. In fact, they were designed to be predictors of school performance. By trial and error, questions that were poor predictors were eliminated, and questions that were better predictors were added. Thus, it is important to realize that an IQ score is not a measure of "innate" intelligence. The test is intentionally culture-bound in the sense that it measures how well one has acquired the symbols, motivations, and values that lead to success in the classroom. The word *intelligence* as usually used is not one, single attribute; it is a complex of many different skills and abilities, and reasonable people may disagree about the importance of these. It is doubtful whether there could be a single test of any sort that could accurately assess all these abilities and assign to each an acceptable measure of importance. The term *intelligence quotient* (IQ) is thus a grave misnomer.

Additional evidence for the culture-bound nature of IQ tests arises from the following observations: (1) When southern whites and northern blacks are compared, the average IQ of northern blacks is higher than that of southern whites; (2) when tests analogous to IQ tests are developed and standardized for blacks raised in ghettos in large cities, the blacks score higher than the whites; (3) black children do better on IQ tests when administered by black examiners than when administered by white examiners; and (4) there is a relatively strong association between IQ score and social class.

In spite of these reservations about what the IQ test actually measures, there has been considerable discussion and disagreement about the interpretation of the data in Figure 14.11. It is in this context that the concepts of quantitative genetics have been introduced—incorrectly. For a trait like IQ, reliable heritability estimates are difficult to obtain because there are many pitfalls in designing a proper study and because certain nongenetic but familial causes of resemblance between relatives are important. The best available estimates put the broad-sense heritability of IQ among whites at about 30 percent.

Recall that the broad-sense heritability of a trait in a population tells how much of the total phenotypic variation in the trait is due to all genetic effects combined. A broad-sense heritability of IQ of 30 percent in the white population thus implies that *within* the white population, about 30 percent in the variation in IQ scores among individuals can be attributed to genetic differences. If a reliable estimate of heritability within the black population were available, a similar statement could be made about variation in IQ *within* the black population.

Unfortunately, the meaning of heritability as applied to IQ score has been completely misinterpreted by some commentators. A heritability of 30 percent does *not* mean that the difference in average IQ between blacks and whites is 30 percent genetic in origin. Heritability has nothing whatever to do with differences between populations. It is concerned exclusively with variation within populations. Early in this chapter we discussed two hypothetical cattle populations, genetically identical, one of which was adequately fed and grew well and the other of which was nutritionally deprived and grew poorly. The point was that these genetically identical populations could have the same heritability, even as high as 100 percent, in spite of their major differences in average growth rate. The reason is that heritability relates only to the resemblance between relatives within populations, whereas the growth-rate difference between the populations is nutritional in origin.

The data in Figure 14.11 have an ambiguity similar to that in the growth-rate example. Here we are comparing two populations of unknown genetic resemblance in two environments of uncertain but probably different effects. Heritability tells us something about genetic versus environmentally caused variation in IQ within each population, but as the growth-rate example shows, it tells us nothing about the causes of the difference between populations. In short, the reasons for the difference in Figure 14.11 are unknown. One may speculate, however, on the basis of the genetic diversity among races with regard to blood groups and allozyme differences. Recall from Chapter 13 that with regard to these traits, the general finding is that genetic variation *within* racial groups is usually much larger than that *among* racial groups. From this point of view, one would expect whatever genetic differences might exist to be rather small. In any case, the application of heritability to differences among races is a misinterpretation of the concept and a misuse of genetics.

SUMMARY

1. **Multifactorial traits** are traits that are determined by "many factors"; the "factors" may be alleles at several or many loci, environmental effects, or both. Multifactorial traits may be classified into three categories. **Quantitative traits** are traits such as height or weight that can be measured on a continuous scale in single individuals. **Meristic traits**, closely related to quantitative traits, are traits in which the phenotype can be determined by counting, such as number of bristles on a fruit fly or number of petals on a flower. **Threshold traits** are traits such as diabetes that are either present or absent in an individual; threshold traits can be interpreted as quantitative traits if there is assumed to be an underlying **liability** (or risk) toward the trait, with all individuals who have a liability greater than a certain level actually developing the trait.

2. Many quantitative traits in populations are distributed approximately according to a smooth, bell-shaped curve known as the **normal distribution**. This distribution is completely characterized by two quantities: the **mean** (average) and the **standard deviation** (a measure of the breadth of the distribution). In a normal distribution, approximately 68 percent of the observations will be within 1 standard deviation from the mean, and approximately 95 percent will be within 2 standard deviations from

the mean. Another measure of the breadth of the distribution is the **variance**, which equals the square of the standard deviation.

3. Without special types of studies that can be carried out only in experimental organisms, one cannot usually determine the number of loci influencing a multifactorial trait, the effects of the loci and the interactions among them, or the allele frequencies at the loci. Yet it is possible to determine what fraction of the total variation in phenotype in a population is due to genetic differences among individuals and to subdivide the genetic variation further into a part that is transmissible and a part that is nontransmissible. The fraction of the total phenotypic variation in a population that is transmissible is called the **narrow-sense heritability** and symbolized h^2; this quantity can be estimated as the slope of the straight line depicting the relationship between the average phenotype of offspring and the average phenotype of their parents. The fraction of the total phenotypic variation in a population that is attributable to genetic differences among individuals is called the **broad-sense heritability** and symbolized H^2, and it includes both transmissible and nontransmissible genetic effects. Broad-sense heritability may be estimated from the resemblance between twins or siblings.

4. For predicting offspring mean from parental mean, the narrow-sense heritability (h^2) is the most important, and it is widely used in plant and animal breeding. If a population has a mean of M and the parents selected as the breeding individuals of the next generation have a mean of P, then the mean of the offspring (C) will be expected to be

$$C = M + (M - P)h^2$$

where h^2 represents the narrow-sense heritability of the trait in the population in question. The standard deviation in phenotype among the offspring is expected to be the same as in the parental generation.

5. Quantitative and threshold traits are often studied by comparisons of twins. **Monozygotic (MZ)** twins arise from a single zygote that separates into two parts at an early stage of embryonic development and goes on to give rise to two genetically identical embryos. MZ twins are often called **identical** twins. **Dizygotic (DZ)** twins arise from two zygotes and are thus genetically related in the same way as ordinary siblings. DZ twins are often called **fraternal** twins. The frequency of twinning varies among populations, being about 1 per 88 births in the United States. Most of the interpopulational variation in twinning rates is due to variation in the rate of DZ twinning. In theory, differences between MZ twins are due entirely to environment, whereas differences between DZ twins are due to genetic and environmental factors. However, MZ twins more often share certain embryonic membranes than do DZ twins, so intrauterine environmental effects may affect MZ and DZ twins differently. In addition, MZ twins are often treated more similarly by parents, teachers, and peers than are DZ twins, so some of the similarities between MZ twins may actually be due to common environmental factors.

6. For threshold traits, twin comparisons are often expressed in terms of **concordance**. The concordance is the proportion of cases in which both twins have a trait when it is known that at least one has it. If the concordance among MZ twins is significantly greater than that among

DZ twins, this suggests there is a genetic component involved in the trait in question. Higher MZ twin concordance is observed for such traits as hypertension, bronchial asthma, and diabetes, but not for such traits as cancer at any site and death from acute infection.

7. Threshold traits are particularly important in **genetic counseling** because many common birth defects (e.g., congenital heart disease, pyloric stenosis, anencephaly, cleft palate) are threshold traits. When a couple has had an affected child, the risk of a subsequent child having the same defect is called the **recurrence risk**. In theory, recurrence risks can be calculated from the population incidence of the trait and the narrow-sense heritability. In practice, many other factors must be taken into account, including the racial background of the parents and the sex of the affected child.

8. Although a large number of rare, simple Mendelian traits is known to affect intellectual performance or behavior (e.g., Tay-Sachs disease, Huntington disease, phenylketonuria, and Lesch-Nyhan syndrome), behavior within the normal range of variation must be considered as a quantitative trait influenced by multiple loci and by environment. The genetic contribution to normal behavioral variation is unknown, possibly small, and doubtless different for different behavioral traits.

9. Investigations in experimental animals such as rodents and monkeys support the view that there are innate sex differences in nonreproductive behavior that result from the prenatal action of certain sex-related hormones, particularly testosterone. The corresponding situation in humans is by no means certain, but such differences may occur in humans as well. However, the sex differences observed are averages, and there is substantial individual variability in behavior and overlap between the sexes.

10. IQ tests were originally developed to identify mildly subnormal children in need of special help in school, but they have since been used to classify normal individuals. IQ tests are not a valid measure of overall "intelligence" in the commonly accepted sense of the term. They are reasonably accurate predictors of success in school, which is what they were designed for. Heritability of IQ has been used to argue for a genetic basis of racial differences in average IQ score. (On standard tests American whites average about 100, American blacks about 85, and Japanese about 111.) This is a wrong application of heritability because heritability relates only to variation within populations, not to differences among populations.

KEY WORDS

Broad-sense heritability	Meristic trait	Recurrence risk
Concordance	Multifactorial trait	Standard deviation
DZ twins	MZ twins	Threshold trait
Genetic counseling	Narrow-sense heritability	Variance
Liability	Normal distribution	
Mean	Quantitative trait	

PROBLEMS

14.1. If the trait in Figure 14-1 were determined by only two unlinked loci instead of three, how many possible phenotypes would there be? What would be the ratio of phenotypes among the offspring of a $\frac{00}{11}$ by $\frac{00}{11}$ cross?

14.2. Why are culturally inherited familial factors a potential source of misinterpretation of resemblance between relatives?

14.3. Why is the concordance in DZ twins 50 percent for an autosomal dominant trait with complete penetrance when only one parent is affected?

14.4. Why is the concordance in DZ twins 25 percent for an autosomal recessive trait with complete penetrance when neither parent is affected?

14.5. Half sibs can share at most one allele at an autosomal locus. Considering this observation, would you expect the resemblance between half sibs to be related to narrow-sense or broad-sense heritability?

14.6. Hybrid corn is produced by crossing highly inbred lines. In a field sown with hybrid seed from the same inbred parents, what is the broad-sense heritability of plant height (or any other quantitative trait)? What does this example illustrate about heritability?

14.7. Two inbred lines differing in a quantitative trait are crossed. One inbred has an average phenotype of 126, the other has an average phenotype of 112. Assuming that there is no dominance of the alleles affecting this trait, what is the expected average phenotype among the offspring?

14.8. The offspring of the members of a monozygotic twin pair are legally related as first cousins. What is their genetic relationship?

14.9. A pair of female monozygotic twins marries a pair of male monozygotic twins. What is the genetic relationship between the children of the two couples?

14.10. If the broad-sense heritability of liability toward a threshold trait is 1.0, could the concordance in MZ twins be predicted?

14.11. Monozygotic twin pairs are studied with respect to a certain disorder, with the following results. What is the MZ twin concordance for the trait?

Both affected	47
One affected	81
Neither affected	206

14.12. A normally distributed quantitative trait has a mean of 50 and a variance of 100. What proportion of the population will have a phenotype between 40 and 60? Between 30 and 70? Less than 30? More than 70?

14.13. In Problem 13, what proportion of the population will have a phenotype between 60 and 70?

14.14. A population has a mean of 75 for a quantitative trait with a narrow-sense heritability of $\frac{1}{3}$ (i.e., 33.3 percent). What is the expected offspring mean from parents whose mean phenotype is 105? From parents whose mean is 45?

14.15. If a trait has a fraction h^2 of its total phenotypic variation that is transmissible and a fraction H^2 of its total phenotypic variation that is genetic (transmissible and nontransmissible combined), then what fraction of the phenotypic variation in the population will be genetic but nontransmissible?

FURTHER READING

Anderson, A. M. 1982. The great Japanese IQ increase. Nature 297: 180–81. How the remarkable increase can be explained.

Blank, R. H. 1981. The Political Implications of Human Genetic Technology. Westview Press, Boulder, Colo. Virtually all aspects of research and intervention are dispassionately discussed.

Cavalli-Sforza, L. L., and W. F. Bodmer. 1971. The Genetics of Human Populations. W. H. Freeman, San Francisco. Excellent textbook.

Ehrhardt, A. A., and H. F. L. Meyer-Bahlburg. 1981. Effects of prenatal sex hormones on gender-related behavior. Science 211: 1312–18. The degree to which prenatal sex hormones influence psychosexual development.

Falconer, D. S. 1981. Introduction to Quantitative Genetics. 2d ed. Longman, New York. A classical textbook well worth the effort to study.

Feldman, M. W., and R. C. Lewontin. 1975. The heritability hang-up. Science 190: 1163–68. Takes the point of view that heritability studies are of no real value in human genetics.

Flynn, J. R. 1984. Japanese IQ. Nature 308: 222. Correction for biases in the sample indicates that the average Japanese IQ may actually be in the range of 102–104, not 111.

Gottesman, I. I., and J. Shields. 1982. Schizophrenia: The Epigenetic Puzzle. Cambridge University Press, Cambridge. A thorough and critical review of the importance of genetic and environmental factors in the causation of this famous form of "madness."

Hunt, E. 1983. On the nature of intelligence. Science 219: 141–46. Confronts a difficult issue, including problems of measurement.

Kempthorne, O. 1978. Logical, epistemological and statistical aspects of nature-nurture data interpretation. Biometrics 34: 1–23. A view that traditional methods of quantitative genetics are relevant to human genetics when properly used.

Loehlin, J. C., G. Lindzey, and J. N. Spuhler. 1975. Race Differences in Intelligence. W. H. Freeman, San Francisco. A cautious, critical, and comprehensive assessment of the race–I.Q. controversy.

McClearn, G. E., and J. C. DeFries. 1973. Introduction to Behavioral Genetics. W. H. Freeman, San Francisco. A short but pungent introduction.

Notkins, A. L. 1979 (November). The causes of diabetes. Scientific American 241: 62–73. On the types of diabetes and their genotype-environmental causation.

Parisi, P., M. Gatti, G. Prinzi, and G. Caperna. 1983. Familial incidence of twinning. Nature 304: 626–28. The extent to which twinning "runs in families."

Plomin, R., J. C. DeFries, and G. E. McClearn. 1980. Behavioral Genetics: A Primer. W. H. Freeman, San Francisco. A useful basic introduction to the study of behavioral genetics.

Rubin, R. T., J. M. Reinisch, and R. F. Haskett. 1981. Postnatal gonadal steroid effects on human behavior. Science 211: 1318–24. Summary of research on testosterone and aggression in men, mood and the menstrual cycle in women, and pubertal sex-role reversal in pseudohermaphrodites.

Tsung, M. X. T., and R. Vandermey. 1980. Genes and the Mind: Inheritance of Mental Illness. Oxford University Press, New York. A brief and readable introduction to the genetics of mental disorders.

Vogel, F. and A. G. Motulsky. 1979. Human Genetics: Problems and Approaches. Springer-Verlag, New York. Contains a good discussion of human behavioral genetics.

Glossary

A (1) Conventional symbol for the deoxyribonucleotide deoxyadenosine phosphate or for the ribonucleotide adenosine phosphate; also used to represent the base adenine. (2) A group of human chromosomes comprising numbers 1 through 3. (3) A blood type in the ABO blood group system. (4) A class of immunoglobulin containing alpha heavy chains.

AB A blood type in the ABO blood group system.

ABO blood groups Refers to various antigens on the surface of red blood cells coded by alleles at a locus denoted *I*.

Abortive infection An unproductive, failed viral infecton.

Acentric A chromosome with no centromere.

Achondroplasia A dominantly inherited form of dwarfism.

Acidic proteins Proteins rich in aspartic and glutamic acids thought to be important in eukaryotic gene regulation.

Acquired immunodeficiency syndrome. See AIDS.

Acridine One of a group of substances that stack between the bases in DNA and so cause small additions or deletions.

Acrocentric A chromosome that has its centromere nearly at one of its ends.

Activation Of enzymes, rendering active or functional.

Acute dose An amount of radiation delivered all at once.

Adenine Purine base in DNA and RNA.

Adrenogenital syndrome Hormone imbalance leading to a buildup of testosterone-like substance and therefore causing masculinization of affected females.

Age of onset The age at which a trait first appears. *Variable age of onset* refers to differing ages of onset of the same trait in different individuals.

Agglutination Clumping or aggregation caused by an antigen-antibody complex.

AIDS   Acquired immunodeficiency syndrome. Partial or complete loss of T-cell functions caused by an infectious agent transmitted by sexual contact or through the blood.

Albinism   Absence of pigment, usually due to homozygosity for an autosomal recessive allele.

Alkylating agent   One of a group of highly reactive chemicals that becomes attached to DNA and thereby causes mutations.

Allele   Alternative forms of a gene that can be present at any particular locus.

Allele frequency   Relative proportion of all alleles at a locus that are of a designated type.

Allelic   The state of being alleles.

Allosteric regulation   Activation or inhibition of an enzyme by a substance chemically unrelated to the substrate of the enzyme.

Allozymes   Enzymes having different electrophoretic mobilities and coded by alternative alleles at the same locus.

Alpha hemoglobin   One of the polypeptide chains in adult and fetal hemoglobin.

Amber $5' - UAG - 3'$ termination codon.

Ames test   A bacterial test for mutagens.

Amino acid   Fundamental chemical subunit of proteins.

Amino acid substitution   Replacement of one amino acid for a different amino acid in a polypeptide.

Aminoacyl-tRNA synthetase   One of 20 or more enzymes that attaches an amino acid to its corresponding tRNA.

Amino end   The end of a polypeptide that has a free amino ($-NH_2$) group; this is the end of the polypeptide at which synthesis begins.

Amniocentesis   Procedure in which fetal cells are obtained from the amniotic fluid for diagnosis.

Amnion   The innermost membrane surrounding the embryo or fetus.

Anaphase   A stage of cell division preceding telephase in which chromosomes begin to move toward opposite poles. In metaphase of mitosis or metaphase II of meiosis, this separation involves splitting of the centromere. In metaphase I of meiosis, it does not involve centromere splitting but rather separation of homologous centromeres.

Ankylosing spondylitis   Joint disease strongly associated with B27 haplotype.

Antibody   Serum protein produced in response to an antigen and capable of binding to the antigen.

Antibody diversity   The variety of antibodies produced or capable of being produced by an organism.

Anticodon   Three-base sequence in tRNA that undergoes base pairing with corresponding codon in mRNA.

Antigen   Any substance able to elicit an immune response.

Antigen-binding site   That part on an antibody molecule that adheres to the corresponding antigen.

Antigenic determinant   That part of an antigen that stimulates an immune response against itself.

Antiparallel   Orientation of strands in duplex DNA, one running $3'$ to $5'$ as the other runs $5'$ to $3'$.

Antisense strand   The DNA strand that is the complement of the sense strand.

Artificial selection   Selection imposed by a breeder in which individuals of only certain phenotypes are permitted to breed.

Asexual inheritance   See Inheritance.

Assortative mating   Mating based on phenotype. In positive assortative mating, like phenotypes mate more frequently than would be expected by chance; in negative assortative mating, the reverse occurs.

Attachment site   The site at which bacteriophage DNA is integrated into the host DNA.

Autoimmune disease   Disorder marked by partial breakdown of immunological tolerance.

Autoradiography   Procedure whereby cellular structures such as chromosomes or DNA photograph themselves by means of incorporated radioactive atoms.

Autosome   Any chromosome that is not a sex chromosome.

Average fitness   The arithmetic mean of the fitnesses of all individuals in a population at a given time.

Avian sarcoma virus   A retrovirus causing connective tissue tumors in birds.

B   A group of human chromosomes comprising numbers 4 and 5. Also, a blood type in the ABO blood type system.

Bacteriophage   A virus that infects bacterial cells.

Balanced   A chromosomal rearrangement in which no genes are lost or present in excess.

Balanced translocation   Translocation in which all genes are represented the normal number of times.

Baldness   See Common baldness.

Barr body   See Sex-chromatin body.

Base   A chemical constituent of nucleotides in DNA and RNA. Five principal bases are of significance in genetics: adenine, guanine, thymine (found only in DNA), cytosine, and uracil (found only in RNA). Pyrimidine bases contain a single carbon-nitrogen ring; purine bases have a fused double-ring structure.

Base analogue   A substance chemically similar to one of the normal bases and incorporated into DNA.

Base substitution   Replacement of one nucleotide with a different nucleotide in DNA.

B cell   Lymphocyte responsible for synthesis of antibody.

B DNA   DNA in its right-handed double-helix configuration.

Beta hemoglobin   One of the polypeptide chains in the predominant form of adult hemoglobin.

Binding   Adherence.

Bivalent   Chromosomal structure occurring only in meiosis I and consisting of two replicated and paired homologous chromosomes held together by chiasmata.

Blastocyst   Hollow ball of cells containing the inner cell mass from which the embryo develops.

Blood group   A collection of alternative red blood cell antigens inherited in simple Mendelian fashion, like ABO, Rh, MN, etc.

bp   Base pairs; a double-stranded DNA molecule containing 100 nucleotides is 100 bp long.

Brachydactyly   A dominantly inherited form of short fingers.

Broad-sense heritability   Fraction of total phenotypic variation in a multifactorial trait that is attributable to genetic differences among individuals, including both transmissible and nontransmissible genetic differences.

Bubble   In DNA replication, an oval structure of DNA with a replication fork at each end.

BUdR   Bromodeoxyuridine, an analogue of thymidine.

C   Conventional symbol for the deoxyribonucleotide deoxycytidine phosphate or for the ribonucleotide cytidine phosphate or for the base cytosine. Also, a group of human chromosomes comprising 6 through 12 plus the X.

cAMP　Cyclic adenosine monophosphate; important regulatory molecule.

Capsid　Part of a virus that encloses the genetic material.

Cap site　Position at which is added an unusual nucleotide in an unusual orientation to the 5′ end of eukaryotic mRNAs forming a cap.

Carboxyl end　The end of a polypeptide that has a free carboxyl (—COOH) group.

Carcinogen　An agent that causes cancer.

Carrier　Heterozygous, as for a chromosome abnormality or mutant gene.

Cell　The smallest unit of any organism that exhibits the properties of living matter.

Cell cooperation　Interaction among cells essential in a normal immune response.

Cell fusion　The joining together of somatic cells from two individuals or species to form a new type of hybrid cell with the chromosomes of both.

Centromere　Part of a chromosome connected to spindle fibers and involved in normal chromosome movement.

Chiasma　Cross-shaped structure connecting nonsister chromatids of homologous chromosomes.

Chimera　See Mosaic.

Chromatid　One of the two strands produced by chromosome replication and sharing a common centromere.

Chromatin　Fundamental structural fiber of the chromosome; visible in interphase as a diffuse granular-appearing substance.

Chromosomal mosaic　An individual composed of two or more chromosomally different types of cells.

Chromosome　Structure containing the genetic material in a highly condensed chemical form and visible with appropriate staining through the light microscope.

Chromosome breakage　Interruption of the physical continuity of a chromosome.

Chromosome fusion　Physical joining of chromosomes or parts of chromosomes.

Chronic dose　An amount of radiation delivered as a series of smaller doses.

Chronic myelogenous leukemia　A type of leukemia frequently found in association with a particular translocation involving chromosomes 9 and 22.

Class　See Immunoglobulin class.

Clonal selection　Process in which an antigenic determinant stimulates mitosis of appropriate B-cell precursor leading to a clone.

Clone　A group of genetically identical cells.

Cloning　See DNA cloning.

Coding sequence　Sequence in DNA or RNA that codes for amino acids.

Codominant　Alleles that are both expressed in heterozygotes.

Codon　Three-base sequence in mRNA specifying an amino acid.

Cohesive ends　Complementary single-stranded ends of lambda DNA.

Colchicine　A drug that inhibits formation of the spindle.

Color blindness　Inability to discriminate certain colors; the common forms are the red and green types, both due to X-linked recessives.

Combinatorial joining　Theory that antibody genes are formed by DNA splicing of sequences corresponding to parts of the complete antibody gene.

Common ancestor　An ancestor of both one's mother and one's father.

Common baldness　Thought to be due to an allele that is dominant in males but recessive in females.

Complementary pairing　Pairing of bases facilitated by hydrogen bonds; in genetics the most important examples of complementary pairing are A-T (or A-U) and G-C.

Complementation　Genetic situation in which an individual carrying two recessive alleles expresses a normal phenotype owing to the alleles being at different loci and therefore doubly heterozygous.

Concordance Among a group of pairs of individuals, the proportion of pairs in which both members have a particular trait when at least one member of each pair has the trait.

Conditional mutation A mutation whose phenotypic effects are expressed under some environmental conditions but not under others, like a temperature-sensitive mutation.

Congenital Present from birth.

Consanguineous mating A mating between relatives.

Constant region The portion of a heavy chain or light chain that varies little in amino acid sequence among antibodies within the same subclass.

Covalent bond Strong chemical bond formed by two atoms that share a common electron.

Crossing-over Physical exchange of parts between nonsister chromatids of homologous chromosomes.

Cultural inheritance See Inheritance.

Curie Unit of radiation equal to 3.7×10^{10} disintegrations per second.

Cystic fibrosis A recessively inherited glandular disorder.

Cytoplasm All cellular substances except the nucleus.

Cytoplasmic inheritance See Inheritance.

Cytosine Pyrimidine base in DNA and RNA.

D (1) A group of human chromosomes comprising 13 through 15. (2) A class of immunoglobulins containing delta heavy chains.

Deamination Removal of an —NH_2 group from a molecule.

Deficiency Having a gene or part of a chromosome missing.

Deletion See Deficiency.

Delta hemoglobin The betalike chain in the minority form of adult hemoglobin.

Denaturation Unfolding of a protein molecule to render it nonfunctional; also separation of complementary strands of duplex DNA.

Deoxyribonucleotide A nucleotide in which the sugar is deoxyribose.

Development Process by which a new organism is formed from the zygote.

Deviation Difference.

Dicentric A chromosome with two centromeres.

Differentiation Occurrence of differing allele frequencies among different populations. See also Development.

Diploid Possessing two complete sets of homologous chromosomes.

Directional selection Selection favoring either phenotypic extreme.

Dispersed repeated gene family Group of identical or similar transposable elements found at scattered sites in the chromosomes.

Disruptive selection Selection simultaneously favoring both extremes of the phenotypic range.

Distribution Statistical description of the phenotypes that occur in a population and their relative proportions.

DNA Deoxyribonucleic acid; composed of two intertwined strands, with each strand being a long sequence of nucleotides.

DNA cloning Production of many identical copies of a defined fragment of DNA.

DNA ligase An enzyme that covalently connects the 3′ end of one DNA strand with the 5′ end of another.

DNA polymerase   An enzyme that catalyzes the production of a DNA strand that is complementary in base sequence to a template strand.

DNA splicing   Excision of certain DNA sequences in the production of an intact antibody gene.

Dominant   An allele that conceals the presence of another allele—the recessive—in heterozygotes.

Dosage   Refers to the relative number of copies of genes or chromosomes.

Dosage compensation   Regulation of X-linked genes in such a way that such genes are equally active in males and females.

Dose-response relationship   Description of how the amount of genetic or biological damage from radiation depends on the dose.

Double helix   A duplex DNA molecule that has a coiled or helical configuration.

Doubling dose   The amount of radiation necessary to increase mutation or biological damage to twice the background level.

Downstream   In RNA or the antisense strand of DNA, toward the 3' end; in the sense strand of DNA, toward the 5' end.

Down syndrome   Trisomy 21—i.e., 47, +21.

Down-syndrome translocation   Robertsonian translocation involving chromosome 21.

Duplex DNA   DNA molecule that has two complementary strands.

Duplication   Occurrence of one excess copy of a gene or region of chromosome. Also used in a more general sense to indicate any number of excess copies of a gene or region of chromosome.

DZ twins   Nonidentical twins; twins arising from two fertilized eggs.

E   (1) A group of human chromosomes comprising 16 through 18. (2) A class of immunoglobulins containing epsilon heavy chains.

Ear cerumen   Waxy substance produced in the ear canal.

Edwards syndrome   Trisomy 18—i.e., 47, +18.

Electrophoresis   Separation of molecules by means of an electric field applied across a jellylike slab.

Embryo   In humans, the developing organism from the second to the seventh week of development.

Emphysema   A lung disorder characterized by distension of air sacs in the lungs. See also Familial emphysema.

Endogenous virus   Chromosomal DNA sequence or sequences obviously similar in structure and characteristics to known integrated proviruses.

Envelope   Part of a virus that encloses the capsid.

Enzyme   A protein able to accelerate certain chemical reactions without itself being altered in the process.

Epidemic   A local outbreak of disease.

Epitope   An antigenic determinant.

Epsilon hemoglobin   One of the polypeptide chains of embryonic hemoglobin.

Equilibrium   A state of stasis; absence of change.

Erythroblastosis fetalis   See Hemolytic disease of the newborn.

Eukaryote   An organism whose cells normally possess a nucleus surrounded by a nuclear envelope and in which true mitosis and/or meiosis occurs.

Evolution   Cumulative change in the genetic characterisics of a species through time.

Excision repair   Process in which abnormal or improper nucleotides in a strand of DNA are cut out of the molecule and the excised portion is resynthesized by enzymes that use the remaining strand as a template.

Exon   A sequence of DNA that codes for amino acids. See also Coding sequence.

Expressivity   The degree or severity with which a trait is expressed; *variable expressivity* refers to the varying severity of a trait among different individuals.

F   A group of human chromosomes comprising 19 and 20. Also, symbol for the fixation index as a measure of genetic divergence.

Familial   Tending to occur in groups of relatives.

Familial emphysema   An inherited lung disease associated with a deficiency of the protein α_1-antitrypsin.

Familial hypercholesterolemia   Inherited high cholesterol levels due to a dominant gene.

Favism   See G6PD deficiency.

Feedback inhibition   The rendering inactive of the first enzyme in a metabolic pathway by the end product of the pathway.

Fetus   In humans, the developing organism from the seventh week to term.

First-division nondisjunction   Nondisjunction occurring in the first meiotic division.

Fitness   Measure of ability to survive and reproduce.

Five-prime (5′) end   The end of a DNA or RNA strand that has a free 5′ phosphate.

Fixation   State at which the allele frequency of a designated allele is 1.0.

Fixation index   Measure of genetic differentiation defined as the proportionate reduction in average heterozygosity as compared with the total heterozygosity that would be expected with random mating.

Fork   Y-shaped structure defining the position of unwinding during DNA replication.

Forward mutation   Change of a normal allele into a mutant form.

Founder effect   Genetic differences between an original population and an offshoot population caused by the alleles in the founders of the offshoot population being nonrepresentative of the original population; a special variety of random genetic drift.

Fragmentation mapping   Production of chromosome fragments by x-rays in somatic cells to enable the localization of genes of interest.

Frameshift mutation   Insertion or deletion of a number of nucleotides in a coding region to alter the translational reading frame downstream of the mutation.

Fraternal twins   See DZ twins.

Frequency   Proportion.

G   (1) Conventional symbol for the deoxyribonucleotide deoxyguanosine phosphate or for the ribonucleotide guanosine phosphate or for the base guanine. (2) A group of human chromosomes comprising numbers 21, 22, and the Y. (3) A class of immunoglobulins containing gamma heavy chains.

Gamete   Reproductive cell, such as sperm or egg in animals.

Gamma hemoglobin   The betalike chain in fetal homoglobin.

G bands   Chromosome bands produced by Giemsa stain.

Gene   The fundamental unit of inheritance; used in several different senses such as unit of function, of mutation, and of recombination.

Gene conversion   The change of one allele to another by means of enzymatic comparison and alteration.

Gene family   Group of genes that has evolved from a common ancestral gene.

Gene frequency   See Allele frequency.

Generalized transduction   Transduction that can involve virtually any genes of the host.

Gene substitution   Replacement of one allele by another in the course of evolution.

Genetic code   Correspondence between three-base codons in mRNA and amino acids incorporated into a polypeptide.

Genetic counseling   Imparting advice and knowledge to individuals with regard to their own genetic constitution or with regard to their offspring or other relatives.

Genetic diversity   Occurrence of alternative genotypes at many loci in a population.

Genetic map   Diagram showing relative positions of genes along a chromosome.

Genome   Refers collectively to all genes carried by a cell.

Genotype   An individual's genetic constitution.

Genotype-environment association   Nonrandom occurrence of genotype-environment combinations.

Genotype-environment interaction   Inability to predict the relative performance of genotypes in one environment from knowledge of their relative performance in a different environment.

Genotype frequency   Relative proportion of individuals who have a designated genotype.

Germinal mutation   A mutation occurring in a germ cell.

Germ-line theory   Theory that antibody diversity has its origin in a large number of distinct genes in the germ line.

Gonadal dose   The amount of radiation reaching the gonads.

Gonads   The sexual organs in which gametes are formed.

Gout   Severe painful swelling of certain joints associated with excess uric acid.

G6PD   Glucose 6-phosphate dehydrogenase enzyme, or the gene coding for the enzyme.

G6PD deficiency   Disease associated with lack or low levels of enzyme glucose 6-phosphate dehydrogenase; symptoms include anemia induced by certain drugs or by fava beans.

H-2   Designation of mouse MHC.

Haploid   Possessing only one set of chromosomes.

Haplotype   The group of particular alleles carried by loci of the MHC on one chromosome.

Hardy-Weinberg rule   Genotype frequencies expected with random mating.

HAT medium   Growth medium containing hypoxanthine, aminopterin, and thymine, which selects for *HGPRT*$^+$, *TK*$^+$ cells.

Heavy chain   The longest type of polypeptide in an antibody.

Hemizygous   Refers to X-linked genes in males to which the terms *heterozygous* and *homozygous* do not apply.

Hemolytic disease of the newborn   Blood disorder caused by breakdown of fetal Rh$^+$ cells by mother's anti-Rh$^+$ antibody crossing the placenta.

Hemophilia   Disorder characterized by abnormally slow or absent blood clotting; the best-known form is an X-linked condition associated with Queen Victoria's descendants and known as Royal hemophilia.

Heritability   Fraction of total phenotypic variance due to genetic factors. Broad-sense heritability refers to all genetic factors; narrow-sense heritability refers only to transmissible genetic factors.

Hermaphroditism   Presence of functional male and female sexual structures in the same individual.

Herpesvirus   Type of virus associated with characteristic runny cold sores.

Heterogeneous nuclear RNA   Refers to large transcripts found in eukaryotic nuclei.

Heterozygosity   Proportion of heterozygotes at a locus, or average proportion of heterozygotes among a group of loci.

Heterozygote inferiority   A situation in which the fitness of a heterozygote is less than that of both homozygotes.

Heterozygous   A genetic situation in which homologous loci in an individual carry different alleles.

HGPRT The enzyme hypoxanthine guanine phosphoribosyl transferase, or the locus that codes for the enzyme.

HGPRT deficiency See Lesch-Nyhan syndrome.

Highly repetitive DNA Highly redundant, simple-sequence DNA tending to be associated with centromeric regions.

Histocompatibility Acceptance by a recipient of transplanted tissue from a donor.

Histone Type of protein rich in lysine and arginine found in association with DNA.

HLA Designation of human MHC.

hnRNA See Heterogeneous nuclear RNA.

Homologous Genetically matched in some way, as two chromosomes matched in size and structure and carrying alleles of the same genes, or loci at corresponding positions on homologous chromosomes.

Homozygous A genetic situation in which homologous loci in an individual carry the same allele.

Host specific A virus or parasite able to live only within one or a restricted range of hosts.

H substance The precursor of A and B antigens in the ABO blood groups.

Huntington disease Autosomal dominant degeneration of the neuromuscular system having an age of onset approximately in midlife. Also called Huntington chorea.

Hybrid cells Cells formed by cell fusion that have the chromosomes of both parental cells.

Hybridization In nucleic acid chemistry, the coming together of nucleic acids that have sufficiently complementary base sequences. In genetics in general, formation of offspring (hybrids) by mating of two species or varieties.

Hydrogen bond Weak chemical bond formed by nearby atoms such as oxygen and/or nitrogen exerting an attraction on a common hydrogen.

Hypercholesterolemia See Familial hypercholesterolemia.

Immune response The body's reaction to viruses, foreign cells, or tissues.

Immune surveillance Theory that one of the important normal functions of the immune system is to screen the body for and destroy cells that have undergone malignant transformation.

Immune tolerance See Tolerance.

Immunodeficiency disease Any disease in which part of the immune response is absent or weak.

Immunoglobulin Antibody molecule.

Immunoglobulin class Type of immunoglobulin defined by the nature of its heavy polypeptide chain.

Inborn errors of metabolism Genetic defects leading to nonfunctional enzymes that upset body chemistry.

Inbreeding Mating between relatives.

Inbreeding coefficient Measure of the genetic effects of inbreeding in terms of the proportionate reduction in heterozygosity in an inbred individual as compared with the heterozygosity expected with random mating.

Incomplete penetrance See Penetrance.

Independent assortment Genetic situation in which the segregation of one pair of alleles has no influence on the segregation of a pair of alleles at a different locus; Mendel's second law.

Inducer A substance that permits genes to be expressed. In prokaryotes, inducers bind with their corresponding repressors, thus permitting transcription and translation.

Influenza Respiratory illness caused by a virus.

Inheritance Passage of genetic elements from one generation to the next. *Sexual inheritance* involves the process of meiosis. *Asexual inheritance* involves mitosis or, in prokaryotes, fission. *Cytoplasmic inheritance* involves synthesis and distribution of cytoplasmic organelles such as mitochondria. *Cultural inheritance* involves learned behavior transmitted by social or cultural means.

Inhibition Of enzymes, rendering inactive or nonfunctional.

Initiation codon First codon translated, typically $5' - AUG - 3'$ in cytoplasmic mRNAs.

Interphase A stage of cell division preceding prophase in which chromosomes are not condensed and characterized by the occurrence of DNA synthesis; conventionally separated into three parts—G_1 (the first part, prior to DNA synthesis), S (the second part, in which DNA synthesis occurs), and G_2 (the third part, after DNA synthesis has ceased). Although DNA synthesis occurs in interphase of mitosis and interphase I of meiosis, interphase II of meiosis is exceptional in that DNA synthesis does not occur.

Intervening sequence Sequence of DNA that interrupts coding sequences.

Intron See Intervening sequence.

Inversion A chromosome with part of its genes in reverse of the normal order.

Inversion loop Circular structure formed by homologous pairing between the inverted region of a chromosome and its normal homologue.

Karyotype Chromosomal constitution.

kb Kilobase pairs; a unit of nucleic acid length consisting of 1000 nucleotides.

Kindred A group of relatives.

Klinefelter syndrome The 47,XXY sex-chromosome abnormality; affected individuals are male.

*Lac*I **gene** Codes for the repressor of the lactose operon in *E. coli*.

Lactose Milk sugar, composed of linked pair of galactose and glucose sugars.

*Lac*Y **gene** Codes for β-galactoside permease in *E. coli*.

*Lac*Z **gene** Codes for β-galactosidase in *E. coli*.

Lambda A bacteriophage that infects *E. coli*.

Left-handed DNA See Z DNA.

Lesch-Nyhan syndrome Severe X-linked nervous disorder due to lack of enzyme HGPRT.

Lethal A condition that is incompatible with life.

Liability Risk or predisposition.

Ligase See DNA ligase.

Light chain The shortest type of polypeptide in an antibody.

Linkage Genetic situation in which alleles at two loci do not exhibit independent assortment owing to their being sufficiently close together on the same chromosome.

Locus Position on a chromosome.

Loss State at which the allele frequency of a designated allele is 0.

Lymphocyte White blood cell.

Lyon hypothesis Dosage compensation in mammals by the inactivation of one X chromosome in each somatic cell of females.

Lysis The bursting open of virus-infected or bacteriophage-infected cells.

Lysogeny Alternative life cycle of temperate bacteriophage in which bacteriophage DNA becomes incorporated into that of the host.

Lytic cycle Viral or bacteriophage life cycle in which progeny viruses are formed and the host cell is killed as these are liberated.

M   A class of immunoglobulins containing μ heavy chains.

Macrophage   Accessory cell important in the immune response.

Major histocompatibility complex   Group of tightly linked loci involved with important cellular antigens and other aspects of the immune response.

Map distance   A measure of the separation between two syntenic loci; obtained by summation of recombination fractions between intervening loci. For loci sufficiently close together, the map distance equals the recombination fraction.

Map unit   The unit of distance on a genetic map; each map unit corresponds to 1 percent recombination if the loci involved are sufficiently close together.

Mean   Arithmetic average.

Meiosis   Process of cell division resulting in production of gametes.

Messenger RNA   See mRNA.

Metabolism   Body chemistry.

Metacentric   A chromosome with its centromere approximately at its center.

Metaphase   A stage of cell division preceding anaphase in which chromosomes align on the metaphase plate. In metaphase of mitosis or metaphase II of meiosis, each chromosome consists of two chromatids as it aligns. In metaphase I of meiosis, the bivalents—paired and replicated homologous chromosomes—align.

MHC   Major histocompatibility complex.

Midparent   Average of the phenotypes of the parents.

Migration   Movement of individuals or gametes among populations.

Missense mutation   A mutation that creates an amino acid substitution.

Mitosis   A mode of cell division in eukaryotes in which a dividing cell produces two genetically identical daughter cells.

Mitotic nondisjunction   Nondisjunction occurring in mitosis.

MN   A blood group system in humans controlled by two codominant alleles.

Moderately repetitive DNA   DNA sequences found tens to hundreds of times per haploid genome and tending to be dispersed throughout the chromosomes.

Modification   Chemical alteration of certain bases in DNA or RNA.

Modifer gene   A gene or allele that influences the expression of a trait associated with a different locus.

Monoclonal   Antibodies produced by a single clone of cells.

Monomorphic locus   Conventionally, a locus at which the most common allele has an allele frequency greater than 0.95.

Monosomic   Having one chromosome missing in an otherwise diploid individual.

Morula   The small clump of cells formed by successive divisions of the zygote.

Mosaic   An individual composed of two or more genetically different types of cells.

mRNA   Fully processed transcript exported to the cytoplasm for translation.

Müllerian ducts   Embryonic ducts associated with internal sexual structures of females.

Multifactorial   A trait whose occurrence or expression is determined by the combined effects of alleles at several or many loci and usually also by environmental factors.

Multiple alleles   The occurrence of more than two alleles at a locus in a population of organisms.

Mutagen   An agent that causes mutations.

Mutagenesis   Creation of mutations.

Mutation   A heritable chemical alteration of the genetic material.

Mutation rate   Probability of occurrence of a new mutation in a gene in one generation.

MZ twins   Monozygotic twins; twins arising from a single fertilized egg.

Narrow-sense heritability   Fraction of total phenotypic variation in a multifactorial trait that is due to transmissible genetic differences among individuals.

Natural selection   Process in which individuals who are best able to survive and reproduce in a particular environment leave a disproportionate share of the offspring and thus gradually increase the overall ability of the population to survive and reproduce in the environment.

Negative control   Manner of regulation in which a gene is transcribed unless it is complexed with a particular protein.

Neutral allele   An allele that has no effects on its carrier's ability to survive or reproduce.

Nitrous acid   A potent deaminating agent and mutagen.

Noncomplementation   Genetic situation in which an individual carrying two recessive alleles expresses a mutant phenotype owing to the alleles being at the same locus and therefore homozygous.

Nondisjunction   Cell division occurring in such a way that daughter cells have extra or missing chromosomes.

Nonsense mutation   A mutation that creates a chain-terminating codon.

Nonsense suppressor   A mutation in a tRNA gene that can read through a termination codon and so suppress nonsense mutations.

Nonsister chromatids   Chromatids that originate by replication of different (though possibly homologous) chromosomes.

Normal distribution   The frequently encountered smooth, bell-shaped distribution curve.

Nucleosome   Tiny spherical particle containing DNA in association with histone molecules.

Nucleosome fiber   Chromosome fiber formed by adjacent nucleosomes.

Nucleotide   A unit of nucleic acid structure consisting of a five-carbon sugar to which is attached a phosphate group (on the $5'$ carbon of the sugar) and a base (on the $1'$ carbon of the sugar). In DNA the sugar is deoxyribose; in RNA it is ribose.

Nucleotide substitution   Replacement of one nucleotide with a different nucleotide in DNA.

Nucleus   The cellular structure containing chromatin.

O   A blood type in the ABO blood group system. Also used to designate the absence of a chromosome, as in the designation XO for 45,X Turner syndrome.

Ochre $5' - UAA - 3'$ termination codon.

Oncogenic   Refers to cancer-causing virus.

One-egg twins   See MZ twins.

Oocyte   Cell destined to undergo meiosis to produce an egg.

Opal $5' - UGA - 3'$ termination codon; also called umber.

Operator   A binding site for repressor in prokaryotic operons.

Operon   In prokaryotes, a group of genes transcribed together and thereby simultaneously expressed.

Overdominance   A situation in which the fitness of a heterozygote is greater than the fitness of both homozygotes.

Pandemic   A worldwide outbreak of disease.

Paracentric inversion   A chromosome with an inverted region that does not include the centromere.

Partial dominance   Any situation in which the phenotype of the heterozygote is between the phenotypes of the corresponding homozygotes, but excluding cases in which the phenotype of the heterozygote is the same as the phenotype of one of the homozygotes.

Patau syndrome   Trisomy 13—i.e., $47, +13$.

Pedigree   A diagram expressing the genetic relationships among individuals.

Penetrance   The proportion of cases in which individuals who have a relevant genotype express the corresponding phenotype. *Complete penetrance* refers to a penetrance of 100 percent. *Incomplete penetrance* refers to any degree of penetrance less than 100 percent.

Peptide bond   Chemical linkage between the carboxyl (—COOH) group of one amino acid and the amino (—NH$_2$) group of another; characteristic linkage in polypeptides.

Pericentric inversion   A chromosome with an inverted region that includes the centromere.

Permissive temperature   See Temperature-sensitive mutation.

Phenocopy   An environmentally caused trait resembling one known to be inherited.

Phenotype   An individual's appearance with respect to a trait, either physical, mental, or biochemical.

Phenotypic variance   The variance in phenotype among individuals in a population.

Phenylketonuria   Recessively inherited defect in phenylalanine metabolism leading to severe mental retardation if untreated by a low-phenylalanine diet.

Phosphate group   Chemical unit with the general molecular formula —O—PO$_3$H$_2$.

Placenta   Organ responsible for the exchange of gases and nutrients between a mother's bloodstream and that of her embryo.

Plasma cell   A B-cell derivative that secretes antibody.

Plasmid   In prokaryotes, a self-replicating molecule of DNA.

Polar body   Tiny product of meiosis I or meiosis II in females. The polar body produced in meiosis I sometimes undergoes meiosis II to produce two even tinier polar bodies.

Poly-A tail   Run of consecutive A's added to the 3′ end of many eukaryotic transcripts.

Polydactyly   Extra fingers or toes; inherited as a dominant in some kindreds, multifactorial in others.

Polygenic   See Multifactorial.

Polymorphic locus   Conventionally, a locus at which the allele frequency of the most common allele is less than 0.95.

Polyoma   A rodent DNA-bearing tumor virus.

Polypeptide   Linear string of amino acids linked by peptide bonds.

Polyploid   Having more than two haploid sets of chromosomes.

Polysomic   A diploid individual who has one or more extra copies of a particular chromosome.

Polytene   Giant chromosomes found in salivary glands and other tissues of certain dipterans formed by several consecutive replications with no intervening cell division.

Population   Group of members of the same species that usually find mates within their own group.

Positive control   Manner of regulation in which a gene is not transcribed unless the appropriate region is complexed with a particular protein.

Primary sex ratio   The sex ratio at the time of fertilization.

Primary structure   The amino acid sequence of a polypeptide.

Primordial germ cells   Cells set aside in early development that are destined to give rise to spermatocytes or oocytes.

Probability   Mathematical expression of the degree of confidence that certain events will or will not occur.

Processing   See RNA processing.

Prokaryote   An organism whose cells lack a well-defined nucleus and in which cells divide by means of fission.

Promoter   Recognition sequence for binding of RNA polymerase.

Prophage   Bacteriophage DNA that has been integrated into the host DNA.

Prophase   A stage of cell division preceding metaphase in which chromosomes first become visible in the light microscope. In prophase of mitosis the chromosomes are visibly double as soon as they appear; each chromosome consists of two sister chromatids sharing a common centromere. In prophase I of meiosis the chromosomes appear as single strands when first visible; prophase I is a relatively long stage characterized by (1) the first appearance of the chromosomes, (2) pairing of homologous chromosomes, (3) paired chromosomes becoming visibly double producing a four-stranded bivalent, (4) increased thickening of chromosomes leading to the clear appearance of chiasmata, and (5) apparent repulsion of homologous centromeres. Prophase II of meiosis is a relatively brief stage in which chromosomes recondense after a slight relaxation in interphase II.

Protein   A molecule composed of one or more amino-acid-containing polypeptide chains.

Protein kinase gene   A gene coding for an enzyme that adds phosphate groups to other proteins or enzymes and thereby influences their ability to function.

Provirus   Circular duplex DNA complementary in sequence to a retrovirus RNA.

Pseudogene   A DNA sequence that has obvious homology with a known gene but is neither transcribed nor translated.

Pseudohermaphroditism   Presence of nonfunctional male and female sexual structures in the same individual.

Punnett square   A cross-multiplication square devised to aid in calculation of types of offspring from a mating and their expected frequencies.

Purine   Base formed from a fused carbon-nitrogen double ring.

Pyrimidine   Base formed from a single carbon-nitrogen ring.

Q bands   Fluorescent chromosome bands produced by such substances as quinacrine.

Quadrivalent   Four-armed structure formed by meiotic pairing involving a reciprocal translocation and its two normal homologues.

Quantitative trait   A trait that can be measured in terms of a relatively continuous scale, such as height or weight.

Race   A population or group of populations that has undergone sufficient genetic differentiation to be regarded as distinctive in some arbitrary sense but not so much genetic differentiation as to be regarded as a subspecies.

Rad   Unit of absorbed radiation dose in terms of the energy imparted per mass of absorbing material; 1 rad equals 100 erg per gram of absorbing material.

Radiation   Emission of energy or subatomic particles from certain substances.

Random genetic drift   Fluctuation in allele frequency from generation to generation due to restricted population size.

Random mating   Mating without regard to genotype or phenotype.

Rate of mutation   See Mutation rate.

Ratio   Relative proportions, such as 1:2:1 ratio of genotypes in a mating of two heterozygotes, or a 3:1 ratio of phenotypes from the same mating when one of the alleles is dominant.

Receptor site   Site on a B cell or T cell that combines with the stimulatory antigenic determinant.

Recessive   An allele that must be homozygous for the corresponding trait to be expressed.

Reciprocal translocation   Interchange of parts between nonhomologous chromosomes.

Recombinant   A chromosome or DNA molecule that carries genetic information from each of two parental chromosomes or DNA molecules.

Recombinant DNA   See DNA cloning.

Recombination   Production of DNA molecule or chromosome carrying hereditary determinants from each of two preexisting DNA molecules or chromosomes.

Recombination fraction   Proportion of gametes produced by an individual that have undergone recombination.

Recurrence risk   Risk that an inherited disorder will occur again in a family.

Relative fitness   Measure of one genotype's fitness as a proportion of another genotype's fitness.

Rem   Unit of radiation dose equivalence used for comparing different types of radiaiton for purposes of radiation protection. The dose in rems is obtained by multiplying dose in rads by a quantity related to the type of radiation and by another quantity related to the distribution of the radiation to different parts of the body. For x-rays or electrons uniformly received, 1 rem equals 1 rad.

Replication   Synthesis of two daughter DNA molecules from a single parental molecule.

Repression   Rendering a gene unexpressible.

Repressor   A molecule, usually a protein, that binds to DNA and so prevents transcription.

Response   Change in population mean in one generation of selection.

Restriction enzyme   An enzyme that cuts duplex DNA at sites that have a particular nucleotide sequence.

Restriction fragment   A DNA molecule produced by cutting a larger molecule with a restriction enzyme.

Restriction map   Diagram of a DNA molecule showing the positions of one or more restriction sites.

Restriction site   The nucleotide sequence recognized by a restriction enzyme.

Restriction-site polymorphism   Refers to genetic variation detected by the presence or absence of a restriction site at particular location in the DNA.

Restrictive temperature   See Temperature-sensitive mutation.

Retinoblastoma: Autosomal dominant disorder characterized by retinal malignancies.

Retrovirus   RNA viruses that produce complementary DNA strands upon infecting cells.

Reverse mutation   Change of a mutant allele back into a normal form.

Reverse transcriptase   An enzyme able to catalyze the synthesis of a complementary DNA strand from an RNA template.

Rh blood groups   Various antigens on the surface of red blood cells coded by alleles at a cluster of tightly linked loci or multiple alleles at a single locus.

Rheumatoid arthritis   A painful joint condition caused by partial breakdown of immunological tolerance.

Ribonucleic acid   See RNA.

Ribonucleotide   Chemical subunit of ribonucleic acid consisting of a ribose sugar to which is attached a phosphate (at its 5' carbon) and a base (at its 1' carbon).

Ribose   The five-carbon sugar found in ribonucleic acid.

Ribosomal RNA   See rRNA.

Ribosome   Cellular organelle that is the structure on which protein synthesis occurs.

RNA   Ribonucleic acid, composed of a linear string of chemical subunits called ribonucleotides; serves several functions in the eukaryotic cell, such as mRNA, rRNA, and tRNA; serves as the genetic material in some viruses.

RNA polymerase   An enzyme capable of catalyzing transcription.

RNA processing   Alteration of transcripts that occurs in the nuclei of eukaryotic cells; includes splicing, capping, and polyadenylation.

RNA splicing   Excision of intervening sequences from eukaryotic transcripts.

RNA transcript   The RNA molecule produced in transcription that is complementary in base sequence to a sense strand of DNA.

Robertsonian translocation   Translocation involving the long arms of two acrocentric chromosomes.

Roentgen   Unit of radiation exposure in terms of the number of ions produced per mass of air; 1 r equals 2.58×10^{-4} coulombs per kilogram of air.

rRNA   The RNA molecules that form part of ribosomes.

Rubella   German measles virus; causes birth defects in embryos.

Satellite   Inconsistently staining blob of chromosomal material attached to the tip of certain chromosomes.

Screening   A survey of phenotypes of a large number of cells or individuals. In genetic counseling, usually refers to widespread tests for individuals with a genetic disorder or for carriers.

Secondary sex ratio   The sex ratio at the time of birth.

Secretor   Autosomal dominant trait associated with ability to secrete A and B antigens of ABO blood groups in body fluids.

Segregation   Separation of homologous chromosomes or of alternative alleles into distinct gametes during meiosis; Mendel's first law.

Selection   In mutation studies, a procedure designed in such a way that only a desired type of cell can survive, as in selection for resistance to an antibiotic. In evolutionary studies, intrinsic differences in the ability of genotypes to survive and reproduce. In animal and plant breeding, choosing individuals with certain phenotypes to be parents of the next generation.

Selection coefficient   The amount by which relative fitness is reduced or increased.

Selfish DNA   DNA sequences that increase in number and spread throughout a population by virtue of their ability to undergo transposition.

Semiconservative   Mode of DNA replication in which each daughter molecule contains one of the parental DNA strands intact.

Sense strand   The DNA strand that is transcribed.

Serum   The fluid part of the blood.

Sex chromosome   An X chromosome or a Y chromosome.

Sex influenced   A trait whose expression or severity differs between the sexes.

Sex limited   A trait that is expressed in only one sex.

Sex linked   A trait determined by genes on a sex chromosome, usually the X.

Sex ratio   Relative proportions of males and females in a population.

Sexual inheritance   See Inheritance.

Sickle cell anemia   A recessively inherited severe anemia due to amino acid substitution in the β-globin chain; heterozygotes tend to be more resistant to falciparum malaria than are normal homozygotes.

Silent mutation   A mutation that creates a synonymous codon and so does not alter the amino acid sequence.

Simple Mendelian   A trait whose occurrence is determined by a major gene such as a dominant or recessive and that thereby occurs in pedigrees according to Mendel's laws.

Single active X principle   The inactivation of all X chromosomes except one in cells of mammals.

Sister chromatid   The partner strand of a chromatid attached to a common centromere.

Somatic cell   Any cell in the body except gametes and cells that give rise to gametes.

Somatic cell genetics   The study of the inherited characteristics of somatic cells grown under laboratory conditions.

Somatic mutation   A mutation occurring in a somatic cell.

Somatic theory   Theory that antibody diversity has its origin in somatic events such as mutation and somatic recombination or gene conversion.

Specialized transduction   Transduction that can involve at most a few genes of the host.

Species   Group of actually or potentially interbreeding organisms that is reproductively isolated from other such groups.

Spermatocyte   Cell destined to undergo meiosis to produce sperm.

Spikes   Projections from surface of influenza virus carrying important antigenic determinants.

Spindle   Football-shaped network of filaments arching through the cell during cell division and serving as the framework for chromosomal and cytoplasmic division.

Splicing   See RNA splicing and DNA splicing.

Split gene   A gene containing one or more intervening sequences.

Spontaneous abortion   Natural premature termination of pregnancy often associated with fetal abnormalities.

Spontaneous mutation   A mutation occurring in the absence of any known mutagen.

Stabilizing selection   Selection favoring intermediate phenotypes.

Stable equilibrium   Of allele frequency, an equilibrium that is automatically and gradually reestablished after a perturbation of allele frequencies.

Standard deviation   The square root of the variance.

Sterile mutation   A mutation that causes infertility.

Subclass   Refers to immunoglobulins within the same class whose heavy chains derive from different heavy-chain genes within the class.

Submetacentric   A chromosome that has its centromere neither approximately at its center nor near one of its ends.

Substrate   Substance acted upon by an enzyme.

SV40   Simian virus 40; related to polyoma, but infects primates.

Synapsis   Pairing of homologous chromosomes; in humans, occurs only in meiosis.

Syndrome   A group of symptoms occurring together with sufficient regularity that it can be recognized as a distinct clinical entity.

Synteny   The state of two loci being on the same chromosome.

T   Conventional symbol for the deoxyribonucleotide thymidine phosphate or for the base thymine.

Tay-Sachs disease   A recessively inherited enzyme defect causing degeneration of the central nervous system; particularly common in certain Jewish populations.

T cell   Lymphocyte responsible for cellular immunity.

Telophase   A final stage of cell division characterized by division of the cytoplasmic contents as the cell pinches into two daughter cells. In telophase of mitosis or telophase I or telophase II of meiosis in males, division of the cytoplasm is approximately equal. In telophase I and telophase II of meiosis in females, cytoplasmic division is very unequal and produces one tiny product called a polar body.

Temperate   A virus or bacteriophage that can infect host cells without killing them.

Temperature-sensitive mutation   A mutation whose phenotypic effects are expressed at one temperature (the restrictive temperature) but not at another (the permissive temperature).

Template   A pattern.

Teratogen   An agent that causes birth defects owing to its effects during development.

Termination codons   Codons signaling the end of synthesis of a polypeptide chain; in cytoplasmic mRNAs, corresponding to $5'-UAA-3'$, $5'-UAG-3'$, or $5'-UGA-3'$.

Terminator codon   See Termination codons.

Tertiary structure   The three-dimensional configuration of a polypeptide.

Testicular-feminization syndrome   X-linked condition characterized by absence of development of male sexual structures in XY individuals.

Testosterone   Male sex hormone.

Tetraploid   Having four haploid sets of chromosomes.

Tetrasomic   An otherwise diploid individual who has two extra copies of a particular chromosome.

Thalidomide   A tranquilizer that is a teratogen because it inhibits normal development of the long bones in the arms and legs.

Three-prime (3′) end   The end of a DNA or RNA strand having a free 3′ hydroxyl.

Threshold trait   A multifactorial trait that is either present or absent in an individual.

Thymine   Pyrimidine base found in DNA.

Thymine dimer   Covalent linkage between adjacent thymines in a strand of DNA.

TK   The enzyme thymidine kinase, or the locus that codes for the enzyme.

Tolerance   Ability to avoid mounting an immune response against one's own antigens.

Total variance   Phenotypic variance.

Trait   A characteristic.

Transcript   See RNA transcript.

Transcription   The process in which an RNA strand complementary in base sequence to a DNA strand is produced.

Transduction   Transfer of genes from one cell to another by means of a virus.

Transfer RNA   See tRNA.

Transformation   (1) Treatment of cells with purified DNA for purposes of altering their genetic material. (2) Production of transformed, tumorlike cells by a viral infection.

Translation   Process of polypeptide synthesis.

Translocation   Attachment of part of a chromosome to a nonhomologous chromosome.

Transposable element   DNA sequence capable of altering its position within a DNA molecule or moving to a different DNA molecule.

Transposition   Physical movement of a DNA sequence from one place to another in the same or a different DNA molecule.

Triploid   Having three haploid sets of chromosomes.

Trisomic   Having one chromosome extra in an otherwise diploid individual.

Trisomy 13   47,+13 karyotype; Patau syndrome.

Trisomy 18   47,+18 karyotype; Edwards syndrome.

Trisomy 21   47,+21 karyotype; Down syndrome.

tRNA   Small RNA molecules including an anticodon (which recognizes a corresponding codon in mRNA) and an amino acid that functions in translation.

Turner syndrome   45,X sex chromosome abnormality; affected individuals are female.

Two-egg twins   See DZ twins.

U   Conventional symbol for the ribonucleotide uridine phosphate or for the base uracil.

Ultraviolet light   Short-wavelength light absorbed by DNA.

Umber   See Opal.

Unbalanced   A chromosome rearrangement in which certain genes are lost or present in excess.

Unbalanced translocation   Translocation in which some genes are missing or present in excess.

Unique-sequence DNA   A DNA sequence found once per haploid genome.

Unstable mutation   A mutation that undergoes further mutation at a high rate.

Unwinding   Separation of strands of duplex DNA during replication.

Upstream   In RNA or antisense strand of DNA, toward the 5′ end; in sense strand of DNA, toward the 3′ end.

Uracil   Base found in ribonucleic acid that serves the role of thymine in deoxyribonucleic acid.

V   Variable region of an antibody polypeptide, or the corresponding DNA sequence.

Variable age of onset   See Age of onset.

Variable expressivity   See Expressivity.

Variable region   The portion of a heavy chain or light chain that varies greatly in amino acid sequence among antibodies in the same subclass.

Variance   Measure of spread of a distribution defined as the mean of the square deviations from the mean.

Variation among populations   Refers to allele frequencies or traits that differ in different populations.

Virulent   A virus or bacteriophage that kills the host cells.

Virus   A tiny internal parasite of cells capable of infectious transmission between cells but incapable of reproduction on its own.

Wolffian ducts   Embryonic ducts associated with internal sexual structures of males.

X chromosome   One of the sex chromosomes; normal human females have two X's, males have only one.

Xeroderma pigmentosum   Autosomal recessive disorder characterized by development of multiple malignant skin tumors; some forms are associated with defective DNA repair processes.

X linked   A gene on the X chromosome.

XO   45,X karyotype; Turner syndrome.

X-ray   Energy emitted as electromagnetic radiation; very penetrating and very mutagenic.

XXX   47,XXX karyotype.

XXY   47,XXY karyotype; Klinefelter syndrome.

XYY   47,XYY karyotype.

Y chromosome   One of the sex chromosomes; normal human males have one Y chromosome, females have none.

Y linked   A gene on the Y chromosome.

Z DNA   DNA in its left-handed configuration.

Zeta hemoglobin   One of the polypeptide chains of embryonic hemoglobin.

Zygote   A fertilized egg.

Answers

Chapter 1

1.1. Because it was thought that these laws would significantly reduce the number of "undesirable" genes in future generations.

1.2. Prokaryotic cells lack a true nucleus and true chromosomes and divide by fission rather than mitosis. Eukaryotic cells do have a true nucleus and chromosomes and divide by mitosis. There are numerous other differences as well.

1.3. DNA, genetic material; ribosomes, sites of protein synthesis; mitochondria, sites of cellular respiration; chloroplasts, sites of photosynthesis; ATP, energy storage molecule; spindle, network for chromosome movement during cell division; cell membrane, semipermeable boundary of cell.

1.4. $1.15 \times 10^6 \ \mu m^3$

1.5. $12.6 \ \mu m^3$ (I.e., the human egg has a volume more than 90,000 times larger than the sperm head.)

1.6. A gamete is a reproductive cell, in humans the sperm or the egg; a zygote is a fertilized egg.

1.7. The sperm and egg, which have 23.

1.8. 69

1.9. 92

1.10. 21; 21

1.11. 86

1.12. Because they are both replicas of the same original chromosome.

1.13. 92; 46

1.14. Anaphase begins at the moment the centromeres split.

1.15. Part of the long arm of the submetacentric chromosome has been broken off and lost.

1.16. TACGGCACGT

CHAPTER 2

2.1. Chromosomal mosaic has two or more chromosomally distinct types of cells. In a chromosomal mosaic, some tissues would have one chromosome constitution, others would have a different one.

2.2. I; II

2.3. I

2.4. 2; 2

2.5. Two gametes will have an extra copy of chromosome 21; two gametes will be missing chromosome 21.

2.6. The mitotic cell would have the diploid number of chromosomes; the meiosis II cell would have the haploid number of chromosomes.

2.7. 18

2.8. The extra chromosome is a Y; the male's chromosome constitution is written 47,XYY.

2.9. The X (drawn upside down of its conventional representation).

2.10. Chromosome 6; it is missing the uppermost band.

2.11. The chromosome consists of the long arm of 21 fused to the long arm of 22. Evidently the short arm of each chromosome was broken off and lost and the remaining parts fused. This type of chromosome abnormality is called a translocation.

2.12. Chromosome 17, but two bands in the long arm are in the reverse of the normal order. Apparently breaks flanking these bands occurred and the segment in the middle became inverted before chromosome restitution.

2.13. *A b* and *a B*.

CHAPTER 3

3.1. It could be due to cultural inheritance (e.g., learning).

3.2. The environment that provokes the trait (e.g., drug sensitivity) may provoke it only in individuals of the appropriate genotype.

3.3. These traits are multifactorial.

3.4. An environmentally induced condition that resembles one known to be inherited.

3.5. A kindred is a group of relatives; a pedigree is a diagram showing how members of a kindred are related.

3.6. $\frac{3}{4}$; $\frac{1}{4}$

3.7. *AA* genotype, *Ww* genotype, woolly hair phenotype, homozygous genotype, bald phenotype, O blood type phenotype, phenocopy phenotype, dry ear cerumen phenotype, heterozygous genotype.

3.8. *Ww, Aa, Bb,* and $I^A I^O$ are heterozygotes; the others are homozygotes.

3.9. $\frac{1}{4}$, $\frac{1}{2}$, $\frac{1}{4}$

3.10. $\frac{3}{4}$; $\frac{2}{3}$

3.11. $\frac{1}{4}$ (i.e., $\frac{1}{2} \times \frac{1}{2}$)

3.12. *Bb*; yes, but it is not possible to state age of onset or degree of expression because these are variable.

3.13. Could not be dominant with complete penetrance because individual III-2 (the second individual from the left in generation 2) is not affected but has an affected offspring. Could be dominant with incomplete penetrance because trait could be unexpressed in III-2.

3.14. $\dfrac{6!}{4!2!}(\frac{1}{2})^6 = \dfrac{720}{(24)(2)}(\frac{1}{64}) = 0.23$

3.15. $(4!/4!0!)(\frac{1}{2})^4 = \frac{1}{16} = 0.0625;$
$(4!/2!2!)(\frac{1}{2})^4 = [24/(2)(2)](\frac{1}{16}) = 0.375$

3.16. $1 - [8!/4!4!](\frac{1}{2})^8 = 1 - [40320/(24)(24)](\frac{1}{256}) = 1 - 0.27 = 0.73$

CHAPTER 4

4.1. Heterozygous

4.2. Falciparum malaria is widespread in west Africa, and carriers of the allele have an enhanced resistance to infection and its complications.

4.3. With 3 alleles, 6 genotypes; 4 alleles, 10 genotypes; 5 alleles, 15 genotypes. In general *n* alleles corresponds to $n(n + 1)/2$ genotypes.

4.4. Consanguinity among the parents of an affected individual. The parents of V-I are second cousins.

4.5. $Aa \times Aa$; $\frac{1}{4}AA$, $\frac{1}{2}Aa$, $\frac{1}{4}aa$

4.6. $Aa \times aa$

4.7. Because formation of a homozygote requires that both sperm and egg carry the rare allele, and these are individually rare events. A more common situation will be to have one or the other gamete (but not both) carrying the rare allele, which will give rise to a heterozygote.

4.8. $-, -, -, -, -, +, -, -$, for chromsome 9; all $+$ for the X.

4.9. A = 6; B = 13; C = 21

4.10. 4 percent

4.11. 50 percent

4.12. Either ABC or BAC; yes, order must be ABC.

4.13. Recombinant (*a*) and (*d*); nonrecombinant (*b*) and (*c*)—because nonrecombinants must always be most frequent classes. Genotype Ab/aB.

4.14. $[4!/(1!)(3!)](\frac{1}{4})^1(\frac{3}{4})^3 = (\frac{24}{6})(\frac{1}{4})(\frac{27}{64}) = 0.42$

4.15. $[4!/(0!)(4!)](\frac{1}{4})^0(\frac{3}{4})^4 = (\frac{81}{256}) = 0.32$

CHAPTER 5

5.1. The cells that give rise to the reproductive cells in the ovaries and testes.

5.2. It interferes with the growth of the long bones of the arms and legs in the developing fetus.

5.3. Primary sex ratio is ratio at fertilization; secondary is ratio at birth. May be unequal because of differential spontaneous abortion of the sexes.

5.4. (*a*) $1048/(1048 + 1000) = 0.5117$

(*b*) $1055/2055 = 0.5134$

(*c*) $1063/2063 = 0.5153$

(*d*) $1070/2070 = 0.5169$

5.5. $(1000/1055)(1000) = 947.87$, or about 948.

5.6. Y-linkage because surnames are inherited from father to son to grandson, and so on. The analogy breaks down because females also receive their father's surname.

5.7. Malelike; up to 3.

5.8. $\frac{2}{3}$; $\frac{1}{3}$; because in each cell only one X chromosome, chosen at random, is active, and $\frac{1}{3}$ of the female's Xs carries $HGPRT^-$.

5.9. Some of the breakdown products of excess progesterone affect the body in a manner similar to testosterone.

5.10. Only males affected, affected males have normal offspring, affected males have affected maternal grandfather.

5.11. Because the X chromosome in a male must be transmitted to his daughters, and in the next generation can pass to his grandson, etc., thereby crisscrossing between the sexes.

5.12. The woman's mother must have been heterozygous; $\frac{1}{2}$.

5.13. $\frac{1}{2}$; makes no difference because the boy receives his X chromosome from his mother.

5.14. Individual III-1; I-1 = CC or Cc; I-2 = c; II-1 = Cc; II-2 = C; III-1 = c; III-2 = Cc; IV-1 = cc.

5.15. The daughter is heterozygous for both color-blind alleles, i.e., $R\ g/r\ G$ where R, r and G, g stand for the red and green alleles. The daughter received the $r\ G$ chromosome from her mother (who received it from the red-color-blind father) and the $R\ g$ chromosome from her father.

5.16. $(1/500)^2 = 1/250,000$ (i.e., 4×10^{-6})

5.17. $[3!/0!3!](\frac{1}{2})^3 = \frac{1}{8}$; $[3!/3!0!](\frac{1}{2})^3 = \frac{1}{8}$; $\frac{1}{8} + \frac{1}{8} = \frac{1}{4}$

5.18. $(4!/0!4!)(\frac{1}{2})^4 = \frac{1}{16}$; $(4!/1!3!)(\frac{1}{2})^4 = \frac{4}{16}$; $(4!/2!2!)(\frac{1}{2})^4 = \frac{6}{16}$;
$(4!/3!1!)(\frac{1}{2})^4 = \frac{4}{16}$; $(4!/4!0!)(\frac{1}{2})^4 = \frac{1}{16}$

5.19. $(3!/1!2!)(.51)^1(.49)^2 = 0.367$
$(3!/2!1!)(.51)^2(.49) = 0.382$

5.20. $(6!/6!0!)(.52)^6(.48)^0 = 0.020$; $(6!/0!6!)(.52)^0(.48)^6 = 0.012$

CHAPTER 6

6.1. Risk is the chance of having an affected child when no previous children are affected; recurrence risk is the chance of having an affected child when one or more previous children are affected.

6.2. Because the incidence of Down syndrome fetuses is greatly elevated in women over 40, and some other chromosomal abnormalities have an elevated incidence as well, but to a smaller extent.

6.3. 47,XXX and 47,XXY.

6.4. It is "inherited" in the sense of "due to the genetic constitution of the affected individual (i.e., genetic)" but it is not inherited because in most cases the recurrence risk is relatively low.

6.5. Routine chromosome study will indicate whether the fetus is XX or XY.

6.6. In cells of an affected fetus the enzyme associated with Tay-Sachs disease will be missing; $\frac{1}{4}$.

6.7. (a) 0; (b) 2; (c) 1; (d) 0.

6.8. Nondisjunction of the Y chromosome during the second meiotic division in the father.

6.9. 46,XX; 46,XY; 45,X; 45,Y in the proportions $\frac{1}{4}, \frac{1}{4}, \frac{1}{4}, \frac{1}{4}$. All of the 45,Y and most of the 45,X would abort.

6.10. $(7500 + 550)/100,000 = 0.080$ (i.e., $\frac{1}{12}$).

6.11. $8/85000 = 9.4 \times 10^{-5}$ (i.e., $1/10625$); $(1350 + 8)/(85000 + 1350) = 0.016$ (i.e., $\frac{1}{64}$).

6.12. The woman must be heterozygous for the color-blind allele, and the X chromosome carrying this allele underwent nondisjunction in the second meiotic division, giving rise to an XX egg in which both Xs carry the color-blind allele.

CHAPTER 7

7.1. A balanced rearrangement changes only the order but not the relative dosage of genes. An unbalanced rearrangement changes relative gene dosage.

7.2. Lacking a centromere, they are lost from cells during cell division and so are quickly eliminated.

7.3. Robertsonian translocation (carriers have 45 chromosomes).

7.4. Breaks flanking the centromere; terminal segments lost, ends of centromere-bearing fragment attach.

7.5. Both abnormal chromosomes have one arm from chromosome 1 and one arm from chromosome 2. This is a translocation originating from breaks very near the centromeres of chromosomes 1 and 2.

7.6. Deletion (one of the small bands in the long arm is missing).

7.7. Chromosome 6; a pericentric inversion.

7.8. A paracentric inversion.

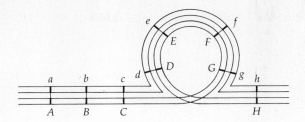

7.9. A paracentric inversion; two moncentrics, an acentric, and a dicentric.

7.10. $\frac{1}{6}$

7.11. It is the "missing half" of the Robertsonian translocation, consisting of the short arms of 14 and 21.

7.12. A Robertsonian translocation involving chromosomes 13 and 21. The trisomy-13 child comes from one mode of segregation, the trisomy-21 child from a second mode, and the carrier child from the third mode. This third mode of segregation could also produce a normal child

7.13. 46; 46

CHAPTER 8

8.1. Because the strands can be separated relatively easily during DNA replication.

8.2. A change in the hereditary characteristics of a living organism by means of the direct introduction of DNA into the cells.

8.3. Physical containment refers to laboratory safety precautions; biological containment refers to the use of host organisms that are so enfeebled by mutation that they cannot survive outside the laboratory.

8.4. $4 \times 4 \times 4 = 4^3 = 64$

8.5. 3'-CGTAATGCTTACG-5'

8.6. Alternating Cs and Gs

8.7. 23% T, 32% A, 19% C, and 26% G; 27.5% A, 27.5% T, 22.5% G, 22.5% C

8.8. One-to-one; one-to-one; one-to-one.

8.9. 17 also

8.10. The 5' end

8.11. 3'-ABABCCGBBBAAC-5'

8.12. 3 fragments

8.13. 5 fragments

8.14. Four ways, resulting in one of the molecules

5'-ATGCGCTA-3'   5'-ATGCTAGC-3'
3'-TACGCGAT-5'   3'-TACGATCG-5'

5'-GCATGCTA-3'   5'-GCATTAGC-3'
3'-CGTACGAT-5'   3'-CGTAATCG-5'

CHAPTER 9

9.1. Metabolism refers to the totality of all biochemical processes (molecular transformations) that occur in the body or, more loosely, body chemistry. An inborn error of metabolism is a genetic defect, usually in a gene coding for an enzyme, resulting in an inability of the body to carry out the corresponding biochemical step.

9.2. 5'-AUG-3' (M) is the "start" codon; 5'-UAA-3', 5'-UAG-3', and 5'-UGA-3' all correspond to "stop" (period).

9.3. Two (M and W) with one each; nine (C, D, E, F, H, K, N, Q, Y) with two codons each; one (I) with three (the termination signal also has three codons); five (A, G, P, T, V) with four codons; none with five; three (L, R, S) with six codons. Altogether the genetic code codes for 20 amino acids.

9.4. The 3' end; the 5' end

9.5. From right to left because RNA polymerase can add ribonucleotides only to the 3' end of a growing chain, hence the 5' end of the RNA (3' end of the DNA template) must be the beginning.

9.6. 5'-AUUGCUGAAUGAC-3'

9.7. The DNA strand would be transcribed from right to left, leading to the RNA transcript 5'-CCCCACAUUUUAGCGGACGAACUCCCGCAUAUAGCU-3'. This would be translated from left to right, leading to the polypeptide PHILADELPHIA.

9.8. SSSSS; that the genetic code has synonyms.

9.9. Alternating As and Gs in the transcript; alternating Rs and Es in the polypeptide.

9.10. ASECRET

9.11. Three, namely: 5'-AUGUGGAUGUGGUAA-3'
　　　　　　　　　　5'-AUGUGGAUGUGGUAG-3'
　　　　　　　　　　5'-AUGUGGAUGUGGUGA-3'

9.12. (a) 5'-ACU-3' (amino acid T); (b) 5'-AAC-3' (N); (c) 5'-GAA-3' (E); (d) 5'-CGA-3' (R); (e) 5'-GUA-3' (V); (f) 5'-CAG-3' (Q)

9.13. PENS; PETS

9.14. FRANCISCRICK; FRNM

9.15. Yes; yes; no; yes.

CHAPTER 10

10.1. To shield the gonads from stray x-rays.

10.2. Estimating mutation rates induced by potential mutagens; uses *Salmonella typhimurium*.

10.3. 5×10^{-5} corresponds to 50 new mutations at the locus per million gametes per generation.

10.4. 1; 2; 2; 1

10.5. Formation of linkage (dimer) between Ts at positions 4 and 5 from the left.

10.6. At 156 rads mutation rate will be 8×10^{-6} (i.e., double the spontaneous rate). Since 156 rads adds 4×10^{-6} to the mutation rate, each 15.6 rads adds 0.4×10^{-6} (assuming proportionality), so expected dose with 15.6 rads will be $0.4 \times 10^{-6} + 4 \times 10^{-6} = 4.4 \times 10^{-6}$.

10.7. (a) 1; (b) 1; (c) 2; (d) 3; (e) 1; (f) 3

10.8. CARDINALS

10.9. CARDINALS; a silent nucleotide substitution

10.10. CARDINAMS; a missense substitution

10.11. CARDINAFS

10.12. CSQRHKCIV; frameshift

10.13. CSQRQNALS

10.14. CA; a chain terminator (nonsense) mutation

10.15. Unmutated tRNA has anticodon 5'-AGC-3' recognizing codon 5'-GCU-3' and inserting A; mutant tRNA has anticodon 5'-AUC-3' recognizing codon 5'-GAU-3'. This codon normally codes for D but when recognized by mutant tRNA will cause insertion of A.

10.16. Three possibilities, namely,

5'-ATTTTAGAT-3'   5'-ATCTTAAAT-3'   5'-ATTTTAAAT-3'
3'-TAAAATCTA-5'   3'-TAGAATTTA-5'   3'-TAAAATTTA-5'

In the first two cases, only one C has been deaminated. In the last case, both have been.

CHAPTER 11

11.1. Cigarette smoking is the single largest risk factor in the occurrence of lung cancer.

11.2. The conversion of a normal cell into one having the characteristics of malignant cell growth.

11.3. Yes, because most mutagens are also carcinogens.

11.4. Influenza

11.5. An epidemic is a local outbreak of infectious disease; a pandemic is a worldwide outbreak of disease.

11.6. Because many individuals will have developed immunity to the prevailing types, and a virus would have to be quite different to be able to circumvent this immunity and cause an epidemic.

11.7. $2^8 = 256$; $10^8 = 100,000,000$

11.8. Abortive

11.9. An RNA-containing virus able to be reverse transcribed into DNA and become incorporated into the chromosomes. No.

11.10. 3'-TAGTAAAGTCA-5'

11.11. These genes are thought to originate ultimately from their normal homologues in the chromosomes of the host.

11.12. Yes, certain cancers such as retinoblastoma (a type of eye cancer) are due to inherited mutations; these simply inherited types of cancer are individually quite rare.

11.13. One part of a reciprocal translocation often found in association with a particular type of leukemia.

11.14. They have characteristic cell-surface antigens that the immune system can attack.

11.15. The closer the genes, the more likely they are to be included in the DNA fragment picked up by the virus, and the greater the frequency of cotransduction.

CHAPTER 12

12.1. The ability to mount a rapid immune response against an invading antigen.

12.2. A disease in which the immune system attacks one's own tissues.

12.3. Any substance capable of eliciting an immune response.

12.4. Amino acid; such small molecules are not normally antigenic.

12.5. Thymus-derived; bone-marrow-derived

12.6. The individual would lack circulating antibodies and would be highly susceptible to infectious disease.

12.7. A T-cell function.

12.8. The patient becomes more susceptible to infection.

12.9. 1, because the son must receive one of his mother's haplotypes.

12.10. $\frac{1}{4}$

12.11. At each locus there are 10 possible genotypes, and there are 10 such loci, so total number of genotypes would be $10^{10} = 10,000,000,000$.

12.12. Antibodies against these antigens are not normally produced in humans.

12.13. Donate to A or AB; receive from A or O.

12.14. $I^A I^O$ and $I^B I^O$

12.15. 7 (A, B, AB, O, C, AC, BC)

12.16. The child could receive the dominant secretor allele and I^O from its father and an I^B allele from its mother.

12.17.

Mother	Father
A	AB or B
B	AB or A
AB	none
O	AB or A or B

12.18. No; the child would have to be A or B.

12.19. Yes; yes; no; yes.

CHAPTER 13

13.1. Missense substitutions; silent mutations are not detected because they do not change the amino acid sequence.

13.2. Mutation rates are normally too small to change significantly allele frequency over the course of just a few generations.

13.3. (1) All populations have more young than can possibly survive and reproduce. (2) There is phenotypic variation in survival and reproduction among individuals. (3) Some of the phenotypic variation in survival and reproduction is due to genotypic differences among individuals.

13.4. That the fitness of both homozygotes is smaller than that of the heterozygote; in the long term, overdominance leads to an equilibrium in which both alleles are maintained in the population, and their allele frequencies do not change.

13.5. A balance between selection against the harmful alleles and mutation creating them.

13.6. Random changes in allele frequency from generation to generation due to relatively small population size preventing all individuals from contributing equally to the progeny in any generation.

13.7. A hunting-and-gathering society, because the traveling groups in such cultures are relatively small in size, thus emphasizing the effects of random genetic drift.

13.8. I^O: $0.450 + (0.363/2) + (0.079)/2 = 0.6710$
I^A: $(0.363/2) + 0.073 + (0.032)/2 = 0.2705$
I^B: $(0.079/2) + 0.003 + (0.032)/2 = 0.0585$

13.9. Allele frequency of $M = (708 + 492)/2000 = 0.60$; allele frequency of $N = (492 + 308)/2000 = 0.40$.

13.10. Expected numbers are
$MM = (0.60)^2(1000) = 360$
$MN = 2(0.60)(0.40)(1000) = 480$
$NN = (0.40)^2(1000) = 160$
The observed numbers are very close to those expected. Conclude that the population would appear to be undergoing random mating for the *MN* locus.

13.11. 0.56; 0.44

13.12. $\frac{1}{2}$, for this reason: Half the time the egg will carry *M* and when it does, the probability that the sperm carries *N* will be 0.44; half the time the egg will carry *N* and when it does, the probability that the sperm carries *M* will be 0.56; the overall probability is thus $(\frac{1}{2})(0.44) + (\frac{1}{2})(0.56) = (\frac{1}{2})(0.44 + 0.56) = \frac{1}{2}$. Note that the answer is $\frac{1}{2}$ irrespective of the allele frequencies.

13.13. Allele frequency of $d = \sqrt{144/10000} = \sqrt{0.0144} = 0.12$; allele frequency of $D = 1 - 0.12 = 0.88$. Expected number of heterozygotes $= 2(0.12)(0.88)(10000) = 2112$.

13.14. Frequency of Rh$^-$ females $= (0.20)^2 = 0.04$; frequency of Rh$^+$ males $= 1 - 0.04 = 0.96$. Frequency of incompatible matings is therefore $(0.04)(0.96) = 0.0384$, or a little less than 4 percent.

13.15. $q = 1/30$, so frequency of affected females $= (1/30)^2 = 1/900 = 0.0011$; heterozygotes $= 2(1/30)(29/30) = 0.064$.

13.16. $q = \sqrt{1/4000} = 0.0158$; frequency of carriers $= 2(0.0158)(1 - 0.0158) = 0.0311$, or about 1 individual in 32.

13.17. $(0.0158)^2(1 - \frac{1}{16}) + (0.0158)(\frac{1}{16}) = 0.0012$, or about 1 in 819.

CHAPTER 14

14.1. Five possible phenotypes (i.e., 0, 1, 2, 3, and 4); ratio would be 1:4:6:4:1.

14.2. Because they might go unrecognized and be interpreted as transmissible genetic factors.

14.3. The concordance rate in DZ twins is the same as the recurrence risk among sibs (although with twins the sibs happen to be born at the same time). For an autosomal dominant this is 50 percent.

14.4. The reasoning is the same as in Problem 14.3; for an autosomal recessive with neither parent affected the recurrence risk in sibs is 25 percent.

14.5. Narrow-sense.

14.6. Broad-sense heritability is 0 because all plants have an identical genotype and therefore none of the variation in phenotype can be due to differences in genotype. This example illustrates that the heritability of a trait depends on the particular population being studied.

14.7. With no dominance, expect hybrid to be the average of the parental inbreds, or $(126 + 112)/2 = 119$.

14.8. Half sibs, because the twins have an identical genotype.

14.9. Full sibs.

14.10. Not predictable from given information because heritability of liability refers to *predisposition* toward the trait, not whether it will actually occur.

14.11. Concordance $= 47/(47 + 81) = 36.7$ percent.

14.12. Standard deviation $= \sqrt{100} = 10$. Consequently, 68 percent of the population will be in the range 50 ± 10 (i.e., 40 to 60), and 95 percent will be in the range 50 ± 20 (i.e., 30 to 70). Also, 5 percent of the population will have a phenotype of less than 30 or more than 70, but these two possibilities must be equally likely because the normal curve is symmetric around the mean. Thus, 2.5 percent of the population has a phenotype of less than 30, and 2.5 percent has a phenotype of greater than 70.

14.13. Since the normal curve is symmetric around the mean, 34 percent have a phenotype between 50 and 60, and 47.5 percent will have a phenotype between 50 and 70. (These numbers correspond to 68/2 and 95/2, respectively.) The proportion between 60 and 70 is therefore $47.5 - 34 = 13.5$ percent.

14.14. $O = 75 + (105 - 75)(\frac{1}{3}) = 85$;

$O = 75 + (45 - 75)(\frac{1}{3}) = 65$

14.15. $H^2 - h^2$

Credits

Chapter 1

1-1 Courtesy of the Library Services Department, the American Museum of Natural History.

1-4 Redrawn from W. DeWitt, Biology of the Cell (New York: Saunders, 1977).

1-5 (a) M. E. Bayer, The Institute for Cancer Research, Philadelphia.
(b) M. E. Doohan.
(c) K. R. Porter, University of Maryland, Baltimore County.

1-6 (a) K. R. Porter, University of Maryland, Baltimore County.
(b) R. A. Dilley, Purdue University.

1-9 B. John, Australian National University, Canberra.

1-10 S. A. Schonberg, Ph.D., Director Cytogenetics Laboratory AC 28, University of California, San Francisco.

1-11 B. E. Schuh, Cytogenetics, Department of Pathology, Monmouth Medical Center, Long Branch, New Jersey.

1-13 G. Albrecht-Buehler, Cold Spring Harbor Laboratory.

1-16 R. J. Green, Center for Disease Control, Department of Health and Human Services, Atlanta, Georgia.

Chapter 2

2-2 S. A. Schonberg, Ph.D., Director Cytogenetics Laboratory AC 28, University of California, San Francisco.

2-3 M. W. Shaw, The University of Texas Health Science Center at Houston.

2-4 Courtesy of A. P. Craig-Holmes, M.D., Department of Medicine, Baylor College of Medicine, Houston.

2-5 C. C. Lin, M.D., University of Calgary.

2-6 Courtesy of A. P. Craig-Holmes, M.D., Department of Medicine, Baylor College of Medicine, Houston.

2-8 (a, c, d) L. F. Meisner, Cytogenetics Unit, State Laboratory of Hygiene, University of Wisconsin, Madison.
(b) H. A. Lubs.

2-9 B. John, Australian National University, Canberra.

2-12 B. John, Australian National University, Canberra.

2-14 Andrew T. L. Chen, Ph.D., Center for Disease Control, Department of Health and Human Services, Atlanta, Georgia.

2-15 B. John, Australian National University, Canberra.

Chapter 3

3-1 V. Orel, Mendelianum of the Moravian Museum, Brno.

3-6, From O. L. Mohr, The Journal of Heredity, 1932,
3-7 23:345–352.

3-9 (a, b) U.P.I./Bettmann Archive
(c) K. Taysi, M.D., Department of Pediatrics, St. Louis Children's Hospital, St. Louis, Missouri.
(d) A. M. Winchester.
(e) Rephotographed from P. R. Dodge, p. 492, fig. XVI. 9. R. D. Adams, L. M. Eaton, and G. M. Shy (eds.), Congenital Neuromuscular Disorders, 1960 38:479–527 (Baltimore: Williams & Wilkins, 1960).
(f) J. Miller, M.D., Department of Ophthalmology, St. Louis Children's Hospital, St. Louis, Missouri.
(g) J. G. Hall, M.D., Department of Medical Genetics, The University of British Columbia, Vancouver.
(h) J. de Framond, Department of Genetics, Washington Univeristy School of Medicine, St. Louis, Missouri.

3-10 R. S. Lees, M.D., New England Deaconess Hospital, Boston, Massachusetts.

3-11 D. L. Rimoin, M.D., Ph.D., Harbor–UCLA Medical Center.

3-12 (a, b) B. O. Hall, University of California, San Francisco.

Chapter 4

4-2 After E. W. Sinnott, L. C. Dunn, and T. Dobzhansky, Principles of Genetics (New York: McGraw-Hill, 1950).

4-3 A. Cerami, Ph.D., The Rockefeller University.

4-4 Adapted from L. L. Cavalli-Sforza, "The Genetics of Human Populations," in Scientific American, September 1974, 231:80–89. Copyright © 1974 by Scientific American, Inc. All rights reserved.

4-6 Photograph by Division of Photography/Courtesy, Field Museum of Natural History, Chicago, FMNH Negative # 118.

4-7 After C. M. Woolf and F. C. Dukepoo, Science, 1969, 164:30–37. Copyright © 1969 by the American Association for the Advancement of Science.

4-13 R. S. Kucherlapati, University of Illinois at Chicago, Center for Genetics, University of Illinois, College of Medicine.

Chapter 5

5-2 Copyright © by Camera M.D. Studios, Inc., New York

5-5 Pedigree after W. Bulloch and P. G. Fildes, The

Treasury of Human Inheritance, 1911, 1, No. 490.

5-7 Culver Pictures.

5-9 (a, b) B. E. Schuh, Cytogenetics, Department of Pathology, Monmouth Medical Center, Long Branch, New Jersey.

5-9 (c, d) March of Dimes Birth Defects Foundation.

5-10 Dr. Pat Calarco, San Francisco Medical Center, University of California.

5-11 (a) Dr. Pat Calarco, San Francisco Medical Center, University of California.
(b) Courtesy of Dr. Arthur T. Hertig and the Regional New England Primate Center.

5-12 (a) Carnegie Institution of Washington, Department of Embryology, Davis Division.
(b, c) Photos by M. Jacobson, D. Yeager, Copyright © by Camera M.D. Studios, Inc., New York.

5-13 Wide World Photos.

5-16 Photograph: J. M. Opitz, M.D., Shodair Children's Hospital, Helena, Montana.
Karyotype: L. F. Meisner, Cytogenetics Unit, State Laboratory of Hygiene, University of Wisconsin, Madison.

Chapter 6

6-3 Photograph: J. M. Opitz, M.D., Shodair Children's Hospital, Helena, Montana.
Karyotype: C. M. Moore, M.D., Department of Pediatrics, University of Texas Medical School, Houston.

6-4 March of Dimes Birth Defects Foundation.

6-5 M. W. Shaw, The University of Texas Health Science Center at Houston.

6-6 (a, b, c) Child Development and Mental Retardation Center, University of Washington.

6-7 Data from E. B. Hook and A. Lindsjö, American Journal of Human Genetics, 1978, 30:19–27 (Chicago: The University of Chicago Press, 1978).

6-8 Photograph: K. Taysi, M.D., Department of Pediatrics, St. Louis Children's Hospital, St. Louis, Missouri.
Karyotype: M. W. Shaw, The University of Texas Health Science Center at Houston.

6-9 Photograph: K. Taysi, M.D., Department of Pediatrics, St. Louis Children's Hospital, St. Louis, Missouri.
Karyotype: M. W. Shaw, The University of Texas Health Science Center at Houston.

6-10 March of Dimes Birth Defects Foundation.

6-11 (a) B. E. Schuh, Cytogenetics, Department of Pathology, Monmouth Medical Center, Long Branch, New Jersey.
(b) M. W. Shaw, The University of Texas Health Science Center at Houston.

Chapter 7

7-2 Micrograph: E. Patau.

7-3 (a, b) K. Taysi, M.D., Department of Pediatrics, St. Louis Children's Hospital, St. Louis, Missouri.

7-6 K. H. Rothfels, Ph.D., University of Toronto.

7-13 M. W. Shaw, The University of Texas Health Science Center at Houston.

7-15 Dr. Jean de Grouchy, Hôpital des Enfants-Malades, Clinique de Génétique Médicale, Paris.

Chapter 8

8-6 A. H. - J. Wang, G. J. Quigley, F. J. Kolpak, J. L. Crawford, J. H. van Boom, G. van der Marel, and A. Rich. Molecular structure of a left-handed double helical DNA fragment at atomic resolution. Nature 282:680–686 (London: Macmillan Journals Ltd., 1979).

8-9 J. A. Huberman, Roswell Park Memorial Institute, Buffalo, New York, and A. D. Riggs, City of Hope Medical Center, Duarte, California.

8-12 Courtesy of Garry Brodeur, M.D., Washington University School of Medicine, St. Louis, Missouri.

8-13 Data from Y. Nishioka and P. Leder, Cell, 1979, 18:875–882. Copyright © 1979, MIT Press, Cambridge.

8-14 T. E. Rucinsky, Director, Center for Basic Cancer Research, Electron Microscope Facility, Washington University School of Medicine, St. Louis, Missouri.

Chapter 9

9-4 W. R. Centerwall, M.D., Professor of Pediatrics and Genetics and Director of Medical Genetic Services, School of Medicine and Medical Center, University of California, Davis.

9-8 O. L. Miller, Jr., and B. R. Beatty, Journal of Cellular Physiology, 1969, 74, Supplement 1:225–282 (New York: Alan R. Liss, Inc., 1969).

9-10 Partially after J. L. Sussman and S. H. Kim, Science, 1976, 192:853–858. Copyright © 1976 by the American Association for the Advancement of Science.

9-11 A. Rich, Massachusetts Institute of Technology, Cambridge.

9-12 Data from Y. Nishioka and P. Leder, Cell, 1979, 18:875-882. Copyright © 1979, MIT Press, Cambridge.

9-13 Based in part on E. R. Huehns, N. Dance, G. H. Beaven, F. Hecht, and A. G. Motulsky, Cold Spring Harbor Symposium on Quantitative Biology, 1964, 29:327–331.

9-14 N. J. Proudfoot.

9-16 After K. Illmensee and P. C. Hoppe, Cell, 1981, 23:9–18. Copyright © 1981, MIT Press, Cambridge.

9-17 B. Mintz, Proceedings of the Natural Academy of Sciences, USA, 1967, 58:344–351.

Chapter 10

10-1 M. W. Strickberger, Genetics, 2d ed. (New York: Macmillan 1976); after J. V. Neel and W. J. Schull, Human Heredity (Chicago: The University of Chicago Press, 1954).

10-7 E. Strobel, Ph.D., Purdue University.

10-8 Data from M. W. Young and H. E. Schwartz, Nomadic gene families in Drosophila, Cold Springs Harbor Symposium on Quantitative Biology, 1980, 45:629–640, and R. Levis, P. Dunsmuir, and G. M. Rubin, Terminal repeats of the Drosophila transposable element copia: Nucleotide sequence and genomic organization, Cell, 1980, 21:581.

10-9 S. A. Schonberg, Ph.D., Director Cytogenetics Laboratory AC 28, University of California, San Francisco.

10-10 Based on J. Schultz, in B. M. Duggar (ed.), Biological Effects of Radiation (New York: McGraw-Hill, 1936).

10-12 L. Bergman and Associates.

10-15 B. N. Ames, Science, 1979, 204:587–593. Copyright © 1979 by the American Association for the Advancement of Science.

Chapter 11

11-1 B. Roizman, The University of Chicago.

11-2 R. C. Williams, Virus Laboratory, University of California, Berkeley.

11-3 J. D. Griffith.

11-4 From H. Dorn, Acta Geneticae Medicae Gemellologiae, 1959, 8:395–408 (New York: Alan R. Liss, Inc., 1959).

11-8 With permission from Natural History, Vol. 82, No. 1; Copyright the American Museum of Natural History, 1973.

11-10 R. C. Williams, Virus Laboratory, University of California, Berkeley.

11-12 Data from J. M. Coffin, in J. R. Stephenson (ed.), Molecular Biology of DNA Tumor Viruses (New York: Academic Press, 1980).

Chapter 12

12-1 Reprinted with the permission of E. Golub, Purdue University.

12-8 Reprinted by permission of the publisher from Comprehensive Immunogenetics, by W. H. Hildemann, E. A. Clark, and R. L. Raison, Fig. 5-10, p. 178. Copyright 1981 by Elsevier Science Publishing Co., Inc.

12-9 Reprinted by permission of the publisher from Comprehensive Immunogenetics, by W. H. Hildemann, E. A. Clark, and R. L. Raison, Fig. 5-9, p. 177. Copyright 1980 by Elsevier Science Publishing Co., Inc.

12-10 (a, b) A. E. Mourant et al., The Distribution of the Human Blood Groups and Other Polymorphisms, 2d ed. (London: Oxford University Press, 1976).

12-11 From D. L. Hartl, 1980. Principles of Population Genetics, Fig. 20, p. 36 (Sunderland, Massachusetts: Sinauer Assoc., 1980).

12-12 L. Bergman and Associates.

Chapter 13

13-1 Courtesy of J. Coyne and D. Hickey.

13-5 Data from Cavalli-Sforza and W. F. Bodmer, The Genetics of Human Populations (New York: W. H. Freeman, 1978).

13-9 From D. L. Hartl, 1980. Principles of Population Genetics, Fig. 20, p. 36 (Sunderland, Massachusetts: Sinauer Assoc., 1980).

13-10 M. Rasmuson, Genetics on the Population Level, p. 103 (Stockholm: Svenska Bokforlaget, 1961).

13-16 A. Jacquard.

13-17 From A. Jacquard, The Structure of Populations (New York: Springer-Verlag, 1974).

Chapter 14

14-2 From Heredity, Evolution and Society, Fig. 11.5, p. 198, by I. M. Lerner and W. J. Libby (San Francisco: W. H. Freeman, Copyright © 1976).

14-3 From S. B. Holt, The Genetics of Dermal Ridges, Fig. 19, p. 57 (Courtesy of Charles C. Thomas, Publisher, Springfield, Illinois, 1968).

14-5 From S. B. Holt, The Genetics of Dermal Ridges, Fig. 20, p. 62 (Courtesy of Charles C. Thomas, Publisher, Springfield, Illinois, 1968).

14-7 © Harvey Stein, 1978.

14-8 Data from H. B. Newcombe in M. Fishbein (ed.), Papers and Discussions of the Second International Conference on Congenital Malformations (New York: The International Medical Congress, Ltd. 1964).

14-9 Adopted from F. Vogel and A. G. Motulsky, Human Genetics: Problems and Approaches (New York: Springer-Verlag, 1979), based on data of R. M. Cooper and J. P. Zubek, Canadian Journal of Psychiatry, 1958, 12:159–164.

14-10 H. F. Harlow, University of Wisconsin Primate Laboratory.

14-11 From W. A. Kennedy et al., Monographs of the Society for Research in Child Development, 1963, 28, ser. 90 © The Society for Research in Child Development, Inc.

Table 6-2 Data from H. A. Witkin et al., 1976, Science 193:547–555. Copyright © 1976 by the American Association for the Advancement of Science.

Table 6-4 Based on data in D. H. Carr and M. Gedeon, 1977, Population cytogenetics of human abortuses, in E. B. Hook and I. H. Porter (eds.), Population Cytogenetics: Studies in Humans, Academic Press, New York, pp. 1–9, and E. B. Hook and J. L. Hamerton, 1977. The frequency of chromosome abnormalities detected in consecutive newborn studies–differences between studies—results by sex and severity of phenotypic involvement, in E. B. Hook and I. H. Porter (eds.), Population Cytogenetics: Studies in Humans, Academic Press, New York, pp. 63–79.

Index